American Short Horn Cattle Importations: Part 1
Containing the pedigrees of all Short Horn Cattle Imported to America

by William Warfield

with an introduction by Jackson Chambers

This work contains material that was originally published in 1884.

This publication is within the Public Domain.

This edition is reprinted for educational purposes
and in accordance with all applicable Federal Laws.

Introduction Copyright 2017 by Jackson Chambers

Self Reliance Books

Get more historic titles on animal and stock breeding, gardening and old fashioned skills by visiting us at:

http://selfreliancebooks.blogspot.com/

Introduction

I am pleased to present another title in the "Cattle" series.

The work is in the Public Domain and is re-printed here in accordance with Federal Laws.

As with all reprinted books of this age that are intended to perfectly reproduce the original edition, considerable pains and effort had to be undertaken to correct fading and sometimes outright damage to existing proofs of this title. At times, this task is quite monumental, requiring an almost total "rebuilding" of some pages from digital proofs of multiple copies. Despite this, imperfections still sometimes exist in the final proof and may detract from the visual appearance of the text.

I hope you enjoy reading this book as much as I enjoyed making it available to readers again.

Jackson Chambers

CONTENTS.

ERRATA.

Page 266: Musketeer, omitted from Index.

The three cows, Rosa Lee, Rose of Dalkeith and Cinderella 2d, on page 298, entered as imported by J. H. Hall, Manitoba, is an error. They were imported by J. J. Hill, Esq., August 30. 1882, and are properly entered in his list on pages 303 and 304.

Page 362: Lovely Boy, omitted from Index.

Page 501: Burletta, called in England Dun Lady, not as printed.

Page 516: Barforth, entered as an imported cow, is the last cross in Devonshire.

PREFACE.

IT was very wisely determined by our ancestors to import representatives of the improved races of cattle already existing in Great Britain, as the foundation of our improved breeding, rather than attempt to form improved types for ourselves. And it has become American Shorn-horn usage to deny the claim of purity to any stock not descended from imported ancestry. There has very naturally arisen a loud call for a complete and handy list of all animals that have been imported, in order that breeders may have a ready means of learning the history and origin of the various tribes that constitute our herds, and to serve as a check against fraud and forgery. The present volume has been prepared at the cost of much labor, in the effort to meet this demand.

This catalogue professes to include the names and pedigrees, so far as still extant, of all Short-horn cattle that were imported into America up to the end of the year 1882. Although a large proportion of those imported in 1883 and the early months of 1884 have been included also, no claim of completeness is made in regard to them. Even during the period up to 1882 no doubt entire completeness and accuracy has not been obtained. It has proved impossible to obtain the details concerning certain importations; concerning others, the information that has been obtained is fragmentary; no doubt concerning others it may be erroneous. Petty errors have probably been unavoidable. Pains have not been spared, however, to present in all cases the fullest and most exact and most authentic information that could be obtained; and I shall be surprised to find that any considerable omissions have occurred or serious errors have been fallen into. I need not say that corrections and additions of all kinds will be most gratefully received, and made public; and breeders are earnestly urged to communicate to me all such criticisms as may tend to perfect the Second Edition.

PREFACE.

The arrangement that has been adopted will be readily understood at a glance. A chronological arrangement would have been, in many aspects, exceedingly valuable; but, on the whole, the alphabetical scheme seemed most likely to be useful. In the body of the book the importers are arranged in alphabetical order, while the index gives an alphabetical list of the animals themselves, with such information concerning them as seemed most needed. It is hoped that by this scheme the matter is presented in the handiest possible shape for actual use. Such animals as seemed, on good evidence, to have been imported, but the history of the importation of which it has been found impossible to obtain, have been classed by themselves at the end of the volume. After them will be found a similar list of such as purport to have been imported, but apparently were not. Some of these pedigrees may have arisen in error, others in fraud; none of them, in the present state of the evidence, can be accepted as true. Their descendants at this late day are, no doubt, Short-horns; but they cannot be accepted as Short-horns descending from imported ancestry.

It would be impracticable to return thanks by name to all who have given me substantial aid in compiling this volume. I must ask the numerous gentlemen who have responded so kindly and promptly to the many questions, and sometimes rather unreasonable demands I have made of them, to permit me to cheaply discharge my debt to them by this general mention. The time may come when I may be able to return their courtesy by a like readiness to give them aid; and I trust I may be enabled then to yield it as readily and as ungrudgingly as they have done to me. Meanwhile, I may be permitted to dedicate this book to them, in company with all lovers of Short-horn history and all friends of the Short-horn race.

WILLIAM WARFIELD.

LEXINGTON, KY., April, 1884.

WARFIELD'S HISTORY

OF

IMPORTED SHORT-HORNS.

1853 to 1856, R. A. ALEXANDER,

Woodburn, Woodford Co., Ky.

COWS.

SWEET MARY—Roan, calved Nov. 11, 1846, bred by Mr. Cattley, got by Rufus (6428), out of Sweet Looks by Prince of Wales (6348)—Vestris by Marton Comet (4409)—Vesta by Plato (2433)—Venus by Bedford Jr. (1701)—Vesta by Isaac (1129)—Rosabella by Northern Light (1281)—Old Rosabella by White Comet (1582)—Rose by Cattley's Bull (1798).

PEERESS (vol. 8, p. 335, E., under dam)—Roan, calved March 23, 1847, bred by Mr. Cattley, got by Lord Marmion (8244), out of Empress by Prince Albert (4791)—Countess by Cyrus (3538)—Empress by Bedford (68)—by Northern Light (1281)—by White Comet (1582).

MISS HUDSON (vol. 23, p. 17980, A. H. B., and vol. 10, p. 493, E.)—Red and white, calved in July, 1847, bred by Mr. Wiley, Brandsby, got by Hermes (8145), out of Mayoress by Carcase (3285)—Matron by Tyro (2781)—Miss Mason by Falstaff (1993)—No. 6 by Dr. Syntax (220)—Charles Cow by Charles (127)—Henry Cow by Henry (301)—Lydia by Favorite (252)—Nell by White Bull (421)—Fortune by Bolingbroke (86)—by Foljambe (263)—by Hubback (319)—bred by Mr. Maynard.

NIGHTINGALE (vol. 8, p. 444, E., under dam)—Roan, calved Dec. 10, 1847, bred by Mr. Kerrich, Arnold's, got by Prince Alfred (8422), out of Molly Bawn by The Lord of Hainault (6588)—Minna by Fergus (3782)—Starlight by Dandy (1902)—Moonshine by Oliver (2386)—by Blyth (797)—by Midas (435)—by Boughton (90)—by Windsor (698)—by R. Colling's son of Favorite (252).

GRACEFUL 2D (vol. 10, p. 389, E., under dam)—Red roan, calved Oct. 9, 1850, bred by Mr. Adkins, got by Earl of Dublin (10178), out of Graceful by Lycurgus (7180)—Marcia by Ranunculus (2479)—Sackbut by William (2840)—Clarion by Childers (1824)—No. 25 by Richard (1376)—by Jupiter (342)—by Charles (127)—by Windsor (698)—by Chilton (136)—by Colonel (152).

PEARLETTE (vol. 10, p. 520, E., under dam)—Red and white, calved Nov. 27, 1851, bred by S. E. Bolden, got by Benedict (7828), out of Pearl 2d by Senator (8551)—Pearl by Homer (2134)—Windermere by Emperor (1974)—Peeress by Snowdrop (2653)—Countess by Sir Charles (593)—Princess by St. Albans (1412)—Blossom by Cupid (177)—by Simon (590)—by Punch (531)—by Bolingbroke (86).

VICTORIA 20TH (vol. 11, p. 741, E.)—Roan, calved April 24, 1851, bred by R. Holmes, Ireland, got by Broken Horn (12500), out of Victoria 14th by Comus (12625)—Victoria 9th by Sir John Sinclair (5165)—Victoria 3d by 2d Comet (5101)—Victoria 2d by Belzoni (783)—Victoria by Satellite (1420)—No. 1 Chilton Sale by Cato (119)—Pope Cow by Pope (514)—by Favorite (252)—by White Bull (421)—by Favorite (252)—by Dalton Duke (188)—by R. Alcock's Bull (19)—by J. Smith's Bull (608)—by Jolly's Bull (337).

BONNY LASS (vol. 11, p. 345, E., under dam)—Roan, calved Jan. 11, 1852, bred by C. W. Goode, got by Earl of Dublin (10178) —Bonny Kate by Ivanhoe (9240)—Heartsease by Saladin (7469) —Pansy by Bloomsbury (3171)—Pasta by Red Rover (4906)—Seraphina by Wharfdale (1578)—by Palemon (479)—by Meteor (432)— by Western Comet (689)—by Favorite (252)—by Cupid (177)—by

Grandson of Bolingbroke (282)—by Foljambe (263)—by R. Alcock's Bull (19).

FILIGREE (vol. 11, p. 442, E., under dam)—White, calved Feb. 18, 1852, bred by Mr. Saunders, got by Abram Parker (9856), out of Fanchette by Petrarch (7329)—Fame by Raspberry (4875)—Farewell by Young Matchem (4422)—Flora by Isaac (1129)—by Young Pilot (497)—by Pilot (496)—by Julius Cæsar (1143).

LADY GULNARE (vol. 11, p. 523, E.)—Roan, calved March 22, 1852, bred by H. Ambler, got by Senator (8548), out of Gulnare by Norfolk (2377)—Medora by Ambo (1636)—Blossom by Memnon (2295)—by Pilot (496)—by Agamemnon (9)—by Burrell's Bull of Burdon (1768).

VALERIA OR VALINA (vol. 11, p. 741, E., under dam, as Valina)—Roan, calved March 29, 1854, bred by R. A. Alexander, gotten in England by Hopewell (10332), out of Victoria 20th by Broken Horn (12500)—Victoria 14th by Comus (12625)—Victoria 9th by Sir John Sinclair (5165)—Victoria 3d by 2d Comet (5101)—Victoria 2d by Belzoni (783)—Victoria by Satellite (1420)—No. 1 Chilton Sale by Cato (119)—Pope Cow by Pope (514)—Flora by Favorite (252)—Nymph by White Bull (421)—Lily by Favorite (252)—Miss Lax by Dalton Duke (188)—Lady Maynard by Alcock's Bull (19)—by J. Smith's Bull (608)—by Jolly's Bull (337).

SCIOTA—Roan, calved April 16, 1854, bred by Mr. Ambler, got by Lancaster Comet (11663), out of Lavender 3d by St. Albans (7464)—Lavender 2d by Queen's Roan (7380)—by Will Honeycomb (5660)—by Spectator (2688)—by Albion (1619)—by Lancaster (360)—by Son of Windsor (698)—by Comet (155).

LAURA—Roan, calved June 17, 1854, bred by Mr. R. A. Alexander, got in England by Zealot (14046), out of Miss Towneley by Brunell (9999)—Lady Laura by Laudable (9282)—Laura by Petrarch (7329)—Fair Spots by Sir Thomas Fairfax (5196)—Spots by Garton (2052)—Latona by Harold (291)—Strawberry by Comet (170)—Venus by Badsworth (47)—by Driffield (223)—bred by Sir G. Strickland.

MINNA—Red, calved Sept. 19, 1854, bred by Mr. Fawkes, Farnley Hall, Otley, got by Bridegroom (11203), out of Moss Rose 3d by Sir Walter (2639)—Moss Rose by Belvedere 2d (3127)—by Waterloo (2816)—by Barmpton (5774)—by Kitt (2179)—by Kitt (2179)—by Page's Bull (6269)—by Middleton's Bull (438).

PRUNELLA (vol. 11, p.650, E., under dam)—Roan, calved Dec. 4, 1854, bred by Mr. Bolden, Hyning, got by Duke of Bolton (12738), out of Prune by Lord Lieutenant (11734)—Pearl 2d by Senator (8551)—Pearl by Homer (2134)—Windermere by Emperor (1974)—Peeress by Snowdrop (2653)—Countess by Sir Charles (593)—Princess by St. Albans (1412)—Blossom by Cupid (177)—by Simon (590)—by Punch (531)—by Bolingbroke (86). Sold at sale of June 5, 1859.

LADY VALENTINE (vol. 11, p. 733, E., under dam)—Red, calved Oct. 8, 1853, bred by H. Smith, Drax Abbey, got by Harbinger (10297), out of Vellum by Abram Parker (9856)—Miss Valentine by Beggarman (3118)—Victoria by Duke (3629)—Venus by Young General (3866)—Maria by Western Comet (689) —by General (272)—by Marquis (407)—by Simon (590)—by Traveler (655)—by Lame Bull (357).

FRANCES FAIRFAX (vol. 11, p. 436, E., under dam)—White, calved Nov. 2, 1853, bred by Mr. Ambler, got by Crusade (7938), out of Fair Frances by Sir Thomas Fairfax (5196)—Feldom by Young Colling (1843)—Lily by Red Bull (2838)—Lily by Son of Hollings (2131)—by Partner (2409)—by R. Alcock's Bull (19).

ZARA—Roan, calved Nov. 27, 1853, bred by Mr. Fawkes, Farnley Hall, Otley, got by Bridegroom (11203), out of Lady Zarifa (vol. 10, p. 440, E.) by Laudable (9282)—Zulieka by Norfolk (2377)—Medora by Ambo (1636)—Blossom by Memnon (2298)—by Pilot (496)—by Agamemnon (9)—by Burrell's Bull of Burdon (1768).

LADY DERBY—Red and white, calved Jan. 29, 1854, bred by R. A. Alexander, begotten in England by Earl Derby (10177), out of Forget-me-not (vol. 10, p. 374, E.) by 2d Cleveland Lad (3408) —Fancy by Duke of Northumberland (1940)—Fanny by Short

Tail (2621)—Fletcher 2d by Belvedere (1706)—Fletcher by son of Young Wynyard (2859)—descended from J. Brown's Red Bull (97).

LENA (vol. 11, p. 526, E., under dam)—White, calved Dec. 6, 1853, bred by Col. Towneley, gotten in England by Jasper (11609), out of Lady Laura by Laudable (9282)—Laura by Petrarch (7329)—Fair Spots by Sir Thomas Fairfax (5196)—Spots by Garton (2052)—Latona by Harold (291)—Strawberry by Comet (170)—Venus by Badsworth (47)—by Driffield (223).—bred by Sir G. Strickland.

CONSTANCE—Red, calved March 11, 1854, bred by Mr. Fawkes, got by Bridegroom (11203), out of Cherry Ripe by Sir Walter (2639)—Young Cherry by Young Waterloo (8757)—Cherry by Waterloo (2816)—Old Cherry by Waterloo (2816)—by Kitt (2179)—by Kitt (2179)—by Page's Bull (6269)—by Middleton's Bull (438).

PRUNE (vol. 11, p. 650, E.)—Roan, calved March 29, 1852, bred by Mr. Bolden, got by Lord Lieutenant (11734), out of Pearl 2d by Senator (8551)—Pearl by Homer (2134)—Windermere by Emperor (1974)—Peeress by Snowdrop (2653)—by Sir Charles (593)—by St. Albans (1412)—by Cupid (177)—by Simon (590)—by Punch (531)—by Bolingbroke (86).

COQUETTE—Roan, calved Jan., 1853, bred by W. Dickerson, got by Monk (11824), out of Jilt by Tom of Lincoln (8714)—Flirt by Purity (8444)—Florence by Berryman (3143)—Fancy by Remus (2524)—White Rosette by Juniper (1114)—Rosette by White Comet (1582)—Young Rose by Wright's Grandson of Favorite (2073)—Rose by Cattley's Gray Bull (1798).

LYDIA LANGUISH (vol. 11, p. 541, E., under dam)—Red and white, calved June 22, 1853, bred by H. Smith, got by Duke of Gloster (11382), out of Lavender by Dan O'Connell (3557)—Lily by Brutus (1752)—Violet by Frederick (1060)—Vestris by Cato (1794)—Verbena by son of Wellington (679)—bred from the herd of Mr. Robertson of Ladykirk.

SUNRISE (vol. 11, p. 712 E., under dam)—White, calved Sept. 14, 1853, bred by Mr. Saunders, got by Abram Parker (9856), out of Sunbeam by Euclid (9097)—Short Tail by 2d Duke of Northumberland (3646)—Spritule by Young Seagull (5100)—by Sultan (1485)—by Crusader (934)—by Chillingham (3374).

SALLY-IN-OUR-ALLEY—Red, calved July 17, 1853, bred by Mr. Fawkes, Farnley Hall, Otley, got by Bridegroom (11203), out of Sally O'Moore 3d by Sir Walter (2639)—Sally O'Moore by Young Remus (2523)—by Remus (550)—Grizzle by Hollings (2131)—Lady by His Honor (2126)—Redley by Partner (2409)—Old Redley by Hutton's Bull (2145)—Lofty, from R. Alcock's stock.

ROSABELLE—Red, calved July 20, 1853, bred by Mr. Fawkes, Farnley Hall, Otley, got by Bridegroom (11203), out of Moss Rose 3d by Sir Walter (2639)—Moss Rose by Belvedere 2d (3127)—by Waterloo (2816)—by Barmpton (5774)—by Kitt (2179)—by Kitt (2179)—by Page's Bull (6269)—by Middleton's Bull (438).

CANNY (vol. 11, p. 379, E., under dam)—Roan, calved Sept. 25, 1853, bred by Mr. Hall, Kiveston Hall, got by Will Watch (12307), out of Comely by Hornby 2d (9223)—Canary by Muley (4519)—Comedy by Kiveton Reformer (4164)—Comical by Topper (2768)—by Wellington (2825)—by Wonderful (700)—by Meteor (431)—by Windsor (698)—by son of Favorite (252).

ALBERTA—Red and white, calved in Feb., 1853, bred by Mr. B. Fuller, got by Holcomb (10324), out of Victoria by Diamond (5918)—Vestris by Young Belshazzar (3122)—Verbena by Noble (4578)—Violet by Monarch (4495)—Julia by Invalid (4076)—Lady Sarah by Satellite (1420)—Portia by Cato (119)—by Jupiter (342)—by George (273)—by Chilton (136)—by Irishman (329)—by B. (45).

CHRISTINE CATTLEY—Roan, calved March 2, 1853, bred by R. Cattley, Brandsby, got by DeGrey (11346), out of Christiana (vol. 10, p. 303, E.) by Lord Marmion (8244)—Vestris by Marton Comet (4409)—Vesta by Plato (2433)—Venus by Bedford Jr. (1701)—Vesta by Isaac (1129)—Rosabella by Northern Light (1281)—Old Rosabella by White Comet (1582)—Rose by Cattley's Gray Bull (1798).

DUCHESS OF AIRDRIE (vol. 20, p. 15598, A., and vol. 11, p. 419, E., under dam)—Red and white, calved Aug. 2, 1853, bred by R. A. Alexander or Col. Towneley, got by 2d Duke of Athol (11376) or Valiant (10989), out of Duchess of Athol by 2d Duke of Oxford (9046)—Duchess 54th by 2d Cleveland Lad (3408)—Duchess 49th by Short Tail (2621)—Duchess 30th by 2d Hubback (1423)—Duchess 20th by 2d Earl (1511)—Duchess 8th by Marske (418)—Duchess 2d by Ketton 1st (709)—Duchess 1st by Comet (155)—Duchess by Favorite (252)—by Daisy Bull (186) —by Favorite (252)—by Hubback (319)—by J. Brown's Red Bull (97).

NOTE.—Mr. R. A. Alexander's catalogue of 1854, on p. 23, No. 73, has Duchess of Airdrie recorded as above. The E. H. B. omits Valiant.

VELLUM (vol. 20, p. 16237 A., and vol. 9, p. 488, E., under dam)—Roan, calved April 1, 1849, bred by Sir C. Tempest, got by Abram Parker (9856), out of Miss Valentine by Beggarman (3118) —Victoria by Duke (3629)—Venus by Young General (3866)— Maria by Western Comet (689)—Lovely by General (272)—Bright Eyes by Marquis (407)—by Simon (590)—by Traveler (655)—by R. Colling's Lame Bull (357).

PRINCESS 4TH (vol. 10, p. 547, under dam)—Roan, calved April 28, 1850, bred by Mr. Malins, got by Revolution (10713), out of Red Duchess 3d by Nonsuch (4581)—Red Duchess by Bachelor (1666)—Duchess by Wellington (683)—Bright Eyes by Admiral (4)—by Sir Harry (5155)—by Colonel (152)—by grandson of Hubback (319)—by son of Hubback (319).

LADY BARRINGTON 13TH (vol. 10, p. 426, E., under dam) —Roan, calved Aug. 18, 1850, bred by R. Bell, got by 4th Duke of York (10167), out of Lady Barrington 5th by 4th Duke of Northumberland (3649)—Lady Barrington 3d by Cleveland Lad (3407)— Lady Barrington 2d by Belvedere (1706)—Lady Barrington by son of Herdsman (304)—Young Alicia by Wonderful (700)—Old Alicia by Alfred (23)—by Young Favorite (6994).

KATHLEEN BAWN—Red and white, calved in Sept., 1853, bred by Mr. Fuller, got by Holcomb (10324), out of Molly Bawn (vol. 10, p. 499, E.) by The Lord of Hainault (6588)—Minna by Fergus (3782)—Starlight by Dandy (1902)—Moonshine by Oliver (2386)—Resplendent by Blyth (797)—by Midas (435)—by Boughton (90)—by Windsor (698)—by son of Favorite (252).

MISS WILEY 2D (vol. 11, p. 589, E., under dam)—Roan, calved Oct. 15, 1852, bred by S. Wiley, Brandsby, got by Prince Royal (8428), out of imp. Miss Hudson by Hermes (8145)—Mayoress by Carcase (3285)—Matron by Tyro (2781)—Miss Mason by Falstaff (1993)—No. 6 by Syntax (220)—Charles Cow by Charles (127)—Henry Cow by Henry (301)—Lydia by Favorite (252)—Nell by White Bull (421)—Fortune by Bolingbroke (86)—by Foljambe (263)—by Hubback (319)—bred by Mr. Maynard.

JESSY 3D—Red and white, calved Oct. 21, 1852, bred by M. Faviell, got by Duke of Albany (10149), out of Jessy (vol. 10, p. 414, E.) by Cramer (6907)—Lady Jane by Plenipo (4724)—Lady Ann by Childers (1824)—Miss Leighton by Umpire (2783)—Young Cowslip by Ratify (2481)—Cowslip by Wellington (680)—by Favorite (252)—by Punch (531).

MISS TOWNELEY (vol. 11, p. 526, E., under dam)—Roan, calved Jan. 7, 1853, bred by Mr. Tanqueray, got by Brunell (9999), out of Lady Laura by Laudable (9282)—Laura by Petrarch (7329)—Fair Spots by Sir Thomas Fairfax (5196)—Spots by Garton (2052)—Latona by Harold (291)—Strawberry by Count (170)—Venus by Badsworth (47)—by Driffield (223)—bred by Sir G. Strickland.

EMMA 2D (vol. 11, p. 431, E., under dam)—Red and white, calved Jan. 30, 1853, bred by R. Cattley, got by De Grey (11346), out of Empress by Prince Albert (4791)—Countess by Cyrus (3538)—Old Empress by Bedford (68)—by Northern Light (1281)—by White Comet (1582).

DORIA PICOLA—Red and white, calved Feb. 13, 1853, bred by M. Faviell, got by Duke of Albany (10149), out of Doria (vol. 10, p. 336, E.) by The Duke (8676)—Clementina by Clementi (3399)—Rosa by Raspberry (4875)—Audrey by Argus (759)—by Admiral (5)—by Ronald (563)—by Cecil (120).

MARY CATTLEY—White, calved Feb. 17, 1853, bred by R. A. Alexander, calved in America but begotten in England by Puritan (9523), out of Sweet Mary by Rufus (6428)—Sweet Looks by Prince of Wales (6348)—Vestris by Marton Comet (4409)—Vesta by Plato (2433)—Venus by Bedford Jr. (1701)—Vesta by Isaac (1129)—Rosabella by Northern Light (1281)—Old Rosabella by White Comet (1582)—Rose by Cattley's Gray Bull (1798).

JOYFUL (vol. 10, p. 406, E., under dam)—Red, calved Sept. 17, 1851, bred by Mr. Tanqueray, got by Lycurgus (7180), out of Jacintha by Fawsley (6004)—Junta by Warden (5595)—Joyance by Javelin (4093)—Joy by Blyth (797)—Janette by Wellington (684)—by Phenomenon (491)—by Favorite (252)—by Favorite (252)—by Favorite (252).

EMMA (vol. 10, p. 350, E., under dam)—Roan, calved Dec. 13, 1851, bred by Mr. Cattley, got by Fair Eclipse (11456), out of Empress by Prince Albert (4791)—Countess by Cyrus (3538)—Old Empress by Bedford (68)—by Northern Light (1281)—by White Comet (1582).

JUBILEE 2D (vol. 11, p. 506, E., under dam)—Roan, calved Jan. 12, 1852, bred by Mr. Tanqueray, got by Marquis of Rockingham (10506), out of Jubilee by Lycurgus (7180)—Jacintha by Fawsley (6004)—Junta by Warden (5595)—Joyance by Javelin (4093)—Joy by Blyth (797)—Janette by Wellington (684)—by Phenomenon (491)—by Favorite (252)—by Favorite (252)—by Favorite (252).

JUNIATA (vol. 11, p. 495, E., under dam)—White, calved June 26, 1852, bred by Mr. Tanqueray, got by Lord Marquis (10459), out of Jardine by Lord Warden (7167)—Jacintha by Fawsley (6004)—Junta by Warden (5595)—Joyance by Javelin (4093)—Joy by Blyth (797)—Janette by Wellington (684)—by Phenomenon (491)—by Favorite (252)—by Favorite (252)—by Favorite (252).

FINELLA (vol. 4, p. 353, A., and vol. 11, p. 449, E., under dam)—Red, calved June 30, 1852, bred by S. E. Bolden, got by

Grand Duke (10284), out of Fay by Foig-a-Ballagh (8082)—Fame by Raspberry (4875)—Farewell by Young Matchem (4422)—Flora by Isaac (1129)—by Young Pilot (497)—by Pilot (496)—by Julius Cæsar (1143). Sold to S. Thorne, N. Y., at sale at Woodburn, June 2, 1858, for $600.

GRISI (vol. 11, p. 733, E., under dam)—Red, calved July 2, 1852, bred by S. E. Bolden, got by Grand Duke (10284), out of Vaudeville by Humber (7102)—Guitar by Zenith (5702)—Serenade by Roman (2561)—Clarion, by Childers (1824)—No. 25 by Richard (1376)—by Jupiter (342)—by Charles (127)—by Windsor (698)—by Chilton (136)—by Colonel (152).

LADY MARY 2D (vol. 11, p. 528, E., under dam)—Red and white, calved Sept. 14, 1852, bred by M. Faviell, got by Duke of Albany (10149), out of Lady Mary by Frederick (9136)—Red Rose by Alexis (1628)—Beeswing by Bellerophon (3119)—Jessy by Matchem 3d (4420)—Clara by Alderman (2976)—by Waterloo (2816)—by Young Wynyard (2859)—by son of Simon (590)—by Styford (629).

DUCHESS OF ATHOL (vol. 20, p. 15600, A., and vol. 10, p. 338, E., under dam)—Red and white, calved Aug. 19, 1850, bred by Mr. Towneley, Towneley Park, Burnley, got by 2d Duke of Oxford (9046), out of Duchess 54th by 2d Cleveland Lad (3408)—Duchess 49th by Short Tail (2621)—Duchess 30th by 2d Hubback (1423)—Duchess 20th by 2d Earl (1511)—Duchess 8th by Marske (418)—Duchess 2d by Ketton 1st (709)—Duchess 1st by Comet (155)—Duchess by Favorite (252)—by Daisy Bull (186)—by Favorite (252)—by Hubback (319)—by J. Brown's Red Bull (97).

BEAUTY (vol. 10, p. 249, E., under dam)—Roan, calved Oct. 27, 1850, bred by Mr. Lowndes, got by Attraction (9912), out of Alice by Colonel (8967)—Moss Rose by Locomotive (4242)—Adelaide by Cleveland (3403)—by Young Eryholme (1981)—by Wonderful (700)—by Merlin (429)—by Alfred (23)—by Cupid (177)—by Suwarrow (636).

WILD EYES JENNY (vol. 10, p. 633, E., under dam)—Red and white, calved Oct. 30, 1850, bred by M. Faviell, got by 4th Duke of York (10167), out of Wild Eyes 17th by 2d Duke of Northumberland (3646)—Wild Eyes 5th by Short Tail (2621)—Wild Eyes by Emperor (1975)—by Wonderful (700)—by Cleveland (145)—by Butterfly (104)—by Hollon's Bull (313)—by Mowbray's Bull (2342) —by Masterman's Bull (422)—descended from the stock of M. Dobison.

ROSE (vol. 10, p. 563, E., under dam)—Roan, calved Dec. 7, 1850, bred by H. Combe, got by Puritan (9523), out of Roseleaf by Earl of Durham (5965)—White Rosalind by Pedestrian (7321)— Red Rosalind by Edrom (1956)—Young Rosalind by Scipio (1421) —Rosalind by Hector (1104)—Rose by Midas (435)—Red Rose by Marquis (407)—by Chilton (136)—Mason's Red Rose by Ben (70)— bought by C. Colling of J. Newby.

BUTTERCUP (vol. 10, p. 264, E., under dam)—Red, calved Dec., 1850, bred by H. Combe, got by Puritan (9523), out of Baroness by Baron Warlaby (7813)—Clematis by Clementi (3399)—Farewell by Young Matchem (4422)—Flora by Isaac (1129)—by Young Pilot (4702)—by Pilot (496)—by Julius Cæsar (1143).

LOBELIA (vol. 10, p. 454, E., under dam)—Roan, calved Feb. 2, 1851, bred by H. Combe, got by Puritan (9523), out of Lily White by Vanish (5546)—White Rose by Orator (2390)—by Lenton (4205) —by a bull of Mr. Wetherell's—Maria by Spanker (1471)—Stella by Snowball (611)—Strawberry by Adonis (1612)—Neesham by son of Phenomenon (491)—by Punch (531).

MAID MARION 2D—Red, calved Aug. 12, 1853, bred by R. A. Alexander, got in England by Lord John (11728), out of Maid Marion by Robin Hood (9555)—Lily by Young Zealot (8797)—Lily by Young Vandyke (8733)—Duchess by Young Spectator (8619)— by Phantassie (8389)—by Young Rockingham (8498).

MISS WILEY 3D (vol. 11, p. 589, E.)—Red and white, calved Sept. 21, 1853, bred by R. A. Alexander, got in England by Gray Friar (9172), out of imp. Miss Hudson by Hermes (8145)—Mayoress by Carcase (3285)—Matron by Tyro (2781)—Miss Mason by Falstaff (1993)—No. 6 by Dr. Syntax (220)—Charles Cow by Charles (127)—Henry Cow by Henry (301)—Lydia by Favorite (252)—Nell by White Bull (421)—Fortune by Bolingbroke (86)—by Foljambe (263)—by Habback (319)—bred by Mr. Maynard.

ABIGAIL—Red and white, calved May 30, 1852. bred by Harvey Combe, Cobham Park, got by Loyalist (10479). out of Albertha (vol. 10, p. 247, E.) by Sulla (7564)—Hernia by Buckingham (3239)—Hypolita by Belshazzar (1704)—Giantess by Mr. Hunter of Gilling's Bull. Sold at sale of June 2, 1858, to E. L. Davison, Springfield, Ky., for $205.

MORLINA (vol. 22, p. 17340, A. H. B.)—Got by Lilyvick (10421), out of Lily (vol. 10, p. 451, E.) by Young Zealot (8797)—Lily by Young Vandyke (8733)—Duchess by Young Spectator (8619)—by Phantassie (8389)—by Young Rockingham (8498).

MINERVA 3D (vol. 11, p. 582, E.)—Red, calved Nov., 1850, bred by Mr. Slade, Kemnal House, got by St. Martin (8525), out of Minerva by Prince Ernest (4818)—Medusa by Mowthorpe (2343)—Magic by Wallace (5586)—by Wellington (2824)—by Marmion (406)—Daphne by Merlin (430)—Nell Gwynne by Layton (366)—Nell Gwynne by Phenomenon (491)—Princess by Favorite (252)—by Favorite (252)—by Hubback (319)—by Snowdon's Bull (612)—by Waistell's Bull (669)—by Masterman's Bull (422)—by Studley Bull (626).

LADY SHERWOOD (vol. 11, p. 582, E., under dam)—Roan, calved Sept. 27, 1853, bred by Mr. Tanqueray, Hendon House, got by 5th Duke of York (10168), out of Minerva 3d by St. Martin (8525)—Minerva by Prince Ernest (4818)—Medusa by Mowthorpe (2343)—Magic by Wallace (5586)—by Wellington (2824)—by Marmion (406)—Daphne by Merlin (430)—Nell Gwynne by Layton (366)—Nell Gwynne by Phenomenon (491)—Princess by Favorite (252)—by Favorite (252)—by Hubback (319)—by Snowdon's Bull (612)—by Waistell's Bull (669)—by Masterman's Bull (422)—by Studley Bull (626).

JUBILEE (vol. 9, p. 402, E., under dam)—Red, calved Nov. 14, 1848, bred by Mr. Beasley, got by Lycurgus (7180), out of Jacinth by Fawsley (6004)—Jacquette by Javelin (4093)—Jonquille by Blyth (797)—Janette by Wellington (684)—by Phenomenon (491)—by Favorite (252)—by Favorite (252)—by Favorite (252).

MAID MARION (vol. 9, p. 449, E., under dam)—Roan, calved Feb. 13, 1849, bred by Mr. B. Fuller, Holcombe, Dorking, got by Robin Hood (9555), out of Lily by Young Zealot (8797)—Lily by Young Vandyke (8733)—Duchess by Young Spectator (8619)—by Phantassie (8389)—by Young Rockingham (8498).

FORGET-ME-NOT (vol. 10, p. 374, E.)—Roan, calved Feb. 5, 1848, bred by R. Bell, got by 2d Cleveland Lad (3408), out of Fancy by Duke of Northumberland (1940)—Fanny by Short Tail (2621)—Fletcher 2d by Belvedere (1706)—Fletcher by son of Young Wynyard (2859)—descended from J. Brown's Red Bull (97).

TIZZY (vol. 10, p. 499, E.)—Roan, calved May, 1850, bred by B. Fuller, got by Robin Hood (9555), out of Molly Bawn by the Lord of Hainault (6588)—Minna by Fergus (3782)—Starlight by Dandy (1902)—Moonshine by Oliver (2386)—Resplendent by Blyth (797)—by Midas (435)—by Boughton (90)—by Windsor (698)—by son of Favorite (252).

BEATRICE (vol. 10, p. 531, E., under dam)—Red and white, calved June 28, 1850, bred by Mr. Lowndes, got by Attraction (9912), out of Primrose by Egremont (9075)—Orinda by Marmion (4383)—Adelaide by Cleveland (3403)—Adelaide by Young Ery-holme (1981)—by Wonderful (700)—by Merlin (429)—by Alfred (23)—by Cupid (177)—by Suwarrow (636). Sold at sale June 2, 1858, for $155.

ROSE 2D—White, calved Nov. 14, 1853, bred by R. A. Alexander, got in England by The Beau (12182), out of Rose by Puritan (9523)—Roseleaf (vol. 10, p. 563, E.) by Earl of Durham (5965) —White Rosalind by Pedestrian (7321)—Roan Rosalind by Edrom (1956)—Young Rosalind by Scipio (1421)—Rosalind by Hector

(1104)—Rose by Midas (435)—Red Rose by Marquis (407)—by Chilton (136)—Mason's Red Rose by Ben (70)—bought by C. Colling of Mr. Newby.

ALICE WILEY (vol. 10, p. 493, E., under dam, as Alice)—Red and white, calved July 25, 1850, bred by Samuel Wiley, Brandsby, got by Rumour (7456), out of imp. Miss Hudson by Hermes (8145)—Mayoress by Carcass (3285)—Matron by Tyro (2781)—Miss Mason by Falstaff (1993)—No. 6 by Dr. Syntax (220) —Charles Cow by Charles (127)—Henry Cow by Henry (301)— Lydia by Favorite (252)—Nell by White Bull (421)—Fortune by Bolingbroke (86)—by Foljambe (263)—by Hubback (319)—bred by Mr. Maynard. Sold at sale of June 2, 1858.

VICTORIA (vol. 9, p. 607, E., under dam)—Roan, calved April 2, 1848, bred by H. Combe, got by Diamond (5918), out of Vestris by Young Belshazzar (3122)—Verbena by Noble (4578)—Violet by Monarch (4495)—Julia by Invalid (4076)—Lady Sarah by Satellite (1420)—Portia by Cato (119)—by Jupiter (342)—by George (273) —by Chilton (136)—by Irishman (329)—by B. (45).

FILBERT (vol. 9, p. 365, E., under dam)—Roan, calved Sept. 4, 1848, bred by R. Bell, got by 2d Cleveland Lad (3408), out of Felicia by 4th Duke of Northumberland (3649)—Fanny by Short Tail (2621)—Fletcher 2d by Belvedere (1706)—Fletcher by a son of Young Wynyard (2859)—descended from James Brown's Red Bull (97).

LADY LAURA (vol. 9, p. 441, E., under dam)—Roan, calved Jan. 25, 1849, bred by Mr. Fawkes, Farnley Hall, got by Laudable (9282), out of Laura by Petrarch (7329)—Fair Spots by Sir Thomas Fairfax (5196)—Spots by Garton (2052)—Latona by Harold (291) —Strawberry by Count (170)—by Badsworth (47)—by Driffield (223)—bred by Sir G. Strickland.

BULLS.

DUKE OF AIRDRIE (12730)—Red and white, calved Aug. 4, 1854, bred by R. A. Alexander, Airdrie House, Airdrie, Scotland, got by Duke of Gloster (11382), out of Duchess of Athol by 2d Duke of Oxford (9046)—Duchess 54th by 2d Cleveland Lad (3408) —Duchess 49th by Short Tail (2621)—Duchess 30th by 2d Hubback (1423)—Duchess 20th by The 2d Earl (1511)—Duchess 8th by Marske (418)—Duchess 2d by Ketton 1st (709)—Duchess by Comet (155)—by Favorite (252)—by Daisy Bull (186)—by Favorite (252)—by Hubback (319)—by J. Brown's Red Bull (97).

NAPIER—Red and white, calved Sept. 15, 1855, bred by R. A. Alexander, got in England by Zealot (14046), out of imp. Nightingale by Prince Alfred (8422)—Molly Bawn by The Lord of Hainault (6588)—Minna by Fergus (3782)—Starlight by Dandy (1902)— Moonshine by Oliver (2386)—Resplendent by Blyth (797)—by Midas (435)—by Boughton (90)—by Windsor (695)—by Mr. Coliing's son of Favorite (252). Sold at public sale Sept. 3, 1856, to Abram Vanmeter, Sr., of Fayette Co., Ky., for $105.

LANGTON 3061—Roan, calved Nov. 6, 1855, bred by R. A. Alexander, got in England by Baron Warlaby (7813), out of Lady Gulnare by Senator (8548)—Gulnare by Norfolk (2377)—Medora by Ambo (1636)—Blossom by Memnon (2295)—by Pilot (496)— by Agamemnon (9)—by Burrell's Bull of Burdon. Sold at public sale Sept. 3, 1856, to Dr. G. W. Perrin, Harrison Co., Ky.

PATRICK—Roan, calved Nov. 10, 1855, bred by R. A. Alexander, got in England by 2d Grand Duke (12961), out of Prune by Lord Lieutenant (11734)—Pearl 2d by Senator (8551)—Pearl by Homer (2134)—Windermere by Emperor (1974)—Peeress by Snowdrop (2653)—by Sir Charles (593)—Princess by St. Albans (1412)— Blossom by Cupid (177)—by Simon (590)—by Punch (531)—by Bolingbroke (86). Sold at public sale Sept. 3, 1856, to J. Allen, Fayette Co., Ky.

RAYMOND 2096—Red, calved Nov. 30, 1855, bred by R. A. Alexander, got in England by Robinson Crusoe (13610), out of

Rosabelle by Bridegroom (11203)—Moss Rose 3d by Sir Walter (2639)—Moss Rose by Belvedere 2d (3127)—by Waterloo (2816)—by Barmpton (5774)—by Kitt (2179)—by Kitt (2179)—by Page's Bull (6269)—by Middleton's Bull (438). Sold at public sale Sept. 3, 1856, to Shakers, Pleasant Hill, Ky., for $360.

SALADIN 2168—Red, calved December 7, 1855, bred by R. A. Alexander, got in England by Robinson Crusoe (13610), out of Sally-in-our-Alley by Bridegroom (11203)—Sally O'Moore 3d by Sir Walter (2639)—Sally O'Moore by Young Remus (2523)—by Remus (550)—Grizzle by Hollings (2131)—Lady by His Honor (2126)—Ridley by Partner (2409)—Old Ridley by Hutton's Bull (2145)—Lofty from Mr. Alcock's stock. Sold at public sale Sept. 3, 1856, to Newton Craig of Scott Co., Ky., for $500.

LORD JOHN (11738)—Roan, calved Jan. 23, 1851, bred by J. H. Downs, Grays, Essex, got by Norfolk (9442), out of Lady Elizabeth by Earl of Essex (6955)—Gainfordine by Gainford (2044)—Mary by Waterloo (2816)—Strawberry by Pyramid (4852)—by a grandson of Cupid (177)—by a son of Cupid (177)—by Cupid (177).

2D DUKE OF ATHOL (11376)—Red roan, calved Sept. 10, 1851, bred by Col. Towneley, Towneley Park, got by Lord George (10439), out of Duchess 54th by 2d Cleveland Lad (3408)—Duchess 49th by Short Tail (2621)—Duchess 30th by 2d Hubback (1423)—Duchess 20th by The 2d Earl (1511)—Duchess 8th by Marske (418)—Duchess 2d by Ketton 1st (709)—Duchess 1st by Comet (155)—by Favorite (252)—by Daisy Bull (186)—by Favorite (252)—by Hubback (319)—by J. Brown's Red Bull (97).

GRAND MASTER (12968)—Roan, calved Jan. 17, 1852, bred by F. H. Fawkes, Farnley Hall, got by Lord Marquis (10459), out of Gretna by The Stuart (7623)—Gulnare by Norfolk (2377)—Medora by Ambo (1636)—Blossom by Memnon (2295)—own sister to Isabella by Pilot (496)—by Agamemnon (9)—by Burrell's Bull of Burdon.

BARON MARTIN (12444)—Roan, calved Feb. 4, 1852, bred by
R. Holmes, Moycashel, Kilbeggan, got by Baron Warlaby (7813),
out of Victoria 3d by 2d Comet (5101)—Victoria 2d by Belzoni
(783)—Victoria by Satellite (1420)—Mason's No. 1 by Cato (119)—
Pope Cow by Pope (514)—Flora by Favorite (252)—Nymph by
White Bull (421)—Lily by Favorite (252)—Miss Lax by Dalton
Duke (188)—Lady Maynard by R. Alcock's Bull (19)—by J. Smith's
Bull (608)—by Jolly's Bull (337).

FANTACHINI (12862)—Red and white, calved June 18, 1852,
bred by F. H. Fawkes, Farnley Hall, got by Beaufort (9943), out of
Lady Fanchette by Laudable (9282)—Fanchette by Austerlitz (3063)
—Fair Sovereign by Sir Thomas Fairfax (5196)—Superb by Young
Colling (1843)—by Young Remus (2523)—by Remus (550)—by
Greathead's Gray Bull (3936)—by Ellerton's Roan Bull (3708).

MICKEY FREE 8626—Roan, calved Aug. 30, 1852, bred by F.
H. Fawkes, Farnley Hall, got by Lord Marquis (10459), out of Moss
Rose 3d by Sir Walter (2639)—Moss Rose by Belvedere 2d (3127)—
by Waterloo (2816)—by Barmpton (5774)—by Kitt (2179)—by
Kitt (2179)—by Page's Bull (6269)—by Middleton's Bull (438).

DOCTOR BUCKINGHAM (14405)—Red, calved March 26,
1853, bred by H. Ambler, Watkinson Hall, got by Hopewell (10332),
out of Bloom by Buckingham (3239)—Hawthorn Blossom by Leonard (4210)—Blossom 3d by Young Red Rover (4905)—Blossom by
Isaac (1129)—by Pilot (496)—by Albion (14).

EL HAKIM (15984)—Red roan, calved Jan. 28, 1853, bred by
Mr. Bolden, got by Grand Duke (10284), out of Fame by Raspberry
(4875)—Farewell by Young Matchem (4422)—Flora by Isaac (1129)
—by Young Pilot (497)—by Pilot (496)—by Julius Cæsar (1143).

SANTON 2174—White, calved Dec. 8, 1855, bred by R. A. Alexander, got in England by Zealot (14046), out of Sweet Mary by
Rufus (6428)—Sweet Looks by Prince of Wales (6348)—Vestris by
Marton Comet (4409)—Vesta by Plato (2433)—Venus by Bedford

Jr. (1701)—Vesta by Isaac (1129)—Rosabella by Northern Light (1281)—Old Rosabella by White Comet (1582)—Rose by Cattley's Gray Bull (1798). Sold Sept. 3, 1856, at public sale to Shakers, Union Co., Ky., for $115.

VICTOR—White, calved Dec. 12, 1855, bred by R. A. Alexander, got in England by Vatican (12260), out of Vellum by Abraham Parker (9856)—Miss Valentine by Beggarman (3118)—Victoria by Duke (3629)—Venus by Young General (3866)—Maria by Western Comet (689)—by General (272)—by Marquis (407)—by Simon (590)—by Traveler (655)—by Lame Bull (357). Sold Sept. 3, 1856, at public sale to J. M. Trimble, Hillsboro, Ohio, for $450.

WOODBURN 2391 — Red, calved Sept., 1855, bred by Mr. Mulins, Thelsford, got by The Prior (13870), out of Princess 4th by Revolution (10713), &c., as in Princess 4th, p. 11. Sold to the Shakers, Pleasant Hill, Ky.

WOODBURN DUKE 2392—Calved Nov. 10, 1855, bred by Mr. Boldin, got by 2d Grand Duke (12961), out of Prune by Lord Lieutenant (11734), &c., as in Prune, p. 9. Sold to Messrs. J. & A. Allen, Fayette Co., Ky.

THE PRIEST 6246—Roan, calved April 19, 1856, bred by R. A. Alexander, got in England by The Prior (13870), out of Graceful 2d by Earl of Dublin (10178)—Graceful by Lycurgus (7180)—Marcia by Ranunculus (2479)—Sackbut by William (2840)—Clarion by Childers (1824)—No. 25 by Richard (1376)—by Jupiter (342)—by Charles (127)—by Windsor (698)—by Chilton (136)—by Colonel (152).

1882. A. J. ALEXANDER AND L. COMBS.

Importation of A. J. Alexander, of Woodburn, Woodford Co., Ky., and Leslie Combs, of Woodford Co., Ky., by ship Palestine. Landed at Boston March 2, 1882. Selected in England by L. Combs. Sold at public auction June 24, 1882, at Woodburn, Woodford Co., Ky.

VISCOUNT OXFORD 5TH (45744)—Red, calved Dec. 1, 1879, bred by Mr. T. Holford, Castle Hill, Cerne, got by 23d Grand Duke (34063), out of Baroness Oxford 3d by Duke of Hillhurst (28401)—Baroness Oxford by 2d Duke of Claro (21576)—Lady Oxford 5th by 3d Duke of Thorndale (17749)—Lady Oxford 4th by 2d Grand Duke (12961)—Maid of Oxford by Lord of Eryholme (12205)—Oxford 13th by 3d Duke of York (10166)—Oxford 5th by Duke of Northumberland (1940)—Oxford 2d by Short Tail (2621)—Matchem Cow by Matchem (2281)—by Young Wynyard (2859). Sold to Palmer & Bowman, Va., for $1,025.

VISCOUNT OXFORD 7TH (48891)—Red, calved Oct. 29, 1881, bred by T. Holford, Castle Hill, Cerne, got by Duke of Leicester (43112), out of Viscountess Oxford by 23d Grand Duke (34063)—Baroness Oxford 3d by Duke of Hillhurst (28401)—Baroness Oxford by 2d Duke of Claro (21576)—Lady Oxford 5th by 3d Duke of Thorndale (17749)—Lady Oxford 4th by 2d Grand Duke (12961), &c., as in Viscount Oxford 5th above. Sold to A. J. Alexander, for $700; resold to W. R. King, Mo.

THORNDALE ROSE 8TH (vol. 26, p. 341, E.)—Roan, calved Sept. 4, 1876, bred by Lord Braybrooke, got by 6th Duke of Oneida (30997), out of Thorndale Rose 3d by 3d Duke of Geneva (23753)—Thorndale Rose 2d by 4th Grand Duke (19874)—Thorndale Rose by 4th Duke of Thorndale (17750)—Cambridge Rose 6th by 3d Duke of York (10166)—Cambridge Rose 5th by 2d Cleveland Lad (3408)—Cambridge Rose 2d by Belvedere (1706)—Cambridge Premium Rose by Belvedere (1706)—Red Rose 9th by 2d Hubback (1423)—Red Rose 2d by His Grace (311)—Red Rose 1st by Yarborough (705)—American Cow by Favorite (252)—by Punch (531)—by Foljambe (263)—by Hubback (319). Sold to A. J. Alexander, for $5,600.

THORNDALE ROSE 16TH (vol. 26, p. 341, E.)—Roan, calved June 1, 1879, bred by Lord Braybrooke, got by 3d Duke of Underley (38196), out of Thorndale Rose 2d by 4th Grand Duke (19874) —Thorndale Rose by 4th Duke of Thorndale (17750), &c., as above. Sold to A. J. Alexander, for $4,000.

DUCHESS OF ROSES—Red and white, calved June 6, 1882 (at Cincinnati, Ohio), bred by Lord Braybrooke, got by Grand Duke 30th (38372), out of Thorndale Rose 8th by 6th Duke of Oneida (30997), &c., as in dam, above. Sold to A. J. Alexander, for $2,025.

HEYDON ROSE 7TH (vol. 25, p. 363, E.)—Red roan, calved May 13, 1878, bred by Lord Braybrooke, got by 3d Duke of Rosedale (33723), out of Heydon Rose 2d by 3d Duke of Geneva (23753)—Heydon Rose by Englishman (19701)—The Beauty by Raritan (9523)—Cambridge Rose 6th by 3d Duke of York (10166), &c., as in Thorndale Rose 8th, above. Sold to A. J. Alexander, for $1,800.

HEYDON ROSE 6TH (vol. 25, p. 363, E.)—Roan, calved June 14, 1877, bred by Lord Braybrooke, got by 6th Duke of Oneida (30997), out of Heydon Rose 2d by 3d Duke of Geneva (23753)— Heydon Rose by Englishman (19701), &c., as in Heydon Rose 7th, above. Sold to A. J. Alexander, for $725.

HEYDON ROSE 12TH (vol. 28, p. 312, E.)—Red and white, calved Nov. 19, 1881, bred by Lord Braybrooke, got by 2d Duke of Sussex (43130), out of Heydon Rose 8th by 3d Duke of Underley (38196)—Heydon Rose 5th by 3d Duke of Rosedale (33723)— Heydon Rose 3d by 17th Grand Duke (24064)—Heydon Rose by Englishman (19701), &c., as above. Sold to A. J. Alexander, for $900.

RED ROSE OF RODIL (vol. 22, p. 405, E.)—Roan, calved Sept. 24, 1875, bred by Earl Dunmore, got by 4th Duke of Geneva (30958), out of Red Rose of Luskentrye by 13th Duke of Airdrie (36459)—Poppy 4th by Airdrie (30365)—Poppy 2d by Duke of Airdrie (12730)—Poppy by Ashland (11122)—Red Rose by Prince

Charles 2d (32113)—Thames by Shakespeare (12062)—Lady of the Lake by Reformer (2505)—Rose of Sharon by Belvedere (1706)—Red Rose 5th by 2d Hubback (1423)—Red Rose 2d by His Grace (311)—Red Rose 1st by Yarborough (705)—American Cow by Favorite (252)—by Punch (531)—by Foljambe (263)—by Hubback (319). Sold to R. C. Estill, for $425.

RED ROSE OF RODIL 2D (vol. 25, p. 847)—Red, calved March 29, 1880, bred by J. J. Hetherington, Brampton, Carlisle, got by 2d Marquis of Oxford (37055), out of Red Rose of Rodil by 4th Duke of Geneva (30958), &c., as in dam, above. Sold to Thomas & Smith, for $525.

RED ROSE OF RODIL 3D (vol. 28, p. 392, E.)—Roan, calved March 2, 1881, bred by George Fox, Elmhurst Hall, got by Lightburne's Duke of Oxford 2d (38564), out of Red Rose of Rodil by 4th Duke of Geneva (30958), &c., as in dam, above. Sold to A. J. Alexander, for $700.

RED ROSE OF PALESTINE (vol. 25, p. 847)—Roan, calved March 2, 1882 (on ship), bred by Geo. Fox, Elmhurst Hall, got by 2d Duke of Elmhurst (43091), out of Red Rose of Rodil by 4th Duke of Geneva (30958), &c., as in dam, above. Sold to T. W. Harvey, Chicago, for $325.

RED ROSE OF SCARRISTA (vol. 25, p. 1042, A., and vol. 24, p. 423, E., and vol. 28, p. 450, E.)—Roan, calved May 17, 1877, bred by Earl of Dunmore, got by 6th Duke of Geneva (30959), out of Red Rose of Luskentrye by 13th Duke of Airdrie (36459)—Poppy 4th by Airdrie (30365), &c., as in Red Rose of Rodil. Sold to H. A. Moran, for $710.

RED ROSE OF THICKET (vol. 28, p. 450, E.)—Roan, calved April 2, 1881, bred by J. I. D. Jefferson, got by Rowfant Duke of Oxford (43926), out of Red Rose of Scarrista by 6th Duke of Geneva (30959), &c., as in dam, above. Sold to Bow Park Association, Canada, for $400.

RED ROSE OF ENNERDALE (vol. 25, p. 960, A., and vol. 27, p. 301, E.)—Red and white, calved June 28, 1880, bred by Earl

3

of Bective, Underley Hall, got by Duke of Underley (33745), out of Red Rose of Eskdale by Airdrie 3d (32919)—Poppy 8th by Joe Johnson (31440)—Poppy 3d by Airdrie (30365)—Poppy 2d by Duke of Airdrie (12730), &c., as in Red Rose of Rodil. Sold to J. S. Latimer & Son, Ill., for $425.

RED ROSE OF DERWENT (vol. 25, p. 846, A., and vol. 25, p. 454, E.)—Red and white, calved Oct. 13, 1878, bred by George Fox, Elmhurst Hall, got by 24th Duke of Airdrie (36460), out of May Rose 5th by Airdrie 3d (32919)—May Rose by Airdrie (30365) —Easterday by Pilot (32066)—Poppy by Ashland (11122), &c., as in Red Rose of Rodil. Sold to T. W. Harvey, Chicago, for $625.

RED ROSE OF THAMES 2D (vol. 25, p. 1091, A., and vol. 22, p. 399, E.)—Roan, calved Sept. 27, 1880, bred by Geo. Fox, Esq., got by 24th Duke of Airdrie (36460), out of Duchess 19th by 4th Duke of Geneva (30958)—Duchess 4th by Airdrie (30365)—Duchess 2d by Pilot (32066)—Duchess by Buena Vista (36623)—Red Rose by Prince Charles 2d (32113)—Thames by Shakespeare (12062), &c., as in Red Rose of Rodil. Sold to Palmer & Bowman, Va., for $550.

ROSE OF MASON 8TH (vol. 25, p. 527, A., and vol. 25, p. 455, E.)—Roan, calved Nov. 16, 1876, bred by Geo. Fox at Vinewood, Ky., and exported to England in 1878, got by 14th Duke of Airdrie (41348), out of Rose of Mason 6th by Double Duke (41334) —Rose of Mason 3d by Duke of Mason (41408)—White Rose by Duke of Mason (41408)—Rose 2d by General Winfield Scott (12936) —White Rose by Young Paragon (11886)—Dorothy by Prince Charles 2d (32113)—Thames by Shakespeare (12062), &c., as in Red Rose of Rodil. Sold to J. M. Bigstaff, for $315.

RED ROSE OF DOVE (vol. 25, p. 527, A., and vol. 26, p. 434, E.)—Roan, calved Aug. 14, 1879, bred by Geo. Fox, Esq., got by 39th Duke of Oxford (38173), out of Rose of Mason 8th by 14th Duke of Airdrie (41348), &c., as in dam, above. Sold to J. M. Bigstaff, for $315.

RED ROSE OF DOVE 4TH (vol. 25, p. 960)—Roan, calved Oct. 16, 1881, bred by Geo. Fox, Esq., got by 39th Duke of Oxford

(38173), out of Rose of Mason 8th by 14th Duke of Airdrie (41348), &c., as in dam, above. Sold to J. S. Latimer & Son, Abingdon, Ill., for $330.

AUDLEY ROSE (vol. 25, p. 455)—Red and white, calved June 25, 1882, bred by Lord Braybrooke, got by Grand Duke 33d (39946), out of Heydon Rose 7th by 3d Duke of Rosedale (33723), &c., as in Heydon Rose 7th, above.

AUDLEY ROSE 2D—Roan, calved Nov. 8, 1882, bred by Lord Braybrooke, got by 11th Duke of Rosedale (44704), out of Heydon Rose 6th (vol 25, p. 455) by 6th Duke of Oneida (30997), &c., as in Heydon Rose 6th, above.

RED ROSE OF TURLINGTON (vol 25, p. 847)—Roan, calved July 7, 1882, bred by Geo. Fox, got by Airdrie's Kirklevington (42652), out of Red Rose of Derwent by 24th Duke of Airdrie (36460), &c., as in dam, above.

DUKE OF RODILL 51281—Red and white, calved Aug. 15, 1882, bred by J. J. Hetherington, got by Underley Prince (45710), out of Red Rose of Rodill 2d by 2d Marquis of Oxford (37055), &c., as in dam, above. Calved the property of Thomas & Smith, Bourbon Co., Ky.

August, 1883. A. J. ALEXANDER,

Of Woodburn, Woodford Co., Ky. Per Steamship Abyssinia, from Liverpool to Quebec.

2D DUKE OF WHITTLEBURY (47789)—Red, calved July 29, 1882, bred by Mr. Loder, Whittlebury, got by 41st Grand Duke (46439), out of 2d Duchess of Whittlebury by Duke of Connaught (33604)—3d Duchess of Hillhurst by 2d Duke of Hillhurst (39748) 10th Duchess of Airdrie by Royal Oxford (18774)—Duchess of Airdrie 7th by Clifton Duke (23580)—by 2d Duke of Athol (11376) —by 2d Duke of Oxford (9046)—by 2d Cleveland Lad (3408)—by Short Tail (2621)—by 2d Hubback (1423)—by 2d Earl (1511)—by Marske (418)—by Ketton 1st (709)—by Comet (155)—by Favorite (252)—by Daisy Bull (186)—by Favorite (252)—by Hubback (319) —by J. Brown's Red Bull (97).

Oct., 1883. A. J. ALEXANDER.
By Steamship Ontario, from Liverpool to Quebec.

BARONESS OXFORD 9TH (vol. 26, p. 402, E.)—Red, calved Sept. 12, 1879, bred by Duke of Devonshire, got by 7th Duke of Gloster (39735), out of Baroness Oxford 4th by 3d Duke of Hillhurst (30975)—Baroness Oxford by 2d Duke of Claro (21576)—Lady Oxford 5th by 3d Duke of Thorndale (17749)—Lady Oxford 4th by 2d Grand Duke (12961)—Maid of Oxford by Lord of Eryholme (12205)—Oxford 13th by 3d Duke of York (10166)—Oxford 5th by Duke of Northumberland (1940)—Oxford 2d by Short Tail (2621)—Matchem Cow by Matchem (2281)—by Young Wynyard (2859).

OXFORD BARON—Red, calved March 30, 1884, bred by the Duke of Devonshire (calved at Woodburn), got by 62d Duke of Oxford (47778), out of Baroness Oxford 9th by 7th Duke of Gloster (39735), &c., as in dam, above.

July, 1872. GEO. ALEXANDER.

WASTELL'S COUNTESS OF KENTORE (vol. 12, p. 1283)—Red, calved Aug., 1870, bred by Geo. Phillips, Aberdeenshire, got by Lord Chatham (26625), out of Kilmeny 6th by The Czar (20947)—Kilmeny 4th by Jeremy (11611)—Anna by Hawthorn (7071)—Kilmeny by The Peer (5455)—Premium by George (2057)—by Togston (5487).

WASTELL'S JENNY LIND 7TH (vol. 12, p. 1284)—Red, calved Feb. 4, 1870, bred by A. E. Hector, Callyhill, got by Lord of the Isles 14784, out of Jenny Lind 6th by Prince Lewis (20560)—Jenny Lind 3d by Kelvinside (14756)—Jenny Lind by The Duke (7593)—Dora by Pasha (7612)—Alice by Mahomet (6170)—Mary Ann by Sillery (5131)—by Fitzwalter (6014).

COUNT BISMARCK 13723—Red, calved Nov. 13, 1872, bred by Geo. Phillips, got by Golden King 14372, out of Wastell's Countess of Kentore by Lord Chatham (26625), &c., as in dam, above.

E. G. ALDEN.

VICTORIA—See vol. 14, p. 10.

Oct. 23, 1879. GEORGE ALLEN & SON,
Of Palermo, Illinois.

From Liverpool—landed at Quebec, Nov. 4, 1879, by Allan Line Steamer Nestorian.

PRINCESS (vol. 22, p. 17382, and vol. 23, p. 309, E., and vol. 25, p. 312, E.)—Roan, calved March 3, 1874, bred by Geo. Allen, Knightly Hall, Eccleshill, England, got by 2d Duke of Wetherby (21618), out of Fennel Duchess 5th by 13th Duke of Oxford (21604)—Fennel 3d by Cherry Duke 2d (14265)—Fennel by 5th Duke of York (10168)—Filbert by 2d Cleveland Lad (3408)—Felicia by 4th Duke of Northumberland (3649)—Fanny by Short Tail (2621)—Fletcher 2d by Belvedere (1706)—Fletcher by son of Young Wynyard (2859)—descended from James Brown's Red Bull (97).

BLOSSOM 3d (vol. 22, p. 17067, A., and vol. 25, p. 310. E.)—Red, calved April 2, 1873, bred by George Allen, got by 2d Duke of Wetherby (21618), out of Blossom by 8th Duke of York (28480)—Peach Blossom 2d by Baron Westbury (19287)—Peach by Gen. Canrobert (12927)—Poppy by 2d Cleveland Lad (3408)—Place 2d by 2d Earl of Darlington (1945)—Place by Son of 2d Hubback (2685)—cow of Mr. Bates'.

LORD ROWLEY 44199—Red and white, calved May 9, 1880, bred by George Allen, got by 2d Duke of Rowley (28441), out of Blossom 3d by 2d Duke of Wetherby (21618), &c., as in dam, above.

PRINCESS 4TH (vol. 22, p. 17384)—Red and white, calved Sept. 2, 1878, bred by George Allen, got by 2d Duke of Rowley (28441), out of Princess 2d by Duke of Wetherby 6th (33756)—Princess by 2d Duke of Wetherby (21618), &c., as in Princess, above.

PRINCESS 2D (vol. 22, p. 17382, A., and vol. 23, p. 309, E., under dam)—Red, calved June 9, 1876, bred by Geo. Allen, got by Duke of Wetherby 6th (33756), out of Princess by 2d Duke of Wetherby (21618), &c., as in dam, above.

PRINCESS 3D (vol. 22, p. 17383, A., and vol. 25, p. 312, E., under dam)—Red and white, calved May 27, 1878, bred by Geo. Allen, got by 2d Duke of Rowley (28441), out of Princess by 2d Duke of Wetherby (21618), &c., as in dam, above.

KIRKLEVINGTON DUKE (41768)—Red, calved April 25, 1877, bred by Geo. Allen, got by Duke of Wetherby 6th (33756), out of Lady Kirklevington 2d by Earl of Oxford (21651)—Lady Kirklevington by Grand Duke of York (12966)—Kirklevington 9th by Gen. Canrobert (12926)—Kirklevington 7th by Earl of Derby (10177)—by Earl of Liverpool (9061)—by Duke of Northumberland (1940)—by Belvedere (1706)—by Son of 2d Hubback (2683)—a cow of Mr. Bates', descended from the stock of Mr. Maynard.

KIRKLEVINGTON LADY 4TH (vol. 22, p. 17211, and vol. 24, p. 294, .E.)—Roan, calved April 28, 1876, bred by Geo. Allen, got by 6th Duke of Wetherby (33756), out of Kirklevington Lady 2d by Earl of Oxford (21651)—Lady Kirklevington by Grand Duke of York (12966), &c., as in Kirklevington Duke (41768), above.

KIRKLEVINGTON LADY 5TH (vol. 22, p. 17211)—Roan, calved April 27, 1877, bred by Geo. Allen, got by 6th Duke of Wetherby (33756), out of Kirklevington Lady by 2d Duke of Wetherby (21618)—Lady Kirklevington 2d by Earl of Oxford (21651)—Lady Kirklevington by Grand Duke of York (12966), &c., as above.

KIRKLEVINGTON LADY (vol. 22, p. 17210, and vol. 25, p. 311, E.)—Roan, calved March 28, 1873, bred by Geo. Allen, got by 2d Duke of Wetherby (21618), out of Lady Kirklevington 2d by Earl of Oxford (21651)—Lady Kirklevington by Grand Duke of York (12966), &c., as in Kirklevington Duke (41768), above.

PRINCESS 5TH (vol. 22, p. 17384)—Roan, calved June 3, 1879, bred by Geo. Allen, got by 2d Duke of Rowley (28441), out of Princess by 2d Duke of Wetherby (21618), &c., as in dam, above.

PRINCESS 6TH (vol. 22, p. 17384)—Red and white, calved April 16, 1880, bred by Geo. Allen, got by 2d Duke of Rowley (28441), out of Princess 2d by 6th Duke of Wetherby (33756), &c., as in dam, above.

PRINCESS 7TH (vol. 22, p. 17384)—Red, calved May 19, 1880, bred by Geo. Allen, got by 2d Duke of Rowley (28441), out of Princess by 2d Duke of Wetherby (21618), &c., as in dam, above.

PRINCESS 8TH (vol. 22, p. 17384)—Red and white, calved May 21, 1880, bred by Geo. Allen, got by 2d Duke of Rowley (28441), out of Princess 3d by 2d Duke of Rowley (28441), &c., as in dam, above.

KIRKLEVINGTON DUCHESS 5TH (vol. 22, p. 17210)—Red, calved Feb. 27, 1880, bred by Geo. Allen, got by 2d Duke of Rowley (28441), out of Kirklevington Lady 4th by 6th Duke of Wetherby (33756), &c., as in dam, above.

KIRKLEVINGTON DUCHESS 6TH (vol. 22, p. 17210)—Red, calved May 4, 1880, bred by Geo. Allen, got by 2d Duke of Rowley (28441), out of Kirklevington Lady by 2d Duke of Wetherby (21618), &c., as in dam, above.

KIRKLEVINGTON LAD 44090—Red roan, calved Feb. 9, 1880, bred by Geo. Allen, got by 2d Duke of Rowley (28441), out of Duchess 7th by Duke of Clarence 5th (36479), &c., as in dam, below.

DUCHESS 7TH (vol. 22, p. 17117)—Red, calved April 11, 1877, bred by Geo. Allen, got by 5th Duke of Clarence (36479), out of Duchess by 2d Duke of Wetherby (21618)—Annette 3d by Lord Oxford 2d (20215)—Annette by Horrox (11591)—Active by 4th Duke of York (10167)—Actress by Duke of Northumberland (1940)—Annie by Short Tail (2621)—Acomb by Belvedere (1706)—a cow of Mr. Bates', of Kirklevington.

DUCHESS ACOMB (vol. 22, p. 17118)—Roan, calved May 13, 1879, bred by Geo. Allen, got by 2d Duke of Rowley (28441), out of Duchess 2d by 2d Duke of Wetherby (21618)—Annette 3d by Lord Oxford 2d (20215), &c., as above.

DUCHESS SURMISE (vol. 22, p. 17125)—Red and white, calved June 29, 1879, bred by Geo. Allen, got by 2d Duke of Rowley (28441), out of Princess Surmise 2d by 4th Duke of Grafton (28396)—Alexandra by Worth (23244)—Princess by May Duke (13320)—Surmise by Duke of Gloster (11382)—Silence by Earl of Derby (10177)—Secret 3d by Duke of Sutherland (6945)—Secret 2d by Locomotive (4242)—Secret by Short Tail (2621)—White Rose by Gambier (2046)—White Rose by Young Wynyard (2859)—by bulls of C. and R. Colling.

DUCHESS 10TH (vol. 22, p. 17117)—Red, calved May 10, 1878, bred by Geo. Allen, got by 2d Duke of Rowley (28441), out of Duchess 2d by 2d Duke of Wetherby (21618)—Annette 3d by Lord Oxford 2d (20215), &c., as in Duchess 7th, above.

DUCHESS 13TH (vol. 22, p. 17117)—Red, calved April 14, 1879, bred by Geo. Allen, got by 2d Duke of Rowley (28441), out of Duchess 6th by Duke of Clarence 6th (36479)—Annette 4th by 8th Duke of York (28480)—Annette 3d by Lord Oxford 2d (20215), &c., as above.

DUCHESS 15TH (vol. 22, p. 17118)—Red, calved May 15, 1880, bred by Geo. Allen, got by 2d Duke of Rowley (28441), out of Duchess 10th by 2d Duke of Rowley (28441), &c., as in Duchess 10th.

DUCHESS 14TH (vol. 22, p. 17118)—Roan, calved May 5, 1880, bred by Geo. Allen, got by 2d Duke of Rowley (28441), out of Duchess 3d by Duke of Wetherby 6th (33756), &c., as in Duchess 3d, below.

DUCHESS 3D (vol. 22, p. 17116)—Roan, calved April 25, 1875, bred by Geo. Allen, got by Duke of Wetherby 6th (33756), out of Duchess (vol. 22, p. 293, E.) by 2d Duke of Wetherby (21618)—Annette 3d by Lord Oxford 2d (20215).

KIRKLEVINGTON DUCHESS 3D (vol. 22, p. 17210)—Red, calved April 30, 1879, bred by Geo. Allen, got by 2d Duke of Rowley (28441), out of Lady Kirklevington 2d by Earl of Oxford

(21651)—Lady Kirklevington by Grand Duke of York (12966), &c., as in Kirklevington Duke (41768).

GEORGIANA 14TH (vol. 21, p. 537, E.)—Red and white, calved Jan. 24, 1867, bred by Geo. Allen, got by 2d Duke of Wetherby (21618), out of Georgiana 9th by 2d Baron Westbury (19288)—Georgiana 5th by Gen. Canrobert (12926)—Georgiana by St. Bernard (15227)—by Lord Geo. Bentinck (10444)—by King Pippin (14769) —by Earl Stanhope (5966). Georgiana 14th died after producing a bull calf.

KIRKLEVINGTON DUCHESS 4TH (vol. 22, p. 17210)—Red, calved Feb. 5, 1880, bred by Geo. Allen, got by 2d Duke of Rowley (28441), out of Kirklevington Lady 5th by Duke of Wetherby 6th (33756), &c., as in dam, above.

RED BULL—Calved April 29, 1880, bred by Geo. Allen, got by 2d Duke of Rowley (28441), out of Georgiana 14th by 2d Duke of Wetherby (21618), &c., as in dam, above.

1835. SAMUEL ALLEN.

RACHEL—Roan, calved April, 1829, bred by Mr. Whitaker. Pedigree lost. See A. H. B., vol. 1, p. 220.

MISS LAWRENCE—Roan, calved May, 1830, bred by Mr. Booth. See A. H. B., vol. 1, p. 205.

MISS MELLON—White, calved 1834. See A. H. B., vol. 1, p. 205.

1870. JOHN S. ARMSTRONG.
Eramosa, Ont.

MISSIE 23D (vol. 2, p. 111, C. H. B.)—Red, calved March 23, 1868, bred by W. S. Marr, got by Young Pacha (20457), out of Missie 5th by Lord of Lorne (18258)—Missie 2d by Augustus (15598)—Miss by a son of Duke 3d (17697)—Countess by The

4

Pacha (7612)—Jessamine by Mahomed (6170)—Rose by Plenipo (4725)—Thorn by Abbot (2899).

LADY ABERDEEN (vol. 2, p. 538, C. H. B.)—Red, calved Feb. 20, 1871, bred by W. S. Marr, got by Royal Briton (29844), out of Missie 23d by Young Pacha (20457), &c., as in Missie 23d, above.

RACHEL 8TH—Red, calved March 10, 1868, bred by W. S. Marr, got by Young Pacha (20457), out of Rachel 3d by Sir Hubert (18844) —Rachel by Clarendon (14280)—Lizzie by Guy Fawkes (12981)— Patience by Duke 7th (7984)—Temperance by Teetotaller (5411)— Fanny by Ivanhoe (1131)—by a son of Blyth Comet (85)—by Atlas (42)—by a son of Favorite (252).

STAMFORD 5TH (vol. 2, p. 799, C. H. B.)—Red, calved April 2, 1868, bred by W. S. Marr, got by Prince Louis (27158), out of Stamford 3d by Young Pacha (20457)—Stamford 2d by Clarendon (14280)—Stamford by Phœnix (10608)—Sprightly by Corporal Trim (7932)—Stamford by Regent (2517)—by Togston (5487)—by a son of Lawnsleeves (365)—by a son of Bolingbroke (86)—by Sultan (633)—by Barnaby (1678)—bred by Messrs. James, of Stamford.

SCOTTISH LASS (vol. 2, p. 787, C. H. B.)—Red, calved March 27, 1871, bred by W. S. Marr, got by Heir of Englishman (24122), out of Rachel 8th by Young Pacha (20457), &c., as in dam, above.

APRICOT 11TH—Red, calved March 16, 1868, bred by Mr. W. S. Marr, Upper Mill, Tarves, Aberdeen, Scotland, got by Prince Louis (27158), out of Apricot 6th by Master Gunner (22316)—Apricot 4th by Clarendon (14280)—Apricot by Jemmie (11611)—Peach by Homer (2134)—Bloom by Sir Launcelot (5166)—Bright Eyes by Viceroy (5561)—by Bulmer (1760)—by Appleton (1647)—by Newtonian (2368)—by a bull of Mr. Cattley's.

MARQUIS OF LORNE 21677—Red, calved Jan. 26, 1871, bred by John S. Armstrong, got by Royal Briton (29844), out of Apricot 11th, &c., as above.

May, 1871, J. S. ARMSTRONG.
From Glasgow to Montreal.

LADY FLORENCE (vol. 2, p. 557, C. H. B.)—Red, calved
Dec. 11, 1869, bred by A. Cruickshank, Sittyton, Aberdeenshire,
got by Julius Cæsar (26486), out of Lady Frances by Prince Imperial (22595)—Lady Fair by Royal Butterfly (18753)—Lady Like
by The Baron (13833)—Lady Isabella by Matadore (11800)—Arabella by Robin O'Day (4973)—Picotee by Premier (6308)—Sunflower by Unicorn (8725)—by Monarch (4495)—by Young Satellite
(8538)—by Valentine (661)—bred by Mr. Rennie, of Phantassie.

GOLDEN BRACELET (vol. 2, p. 495, C. H. B.)—Red, calved
Feb. 26, 1870, bred by Wm. Duthie, Collynie, Aberdeen, Scotland, got by Bulwark (25696), out of Bracelet by Hand-and-Glove
(24099)—Velvet by Champion of England (17526)—Queen's Anni-
versary by Master Butterfly 2d (14918)—May Dew by Matadore
(11800)—Fidelity by Prince Edward Fairfax (9506)—by Premier
(6308)—by Alamode (725)—by Slater (2643)—by Slater (2643)—
by son of Phenomenon (1318).

Aug. 6, 1873. J. S. ARMSTRONG.
Guelph, Ont.
Ship Italia from Glasgow to Montreal.

PRINCESS JOSEPHINE (vol. 3, p. 702, C. H. B.)—Red,
calved Dec. 29, 1870, bred by Mr. W. Duthie, Collynie, Aberdeenshire, got by Grand Knight (26303), out of Josephine by Prince
Alfred (22567)—Jenny Lind by Guy Fawkes (12981)—Red Lady by
Van Dunck (10992)—Patience by Duke 7th (7984)—Temperance
by Teetotaller (5411)—Fanny by Ivanhoe (1131)—by a son of Blyth
Comet (85)—by Atlas (42)—by a son of Favorite (252).

PRINCESS JOSEPHINE 2D—Red, calved Dec. 28, 1874, bred
by Wm. Duthie, got by Duke of Cambridge (33586), out of Princess
Josephine by Grand Knight (26303), &c., as in dam, above.

SALVIA 14TH (vol. 3, p. 768, C. H. B.)—Red, calved March 29,
1871, bred by W. S. Marr, Upper Mill, Tarves, Aberdeenshire,
Scotland, got by Gold Digger (24044), out of Salvia 2d by Baron

Renfrew (15624)—Salvia by The Baron (13833)—Stephania by Procurator (10657)—Sympathy by Prince Edward Fairfax (9506)—Fancy by Billy (3151)—Jessie by Sovereign (7539)—Rose by Satellite (1420)—by Baronet (60)—by Cleveland (144)—by Symmetry (641).

BEAUTY 15TH (vol. 3, p. 363, C. H. B.)—Red, calved March 26, 1872, bred by W. S. Marr, Upper Mill, Tarves, Aberdeenshire, Scotland, got by Heir of Englishman (24122), out of Beauty 11th by Young Pacha (20457)—Beauty 7th by Sir Hubert (18844)—Beauty 4th by Clarendon (14280)—Beauty 2d by Sir Arthur (12072)—Roan Beauty by Robin O'Day (4973)—by Emperor (3716).

KINDNESS 7TH (vol. 3, p. 528, C. H. B.)—Red and white, calved Oct. 22, 1872, bred by W. S. Marr, got by Gold Digger (24044), out of Kindness 5th by Young Pacha (20457)—Kindness by Clarendon (14280)—Patience by Duke 7th (7984)—Temperance by Teetotaller (5411)—Fanny by Ivanhoe (1131)—by a son of Blyth Comet (85)—by Atlas (42)—by a son of Favorite (252).

EARL OF MARR [4324]—Red, calved April 10, 1874, bred by W. S. Marr, Upper Mill, got by Young Heir (31351), out of Beauty 15th by Heir of Englishman (24122). See Beauty 15th, above.

Aug. 9, 1873. J. S. ARMSTRONG.
Ship Italia.

BRIDE 2D (vol. 3, p. 380, C. H. B.)—Roan, calved May 13, 1871, bred by W. S. Marr, Upper Mill, Tarves, Aberdeenshire, Scotland, got by Gold Digger (24044), out of Bride by Baron Alfred (23351)—Bridesmaid 4th by Prince (16715)—Bridesmaid 2d by Guy Fawkes (12981)—Bridesmaid by Sir Arthur (12072)—Crescent by The Pacha (7612)—Dainty by 2d Duke of Northumberland (3646)—by Emperor (3716)—by Invalid (4076)—by Magnet (392)—by Palmflower (480).

PRINCESS ROYAL 13TH (vol. 3, p. 706, C. H. B.)—Red, calved February, 1872, bred by Jas. Whyte, Clinterty, Blackburn, Aberdeen, Scotland, got by Lord Charles (31634), out of Princess

Royal 9th by Sir Charles (16945)—Princess Royal 8th by Bosquet (14183)—Princess Royal 3d by Grand Duke (10284)—Princess Royal by Robin O'Day (4973)—Vesta by Leander (4199)—Varna by Matchem (2281)—by Sir Henry (1446)—by Young Neswick (1268)—by son of Rose's Red Bull (5009)—by Southampton—by Prince (521).

ENGLISH LADY (vol. 3, p. 455, C. H. B.)—Red and white, calved May 25, 1872, bred by W. S. Marr, Upper Mill, Tarves, Aberdeenshire, Scotland, got by Heir of Englishman (24122), out of Red Lady by Young Pacha (20457)—Roan Lady by a son of Ury (10984)—Red Lady by Van Dunck (10992)—Patience by Duke 7th (7984)—Temperance by Teetotaller (5411)—Fanny by Ivanhoe (1131)—by a son of Blyth Comet (85)—by Atlas (42)—by a son of Favorite (252).

YOUNG HEIR (31351)—Roan, calved April 17, 1872, bred by W. S. Marr, Upper Mill, got by Heir of Englishman (24122), out of Kindness 2d by Master Gunner (22316)—Kindness by Clarendon (14280)—Patience by Duke 7th (7984)—Temperance by Teetotaller (5411)—Fanny by Ivanhoe (1131)—by a son of Blyth Comet (85)—by Atlas (42)—by a son of Favorite (252).

BRIDESMAN [2794]—Red, calved Dec. 21, 1873, bred by W. S. Marr, got by Young Englishman (31113), out of Bride 2d by Gold Digger (24044), &c., as in Bride 2d, above.

SCOTTISH HEIR [4439]—Red and white, calved April 7, 1874, bred by W. S. Marr, got by Young Heir (31351), out of English Lady by Heir of Englishman (24122), &c., as in English Lady, above.

RACHEL 10TH (vol. 3, p. 716, C. H. B.)—Roan, calved April 23, 1871, bred by W. S. Marr, Upper Mill, Tarves, Aberdeenshire, Scotland, got by Heir of Englishman (24122), out of Rachel 6th by Lord Lyons (22173)—Rachel 5th by Lord of Lorne (18258)—Rachel 2d by Baron Renfrew (15624)—Rachel by Clarendon (14280)—Lizzie by Guy Fawkes (12981)—Patience by Duke 7th

(7984)—Temperance by Teetotaller (5411)—Fanny by Ivanhoe (1131)—by a son of Blyth Comet (85)—by Atlas (42)—by a son of Favorite (252).

WHITE RACHEL (vol. 4, p. 592, C. H. B.)—White, calved Jan. 16, 1874, bred by W. S. Marr, got by Gladstone (31253), out of Rachel 10th by Heir of Englishman (24122), &c., as in Rachel 10th, above.

June 8, 1876. J. S. ARMSTRONG.
Cranberry Farm, Guelph, Ont.
By Ship Phœnician.

HELEN 11TH (vol. 4, p. 197, C. H. B.)—Roan, calved April 26, 1872, bred by W. S. Marr, Upper Mill, Aberdeenshire, Scotland, got by Heir of Englishman (24122), out of Helen 5th by Gold Digger (24044)—Helen 4th by Young Pacha (20457)—Helen 2d by Sir Hubert (18844)—Helen by Sir Arthur (12072)—Bonny Lass by Young Ury (10984)—Likely by The Pacha (7612)—by 2d Duke of Northumberland (3646)—by Sillery (5131)—by Carleton (843)—by Diamond (205)—by Diamond (205).

MARY ANNE 10TH (vol. 4, p. 361, C. H. B.)—Roan, calved March 19, 1873, bred by W. S. Marr, got by Heir of Englishman (24122), out of Mary Anne 7th by Young Pacha (20457)—Mary Anne 2d by Clarendon (14280)—Mary Anne by Mosstrooper (11827)—Cressy by Vice-President (11002)—Geraldine by Robin O'Day (4973)—Angelica by Holkar (4041)—by Jopp's Bull (9256)—Kate of Darlington.

MISSIE 46TH (vol. 4, p. 391, C. H. B.)—Roan, calved Feb. 2, 1874, bred by W. S. Marr, got by Young Englishman (31113), out of Missie 35th by Prince of Stokesley (27177)—Missie 5th by Lord of Lorne (18258)—Missie 2d by Augustus (15598)—Missie by a son of Duke 3d (17697)—Countess by The Pacha (7612)—Jessamine by Mahomed (6170)—by Plenipo (4725)—by Abbott (2899).

RASPBERRY 7TH—Red, calved Feb. 17, 1875, bred by W. S. Marr, got by Young Englishman (31113), out of Raspberry 5th by

Heir of Englishman (24122)—Raspberry 2d by Prince Arthur 2d (22571)—Raspberry by Vladimir (21043)—Matilda 5th by Cunningham (11323)—Matilda by Robin Hood (8494)—Ruby by Mahomed 2d (10492)—Daisy by Billy (3151)—Maria by Belshazzar (1703)—by Abraham (2905)—by Simon (5134).

LOVELY 19TH—Red, calved Dec. 19, 1874, bred by A. Cruickshank, got by Millionaire (31917), out of Lovely 12th by Scotch Rose (25099)—Lovely 9th by Windsor Augustus (19157)—Lovely 8th by Bosquet (14183)—Lovely by Kelly 2d (9265)—Lady Ythan by Robin O'Day (4793)—Lady by Favorite (9116)—Marion by Anthony (1640)—Merino by Edgcott (1953)—Matilda by Son of Merlin (6522)—White Cow by Acton (1607).

BRITISH PRINCE [4687]—Roan, calved Feb. 7, 1876, bred by S. W. Marr, got by Royal Prince (35398), out of Helen 11th by Heir of Englishman (24122), &c., as in Helen 11th, above.

STARLIGHT (vol. 4, p. 556, C.)—Red and white, calved April 11, 1875, bred by Jas. White, Clinterty, Scotland, got by Star (32601), out of Ury Lass by Prince of Warlaby (20593)—Clipper by Old England (24681)—Mary Rose by Master Gunner (23316)—Daisy by Canrobert (17493)—Legacy by The Pacha (7612)—Crocus by 2d Duke of Northumberland (3646)—Norah by Sillery (5131)—Emily by Sillery (5131)—Eliza by Young Western Comet (1575)—Lady Betsy by Diamond (205)—Betty by Favorite (256)—by Charge's Red Bull (1810).

1857. W. & R. ARMSTRONG.
Markham, Ont.

FAWKES [249] (14539)—Light roan, calved Feb. 26, 1855, bred by Robt. Syme, Red Kirk, Dumfries, Scotland, got by Sir John (13735), out of Lady Bird 3d by Remus (11987)—Red Rose by Strathmore (6547)—Lady Bird 2d by Playfellow (6297)—Lady Bird by Scrip (2604)—by Thornington (5472).

Oct. 20, 1875. R. ASHBURNER.

By Ship Erin, from Liverpool for California.

KIRKLEVINGTON DUKE 2D (34364)—Roan, calved May 1, 1874, bred by W. Ashburner, Netherhouse, Ulverston, got by 23d Duke of Oxford (31001), out of Kirklevington Duchess 7th by Duke Kirklevington (25982)—Kirklevington 18th by 3d Lord Oxford (22200)—Kirklevington 10th by Delhi (15865)—Kirklevington 8th by Gen. Canrobert (12926)—Kirklevington 7th by Earl of Derby (10177)—Kirklevington 4th by Earl of Liverpool (9061)—Kirklevington 1st by Duke of Northumberland (1940)—by Belvedere (1706)—by Son of 2d Hubback (2683)—a cow of Mr. Bates', descended from the stock of Mr. Maynard, of Eryholme.

GRAND PRINCE OF LIGHTBURNE (36730)—Red and white, calved June 26, 1875, bred by Mr. A. Brogden, Ulverston, by 2d Duke of Gloster (28392)—Grand Princess of Lightburne by Grand Duke 10th (21848)—Princess 2d by 3d Duke of Thorndale (17749)—Lady Sale by Old Rowley (15020)—Lady Coke by Jethro Tull (11616)—by General Sale (8099)—by Napier (6238)—by Mameluke (2258)—by Waterloo (2816)—by Baron (58)—by Phenomenon (491)—by Favorite (252)—by Favorite (252)—by Favorite (252)—by Hubback (319)—by Snowdon's Bull (612)—by Waistell's Bull (669)—by Masterman's Bull (422)—by Studley Bull (626).

COWS.

OXFORD MINSTREL 2D (vol. 22, p. 300, E., under dam)—Red and white, calved May 11, 1874, bred by G. Ashburner, got by Duke of Oxford (31004), out of Park Minstrel by Hematite (21917)—Minstrel 4th by 10th Duke of Oxford (17739)—Minstrel 2d by Prince of Gloster (13517)—Minstrel by Count Conrad (3510)—Magic by Wallace (5586)—by Wellington (2824)—by Marmion (406)—by Merlin (430)—by Layton (366)—by Phenomenon (491)—by Favorite (252)—by Favorite (252)—by Hubback (319)—by Snowdon's Bull (612)—by Waistell's Bull (669)—by Masterman's Bull (422)—by Studley Bull (626).

OXFORD'S ELVIRA (vol. 22, p. 300, E., under dam)—Red and white, calved March 5, 1875, bred by G. Ashburner, Low Hall, got by Duke of Oxford (31004), out of Elvira 8th by 10th Grand Duke (21848)—Elvira 3d by 8th Duke of Oxford (15939)—Ruby Rose 2d by The Baronet (10918)—by Burgundy (7861)—by Pilot (4707)—by Navarino (2352)—by Favorite (256)—by Phenomenon (491)—by Favorite (252)—by Favorite (252)—by Hubback (319)—by Snowdon's Bull (612)—by Waistell's Bull (669)—by Masterman's Bull (422)—by Studley Bull (626).

LIGHTBURNE GWYNNE (vol. 22, p. 300, E., under dam)—Red and white, calved May 5, 1875, bred by G. Ashburner, got by Grand Duke of Lightburne 3d (28761), out of Park Minstrel by Hematite (21917)—Minstrel 4th by 10th Duke of Oxford (17739)—Minstrel 2d by Prince of Gloster (13517)—Minstrel by Count Conrad (3510)—Magic by Wallace (5586)—by Wellington (2824)—by Marmion (406)—by Merlin (430)—by Layton (366)—by Phenomenon (491)—by Favorite (252)—by Favorite (252)—by Hubback (319)—by Snowdon's Bull (612)—by Waistell's Bull (669)—by Masterman's Bull (422)—by Studley Bull (626).

DAME GWYNNE (vol. 22, p. 300, E.)—Red and white, calved June 24, 1871, bred by Mr. H. Caddy, Roughholm, Bootle, Cumberland, got by Waterloo Cherry (27763), out of Daisy Gwynne by Knight of Distington (18158)—Dolly Gwynne by Duke of York (14461)—Young Dowager Gwynne by St. Thomas (10777)—Dowager Gwynne by Prime Minister (2456)—by Wallace (5586)—by Marmion (406)—by Merlin (430)—by Layton (366)—by Phenomenon (491)—by Favorite (252)—by Favorite (252)—by Hubback (319)—by Snowdon's Bull (612)—by Waistell's Bull (669)—by Masterman's Bull (422)—by Studley Bull (626).

CHERRY OXFORD 2D—Roan, calved Feb. 2, 1873, bred by W. Ashburner, Netherhouse, Ulverston, got by Baron Bates (30421), out of Cherry Princess 2d by Lord Oxford 2d (20215)—Southwick Cherry (vol. 18) by 3d Grand Duke (16182)—Cherry Bloom by Heir-at-Law (13005)—Cherry Blossom by Roland (2556)

—Old Cherry by Pirate (2430)—by Houghton (318)—by Marshal Blucher (416)—from the stock of Messrs. Wright & Charge.

ROSE OF RABY 2D (vol. 22, p. 301, E.)—Roan, calved March 29, 1874, bred by Mr. W. Ashburner, got by Baron Wellington (30495), out of Maid of Lorne by 6th Duke of Airdrie (19602)—Rose of Raby by Lumley (16478)—Peace by 2d Grand Duke (12961)—Dolly by 2d Cleveland Lad (3408)—by 4th Duke of Northumberland (3649)—by 2d Earl of Darlington (1945)—a cow of Mr. Bates'.

1854. WILL ASHTON.
Galt, Canada.

RATTLER 887 [595]—White, calved Feb. 24, 1854, bred (supposed) by R. Ashton, Limefield, got by Gilliver (11529), out of Rosebud by Earl of Durham (5965)—White Rosalind by Pedestrian (7321)—Roan Rosalind by Edsom (1956)—Young Rosalind by Scipio (1421)—Rosalind by Hector (1104)—Rose by Midas (435)—Red Rose by Marquis (407)—by Chilton (136)—by Ben (70).

MELODY (vol. 2, p. 475)—Red and white, calved Feb. 11, 1852, bred (supposed) by R. Ashton, Limefield, England, got by Valiant (10989), out of Mi by Tom o' Lincoln (8714)—Mirth by Buchan Hero (3238)—Merry by Rockingham (2550)—Moral by Norfolk (2377)—Vesta by Frederick (1060)—Vestris by Cato (1794)—Verbena by Son of Old Wellington (679)—bred by and from the herd of Mr. Robertson, of Ladykirk.

THE OCEAN 1032—Red, calved June 5, 1854, bred (supposed) by R. Ashton, Limefield, got by Rivington (13600), out of Melody by Valiant (10989), &c., as above.

LADY EVELYN (vol. 2, p. 425)—Red and white, calved June 1, 1853, bred (supposed) by R. Ashton, Limefield, got by Valiant (10989), out of Etiquette by Robin Hood (9555)—Eudine by Lord John (4257)—Eudine by Sandy (5088)—by Blyth Eclipse (1728)—by Blyth Comet (85).

May, 1863. JOHN ASHWORTH.
Belmont, Ottawa, Canada.

SWEETMEAT (20924) [868]—Roan, calved Nov. 12, 1861, bred by J. Robinson, Clifton Pastures, Bucks, got by Duke of Leinster (17724), out of Sweetheart 2d by Earl of Dublin (10178)—Sweetheart by Accordion (5708)—Charmer by Little John (4232)—Graceful by Caliph (1774)—Sylph by Sir Walter (2637)—by Hotspur (1117)—by Coxcomb (928)—by Midas (435)—Rachel by Comet (155)—Russell by R. Colling's son of Favorite (252)—by same son of Favorite (252)—by Hubback (319).

1864. JOHN ASHWORTH.
Belmont, Canada.

RED DUCHESS (vol. 1, p. 442, C. H. B., and vol. 16, p. 648, E. H. B.)—Red, calved Jan. 9, 1860, bred by Edward Lawferd, Southcott, near Leighton Buzzard, by John O'Gaunt (16322), out of Duchess by Clarendon (12605)—Lily by Honeycomb (10330)—Old Moss Rose by Bower's Bull (19332)—by May Duke (424).

TURK'S DELIGHT (vol. 1, p. 487, C. H. B., and vol. 15, p. 451, E. H. B.)—Red and white, calved Feb. 8, 1861, bred by James H. Langston, Sarsden House, Chepping Norton, got by Royal Turk (16875), out of Delightful by Field Marshal (14545)—Dinah by Lord Milton (10461)—Roan Daisy by Prince of Wales (8432)—Daisy by Bucephalus (6816)—by Stanhope (5315)—by Blyth Favorite (801)—by a son of Wellington (683).

August, 1881. H. Y. ATTRILL.
Goderich, Ont., Canada.

5TH DUKE OF TREGUNTER (33743)—Roan, calved April 6, 1874, bred by Col. Gunter, Wetherby Grange, Yorkshire, England, got by 4th Baron Oxford (25580), out of Duchess 94th by 2d Duke of Wharfdale (19649)—Duchess 84th by Archduke (14099)—Duchess 72d by 4th Duke of Oxford (11387)—Duchess 67th by

Usurer (9763)—Duchess 59th by 2d Duke of Oxford (9046)—Duchess 56th by 2d Duke of Northumberland (3646)—Duchess 51st by Cleveland Lad (3407)—Duchess 41st by Belvedere (1706)—Duchess 32d by 2d Hubback (1423)—Duchess 19th by 2d Hubback (1423)—Duchess 12th by The Earl (646)—Duchess 4th by Ketton 2d (710)—Duchess 1st by Comet (155)—by Favorite (252)—by Daisy Bull (186)—by Favorite (252)—by Hubback (319)—by J. Brown's Red Bull (97).

GRAND DUCHESS 28TH (vol. 26, p. 574, E.)—Roan, calved Aug. 5, 1874, bred by R. E. Oliver, Sholebroke Lodge, England, got by 3d Duke of Clarence (23727), out of Grand Duchess 17th by Imperial Oxford (18084)—Grand Duchess 10th by Grand Duke 3d (16182)—Grand Duchess 5th by Prince Imperial (15095)—Grand Duchess 2d by Grand Duke (10284)—Duchess 51st by Cleveland Lad (3407)—Duchess 41st by Belvedere (1706)—Duchess 32d by 2d Hubback (1423)—Duchess 19th by 2d Hubback (1423)—Duchess 12th by The Earl (646)—Duchess 4th by Ketton 2d (710)—Duchess 1st by Comet (155)—by Favorite (252)—by Daisy Bull (186)—by Favorite (252)—by Hubback (319)—by J. Brown's Red Bull (97).

GRAND DUKE OF CONNAUGHT AND RIDGEWOOD 46202—Roan, calved Sept. 6, 1881, bred by R. E. Oliver, calved the property of Mr. Attrill, after dam's landing in Canada, got by Duke of Connaught (33604), out of Grand Duchess 28th by 3d Duke of Clarence (23727), &c., as in dam, next above.

GRAND DUCHESS 35TH (vol. 25, p. 607, E.)—Roan, calved March 30, 1878, bred by R. E. Oliver, got by Duke of Underley (33745), out of Grand Duchess 25th by 2d Duke of Tregunter (26022)—Grand Duchess 23d by Grand Duke 7th (19877)—Grand Duchess 17th by Imperial Oxford (18084)—Grand Duchess 10th by Grand Duke 3d (16182)—Grand Duchess 5th by Prince Imperial (15095)—Grand Duchess 2d by Grand Duke (10284)—Duchess 51st by Cleveland Lad (3407)—Duchess 41st by Belvedere (1706)—Duchess 32d by 2d Hubback (1423)—Duchess 19th by 2d Hubback

(1423)—Duchess 12th by The Earl (446), &c., as in Grand Duchess 28th, above.

GRAND DUCHESS OF RIDGEWOOD (vol. 23, p. 17861)— Roan, calved May 26, 1881, bred by R. E. Oliver, got by Cherry Grand Duke 8th (39515), out of Grand Duchess 35th by Duke of Underley (33745), &c., as in dam, above.

Nov., 1883. H. Y. ATTRILL.

BARONESS OXFORD 12TH (vol. 27, p. 369, E.)—Red, calved Sept. 19, 1880, bred by Duke of Devonshire, Holkar Hall, got by Duke of Gloster 7th (39735), out of Baroness Oxford 4th by 3d Duke of Hillhurst (30975)—Baroness Oxford by 2d Duke of Claro (21576)—Lady Oxford 5th by 3d Duke of Thorndale (17749)—Lady Oxford 4th by Grand Duke 2d (12961)—by The Lord of Eryholme (12205)—by 3d Duke of York (10166)—by Duke of Northumberland (1940)—by Short Tail (2621)—by Matchem (2281)—by Young Wynyard (2859).

GRAND DUCHESS OF OXFORD 60TH (vol. 28, p. 362, E.) —Red and white, calved Nov. 8, 1881, bred by Duke of Devonshire, Holkar Hall, got by Baron Oxford 8th (41057), out of Grand Duchess of Oxford 26th by Baron Oxford 4th (25580)— Grand Duchess of Oxford 11th by Grand Duke 10th (21848)— Grand Duchess of Oxford 5th by Priam (18567)—Countess of Oxford by Earl of Warwick (11412)—by 4th Duke of York (10167) —by 2d Duke of Northumberland (3646)—by Short Tail (2621) —by Matchem (2281)—by Young Wynyard (2859).

GRAND DUCHESS OF OXFORD 63D (vol. 29, p. 413, E.) —Red, calved Jan. 23, 1882, bred by Duke of Devonshire, got by Baron Oxford 8th (41057), out of Grand Duchess of Oxford 45th by 5th Duke of Wetherby (31033)—Grand Duchess of Oxford 23d by Baron Oxford 4th (25580)—Grand Duchess of Oxford 14th by Grand Duke 10th (21848)—Grand Duchess of Oxford 7th by Lord Oxford (20214)—Grand Duchess of Oxford by Grand Duke 3d (16182)—by Earl of Warwick (11412)—by 4th Duke of York

(10167)—by 2d Duke of Northumberland (3646)—by Short Tail (2621)—by Matchem (2281)—by Young Wynyard (2859).

MAID OF MOWBRAY (vol. 27, p. 622, E.)—Roan, calved Nov. 18, 1880, bred by R. Welsted, Ballywalter, Castle Town Roche, Ireland, got by Royal Mowbray (42330), out of Maid of the Vale by Master of Arts (34816)—Maid of the Mulla by England's Glory (23889)—Matchless by Uncle Ned (19026)—Mabel by Sir James (16980)—Marchioness by Prince Regent (18637)—Maid Royal by Blood Royal (14169)—Memento by Vanguard (10994)—Magnet by Baron Warlaby (7813)—Mosaic by Leonard (4210)—by Prince Comet (1342)—by Constellation (163)—by Prince of Waterloo (528)—by Young Favorite (255).

CHERRY GRAND DUCHESS OF RIDGEWOOD—Red roan, calved Nov. 26, 1883, at Quebec, Can., bred by L. H. Wraith, got by Lord Turncroft of Oxford 4th (46708), out of Cherry Queen 2d by Duke of Tregunter 7th (38194), &c., as in dam, below.

WINSOME TWENTY-SECOND (vol. 26, p. 404, E.)—Red, calved March 15, 1879, bred by Duke of Devonshire, got by 7th Duke of Gloster (39735), out of Winsome 4th by Grand Duke 10th (21848)—Winsome by Oxford 2d (18507)—Beauty by Crusade (7938)—Bright Eyes by 3d Duke of York (10166)—by 2d Cleveland Lad (3408)—by Duke of Northumberland (1940)—by Belvedere (1706)—by Emperor (1975)—by Wonderful (700)—by Cleveland (145)—by Butterfly (104)—by Hollon's Bull (313)—by Mowbray's Bull (2342)—by Masterman's Bull (422)—descended from M. Dobison's stock.

COUNTESS OF BARRINGTON TENTH (vol. 26, p. 402, E.)—Red and white, calved Oct. 17, 1879, bred by Duke of Devonshire, got by Duke of Gloster 7th (39735), out of Countess of Barrington 8th by 2d Duke of Tregunter (26022)—Lady Laura Barrington by Baron Oxford 4th (25580)—Lady Ellen Barrington by Lord Stanley (24467)—Grand Duchess of Barrington by Grand Duke 7th (19877)—Countess of Barrington 2d by 9th Duke of Oxford (17738)—by Grand Duke 3d (16182)—by Grand Turk

(12969)—by Earl of Derby (10177)—by Earl of Liverpool (9061)—by 2d Duke of Cambridge (3638)—by Belvedere (1706)—by son of Herdsman (304)—by Wonderful (700)—by Alfred (23)—by Young Favorite (6994).

CHERRY QUEEN SECOND (vol. 27, p. 632, E.)—Roan, calved April 16, 1880, bred by L. H. Wraith, Newfield, Lower Darwin, Lancashire, got by Duke of Tregunter 7th (38194), out of Cherry Queen by Baron Oxford 5th (27958)—Cherry Princess by Gen. Napier (24023)—Cherry Duchess 8th by Grand Duke 3d (16182)—Cherry Duchess 6th by Grand Duke 3d (16182)—by Grand Duke 2d (12961)—by Grand Duke (10284)—by Sheldon (8557)—by The Colonel (5428)—by Thorp (2757)—by Pirate (2430)—by Houghton (318)—by Marshal Blucher (416)—from the stock of Messrs. Wright & Charge.

JAMES BAGG and S. WAIT.

COWS.

NONSUCH (vol. 5, p. 741, E., under dam)—Roan, calved May 5, 1836, bred by Mr. Cattley, got by Luck's-All (2230), out of Nonpareil by Bedford (68)—Palmflower by Barrister (776)—Rosanna by Whitworth (1584)—by Young Jupiter (1148)—by White Comet (1582)—by a son of Charge's Gray Bull (872).

EMPRESS (vol. 5, p. 832, E., under dam)—Roan, calved 1837, bred by Mr. Cattley, got by Cyrus (3538), out of Raspberry by Bedford Jr. (1701)—Mulberry by Isaac (1129)—by Whitworth (1584)—by White Comet (1582)—by a son of Charge's Gray Bull (872). Sold to Dr. Bradford, Tenn.

COUNTESS (vol. 5, p. 741, E.)—Roan, calved in Oct., 1837, bred by Robt. Cattley, Bransby, got by Cyrus (3538), out of Nonpareil by Bedford (68)—Palmflower by Barrister (776)—Rosanna by Whitworth (1584)—by Young Jupiter (1148)—by White Comet (1582)—by a son of Charge's Gray Bull (872).

BULLS.

HARKAWAY (vol. 5, p. 389, E.)—Roan, calved 1839, bred by Mr. Cattley, got by Espersykes (3738), out of Fortune by Luck's-All (2230)—Fanny by Remus (2524)—Flora by Freeman (1062)—by White Comet (1582).

RALPH (4862)—White, calved May 5, 1839, bred by Mr. J. Colling, got by Borderer (3191), out of Ruby by The Monk (2752)—Rally by Magnum Bonum (2243)—Red Rose by Forester (1055)—Rachel by Frederick (1060)—by Planet (502).

YORKSHIREMAN (5701)—Roan, calved in 1839, bred by Mr. Cattley, got by Plato (2433), out of Nonsuch by Luck's-All (2230)—Nonpareil by Bedford (68)—Palmflower by Barrister (776)—Rosanna by Whitworth (1584)—by Young Jupiter (1148)—by White Comet (1582)—by son of Charge's Gray Bull (872).

MILO 711—Bred by W. F. Paley, got by ——————, out of Sprightly by Fitz Roslyn (2026)—Clarinda by Buckingham (1755)—Clara by Election (1961)—Young Charlotte by Pilot (1319)—Charlotte by Clarence (888)—by Lame Bull (358)—by Punch (531)—by Hubback (319).

NOTE.—Mr. Bagg says he never owned nor imported such a bull. See vol. 3, p. 26. He appears as sire of Miss Kerr, vol. 2.

————————

1839-40. JAMES BAGG and MR. WAIT.
Into Kentucky.

COWS.

POMONA (vol. 3, p. 564, E.)—White, calved Jan. 14, 1834, bred by Mr. Cattley, got by Bedford Jr. (1701), out of Mulberry by Isaac (1129)—by Whitworth (1584)—by White Comet (1582)—by a son of Mr. Charge's Kitt (7127).

LAURENTIA (vol. 5, p. 543, E.)—Roan, calved in 1835, bred by Mr. Cattley, got by Bedford Jr. (1701), out of Columbine by Baronet (775)—Carnation by Young Jupiter (1148)—by White Comet (1582)

—by a grandson of Favorite (252)—by a son of Charge's Gray Bull (872).

Vol. 3, p. 319, E. H. B., gives Laurentia's birth as in 1834; vol. 5, p. 543 gives it as 1835. Sold to Dr. S. Bradford, of Nashville, Tenn.

AMELIA (vol. 5, p. 543, E., under dam)—Roan, calved in 1838, bred by Mr. Cattley, got by Plato (2433), out of Laurentia by Bedford Jr. (1701)—Columbine by Baronet (775)—Carnation by Young Jupiter (1148)—White Comet (1582)—by a grandson of Favorite (252)—by a son of Charge's Gray Bull (872).

YOUNG AMELIA (vol. 2, p. 600)—Bred by Mr. Cattley, got by Marton Comet (4409), out of Amelia by Plato (2433)—Laurentia by Bedford Jr. (1701)—Columbine by Baronet (775)—Carnation by Young Jupiter (1148)—by White Comet (1582)—by a grandson of Favorite (252)—by a son of Charge's Gray Bull (872).

SYLVIA (vol. 3, p. 639, E.)—White, calved April 4, 1835, bred by Mr. Cattley, got by Luck's-All (2230), out of Stately by Romulus (1403)—by Juniper (1144)—by Snowdrop (614)—by White Comet (1582)—by a son of Charge's Gray Bull (872).

CLAID—Got by Brandon (3206), dam by Sir Charles (1440)—by Wellington (683)—by Phenomenon (491)—by Pansy (186)—by a son of Hubback (319)—by Mr. Hill's Red Bull (310).
Owned by Mr. R. Cockrell, Tennessee.

BLOOM (vol. 5, p. 94, E.)—Roan, calved in 1837, bred by Mr. Cattley, Bransby, got by Luck's-All (2230), dam by Baron (1681)—by Emperor (1013).

YOUNG CARNATION (vol. 5, p. 136, E.)—Roan, calved March 10, 1833, got by Bedford Jr. (1701), out of Pink by Isaac (1129)—Carnation by Young Jupiter (1148)—by White Comet (1582)—by a grandson of Favorite (252)—by a son of Charge's Gray Bull (872).

6

RASPBERRY (vol. 5, p. 708, E.)—Roan, calved Jan. 18, 1833, bred by Mr. Cattley, got by Bedford Jr. (1701), out of Mulberry by Isaac (1129)—by Whitworth (1584)—by White Comet (1582)—by a son of Charge's Gray Bull (872).

SPLENDISSIMA (vol. 5, p. 136, under dam)—Roan, calved in 1839, bred by Mr. Cattley, got by Marton Comet (4409), out of Young Carnation by Bedford Jr. (1701), &c., as in dam, above.

MAJOR (4345)—White, calved in 1838, bred by Lord Hunting-field, got by Cramer (3515), out of Vesta by Colton (1849)—Virtue by Amor (747)—Virtue by Candour (107)—Violet by Petrarch (488)—by an own brother to Mr. R. Colling's White Heifer—by Butterfly (104)—by Globe (278).

JAKE 593—Twin to Esau—Roan, calved June, 1841, bred by M. R. Cockrill, got by Baronet (1688), out of Claid by Brandon (3206)—by Sir Charles (1440), &c., as in dam Claid.

ESAU—Twin to Jake, above.

1857. SAMPSON BAKER.
Charlottsville, Can.

LADY KINGSCOTE—Bred by Col. Kingscote, got by 6th Duke of York (15950), out of Lofty by Helicon (2107).

S. BANCROFT.

FAVORITE (bull)—See Red Crump 2d (vol. 2, p. 528).

1856. ALEX. BARRETT,
Of Henderson, Ky. To New Orleans.

SIR DAVID (16969)—Roan, calved Feb. 1, 1858, bred by J. Douglas, Athelstaneford, Scotland, got by Sir James the Rose (15290), out of Bonna Fortuna by Lord Marquis (10459)—Fanchette by Austerlitz (3063)—Fair Sovereign by Sir Thomas Fairfax

(5196)—Superb by Young Colling (1843)—by Young Remus (2523) —by Remus (550)—by Greathead's Gray Bull (3936)—by Ellerton's Roan Bull (3708).

LORD MORPETH—

QUEEN OF TRUMPS (vol. 11, p. 655, E.)—Died on passage. Roan, calved Feb., 1849, bred by Mr. Unthank, Netherscales, got by Belleville (6778), out of Queen of Trumps by Capt. Shaftoe (6833)—Cherry by Pirate (2430)—by Houghton (318)—by Marshal Blucher (416)—Colling bred from the stock of Messrs. Wright & Charge.

NOTE.—Mr. Barrett shipped other short-horns, but lost them on the voyage, which proved very long and disastrous.

SIMON BEATTIE.

BESSIE BELL (vol. 7, p. 358, C. H. B.)—Roan, calved April 28, 1877, bred by Jas. Beattie, Annam, Scotland, got by Titan (35805), out of Bessie Lee (vol. 21, p. 562, E.) by Bentinck (28016) —Lady Towneley by Baronet (27930)—Miss Beverley 28th by Royal Butterfly 16th (20724)—Miss Beverley 3d by Voltigeur (13964) —Miss Beverley by Disraeli (10125)—Lady Beverley by Geo. Bentinck (9317)—by 2d Earl of Beverley (5963)—by 2d Cleveland Lad (3408)—by Red Darlington.

PEACH (vol. 7, p. 358, C. H. B.)—Roan, calved March 15, 1878, bred by J. Johnson, Halleath, Dumfrieshire, Scotland, got by Magician (34720), out of Primrose (vol. 23, p. 515, E.) by Keir Butterfly 8th (28950)—Pansy by Union Jack (30193)—Miss Pigot by Baron Cherry (19268)—Lady Pigot by Mac Turk (14872)—Vanity by Heir-at-Law (13005)—Haddam by Baron of Ravensworth (7811) —Abbess of St. Mary's by Hudibras (10339)—Crucifix by Baronet (1686)—by Spectator (2688)—by Fitz Remus (2025)—by Whitworth (695)—a cow, bought of Mr. Mason, of Chilton.

BUD (vol. 7; p. 358, C. H. B.)—White, calved Oct. 20, 1880, bred by J. Johnson, got by Wallaby Warrior (44216), out of Peach by Magician (34720), &c., as in Peach, above.

1861. SIMON BEATTIE.

BARON SOLWAY [45]—Roan, calved Oct. 9, 1860, bred by R. Syme, Red Kirk, Dumfrieshire, Scotland, got by General Havelock (16130), out of Snowdrop by Strathmore (6547)—Catherine 3d by Playfellow (6297)—Young Catherine by Sir William (12102)—Catherine by Emperor (1974).

FASHION ALIAS SNELL (vol. 1, p. 296, C. H. B.)—Red and white, calved March 30, 1861, bred by R. Syme, got by Gen. Havelock (16130), out of Snip by Tweedside (12246)—Spoors by Remus (11987)—by Strathmore (6547)—by Sir William (12102)—by Togston (5487).

FAVORITE (vol. 1, p. 298, C. H. B.)—Roan, calved March, 1861, bred by R. Syme, got by General Havelock (16130), out of Silky by Baron of Kidsdale (11156)—Lady by Remus (11987)—Young Lady Bird by Strathmore (6547)—Lady Bird 2d by Playfellow (6227)—Lady Bird by Scrip (2604)—by Thornington (5472).

PRESIDENT 15173 [536]—Roan, calved Dec. 2, 1857, bred by R. Syme, got by Tweedside (12246), out of Syren by Petrarch (7329)—Countess by Ethelred (5990)—Sarah by Hecatomb (2102)—by Belvedere 2d (3127)—by Barmpton (5774)—by Kitt (2179)—by Middleton's Bull (438).

YOUNG SNOWDROP—Roan, calved Nov. 15, 1858, bred by R. Syme, got by Tweedside (12246), out of Jane by Lanercost (11665)—Mary by Sir William (12102)—Young Rose by Remus (11987)—by Togston (5487). C. H. B., vol. 1, p. 478.

LADY ANNE (vol. 1, p. 333, C. H. B.)—Red and white, calved Aug. 11, 1861, bred by R. Syme, got by General Havelock (16130), out of Young Snowdrop by Tweedside (12246), &c., as in dam, above.

1870. SIMON BEATTIE.

BISMARCK [983]—Roan, calved Jan. 28, 1871, bred by James Beattie, Newbie House, Annan, Scotland, got by Bentinck [957] (28016), out of British Beauty by Elegant Prince (21676)—Beauty by Duke of Cumberland (15927)—Roan Lady by Sir Humphrey (9640)—White Lady by Major (4356)—Lady Lowther by Prime Minister (2456)—Strawberry by Hartley's Bull (8134)—by Magician (1184)—by Western Comet (689).

HER HIGHNESS (vol. 2, p. 504, C. H. B.)—Roan, calved Dec. 9, 1868, bred by J. B. Booth, Yorkshire, got by Major (21312), out of Request by Percival (20486)—Romance by Hartforth (16224)—Lady Day by Man Friday (13290)—Anna by The Irishman (5446) —by Sir Richard (5175)—by a son of Booth's White Bull—by Seymour's Red Bull.

MARY BOOTH (vol. 2, p. 641, C. H. B.)—Roan, calved Nov. 15, 1870, bred by J. B. Booth, got by K. C. B. (26492), out of Her Highness by Major (21312), &c., as above.

MARQUIS OF WORCESTER 3D (vol. 23, p. 429, E., under dam)—Red, calved March 17, 1875, bred by Earl of Dunmore, got by 3d Duke of Hillhurst (30975), out of Lady Worcester 2d by Charleston (21400)—Clear Star by Marton Duke (22307)—Bright Star by Red Duke (18676).

Oct. 24, 1870. SIMON BEATTIE.
By Ship European.

INNOCENT—Red and white cow, calved Nov. 1. 1863, bred by C. Barnard, Harlowbury, Essex, got by 5th Duke of York (19652), out of Jessamine by Nelson (14989)—Jessie by Lottery (10472)—Jenny Lind by Fame (10221)—Jessy by Cramer (6907)—Lady Jane by Plenipo (4724)—by Childers (1824)—by Umpire (2783)—by Ratify (2481)—by Wellington (680)—by Favorite (252)—by Punch (531). E. H. B., vol. 17, p. 532. C. H. B., vol. 2, p. 510.

BOTHWELL (25661) [1008]—Roan, calved May 29, 1868, bred by W. Torr, Aylesby Manor, got by Lord Blithe (22126), out of Blink Bonny by Booth Royal (15673)—Bright Dawn by Vanguard (10994) —Bright Phœbus by Crown Prince (10087)—Blanche 2d by Zadig (8796)—Blanche by Auld Robin Gray (6753)—White Rose by a bull of Mr. Crisp's—Rose by Burley (1766)—Young Anna by Isaac (1129)

—Anna by Pilot (496)—Ariadne by Albion (14)—Bright Eyes by
Lame Bull (359)—by Shipton (587)—by a son of Suwarrow (636)
—by a Son of Twin Brother to Ben (88)—by Twin Brother to
Ben (660).

LORD YORK (26766) [1714]—Roan, calved Jan. 17, 1868,
bred by Mr. Harward, Kidderminster, got by 3d Duke of Wharfdale
(21619), out of Duchess of York by 7th Duke of York (17754)—
Grand Duchess by 4th Duke of Oxford (11387)—Juliet by Sol
(8608)—Kate by Leo (4208)—by Treasurer (5513)—by Rupert
(2580)—by North Star (460)—by Cripple (173)—by Minor (441)—
by Freeman (269)—by Danby (190).

1872. SIMON BEATTIE.

DUCHESS OF WETHERBY—Roan heifer, calved June 20,
1870, bred by Hon. Col. Duncombe, Waresley Park, St. Neots,
got by Gen. Wetherby (24026), out of Girl of the Mist by Hypocrite
(19996)—Maid of the Mist by Admiral (14064)—Heather Bell by
Hero (18055)—Fanny by Rubens (5057)—by Young Matchem (4422)
—by Isaac (1129)—by Young Pilot (4702)—by Pilot (496)—by
Julius Cæsar (1143).

AZALEA—Roan heifer, calved Dec., 1870, bred by W. Desham,
Palmer's Green, Southgate, Middlesex, got by Bismarck (25637),
out of Acacia by Knight Errant (18154)—Amethyst by Magna
Charta (16486)—Applin by Lord Raglan (14849)—Amaryllis by
Burgomaster (12513)—by Baron Ravensworth (7811)—by Lycurgus
(7180)—by Zenith (5702)—by Guardian (3947)—by Firby (1040)—
by Ivanhoe (1131)—by Regent (544)—by Blyth Comet (85).

ROYAL DUKE 18798 (vol. 20, p. 616, E., under dam)—Calved
Feb. 25, 1872, bred by J. Beattie, Newbie House, Annan, Scotland,
got by Blood Royal (28047), out of Lady Towneley by Baronet
(27930)—Miss Beverley 28th by Royal Butterfly 16th (20724)—Miss
Beverley 3d by Voltigeur (13964)—Miss Beverley by Disraeli (10125)
—Lady Beverley by Lord George Bentinck (9317)—Blossom 2d
by 2d Earl of Beverley (5963)—by 2d Cleveland Lad (3408)—
Darlington.

1873. SIMON BEATTIE.

LADY KNOWLMERE (vol. 14, p. 624)—Roan, calved Oct. 19, 1873, bred by J. Beattie, got by Knight of Knowlmere (31542), out of Lady Gunter by Lord York (26766), &c., as in Lady Gunter.

1872. SIMON BEATTIE.

JUSTINE (vol. 17, p. 691, E., under dam)—Roan, calved Dec. 4, 1866, bréd by J. N. Beasley, Pittsford Hall, Northampton, got by Sir Launcelot (25159), out of Queen Janette by Royal Butterfly 5th (18756)—Countess Janette by Lilyvick (10421)—Young Jocund by Japetus (10359)—Jocund 2d by Monarch (7249)—Jocund by McIvor (2237)—by Northern Light (1280)—by Wellington (684)—by Phenomenon (491)—by Favorite (252)—by Favorite (252)—by Favorite (252)—by Hubback (319)—by Snowdon's Bull (612)—by Waistell's Bull (669)—by Masterman's Bull (422)—by Studley Bull (626).

WELCOME (vol. 12, p. 1287, and vol. 19, p. 789, E., under dam)—Roan heifer, calved May 31, 1870, bred by J. N. Beasley, got by 4th Grand Duke (19874), out of Winsome by Lycidas (20249)—Wildflower by Wolfsbane (15518)—Jessy by John Ford (9253)—Flower by Lord Warden (7167)—Farce by Zenith (5702)—by Orontes (4623)—by William (2840)—by Childers (1824)—by Richard (1376)—by Jupiter (342)—by Charles (127)—by Windsor (698)—by Chilton (136)—by Colonel (152).

1873. SIMON BEATTIE.

White Vale, Pickering, Ontario. By Ship Canadian, from Liverpool, August, 1873.

LADY GUNTER (vol. 14, p. 620)—Red heifer, calved in March, 1871, bred by James Beattie, Newbie House, Annan. N. B., got by Lord York (26766), out of Eliza by Mac Turk (14872)—Edith by Pride (10631)—Brilliant 3d by Baron of Ravensworth (7811)—Brilliant by Sir Thomas (5194)—by Noble Henry (2374)—

by Abraham (2905)—by Mustachios (4527)—by Simon (5134)—by
Young George (3885)—by George (276).

Sold to Geo. Murray, Racine, Wis.

MALMSEY (vol. 13, p. 767)—Roan heifer, calved April 1, 1870,
bred by Mr. T. Garne, Broadmoor, Gloucestershire, got by Royal
Benedict (27348), out of Moselle by Gondomar (19867)—Moss
Rose by Royal Oak (16870)—Young Moss Rose by Bashaw
(12449)—Moss Rose by Marchmont (9367)—Tortworth Rose by Fitz
Hardinge (8073)—Moss Rose by Augustus (6751)—by son of Anthony (1640)—by a bull of Mr. Champion's, of Blyth.

Sold to C. C. Parks, Waukegan, Ill.

MAID OF HONOR (vol. 20, p. 635, E., under dam, and vol. 14,
p. 692)—Roan heifer, calved Aug. 7, 1871, bred by Mr. T. Garne,
got by Royal Benedict (27348), out of Madeira by Monk (24616)
—Moselle by Gondomar (19867)—Moss Rose by Royal Oak
(16870)—Young Moss Rose by Bashaw (12449)—Moss Rose by
Marchmont (9367)—Tortworth Rose by Fitz Hardinge (8073)—
Moss Rose by Augustus (6751)—by son of Anthony (1640)—by a bull
of Mr. Champion's, of Blyth.

Sold to Mr. Geo. Murray, Racine, Wis.

LORD EGLINTOUN (31652)—Red bull, calved April 13,
1870, bred by C. A. Barnes, Charleywood House, Rickmansworth,
got by Royal Duke (32375), out of Evangelina by Lord Chancellor (20160)—English Emily by Englishman (19701)—Miss Emily
by Young Duke of Cambridge (14433)—Young Celia 2d by Lord
of the South (13216)—by Lord of the North (11743)—by 3d Duke
of Northumberland (3647)—by Bashaw (1692)—by Helmsman
(2109)—by Columella (904)—by Regent (544)—by Palatine (478)—
by Palmflower (480)—by Patriot (486)—by Driffield (223)—by C.
Holmes' Bull (314).

1881. SIMON BEATTIE.

LORD MONTRATH (46681)—Red and white, calved Dec. 28,
1880, bred by Hugh Aylmer, West Dereham Abbey, got by Lord
President (41908), out of Phillis 22d by High Sheriff (26392)—

Phillis 14th by Royal Broughton (27352)—Phillis 6th by General Hopewell (17953)—Phillis 2d by Red Knight (16809)—Phillis by Homer (14714)—Young Polly by Cardigan (12556)—Polly by Young Rufus (13649)—by Constitution (12634)—by Young Comet (1853)—descended from Jolly's Bull (4115).

Aug. 13, 1874. S. BEATTIE and W. M. MILLER.

By Ship Phœnician, from Glasgow to Canada.

ROYAL DEREHAM 24715 (35353)—Red, little white, calved May 18, 1873, bred by Hugh Aylmer, West Dereham Abbey, got by Royal Broughton (27352), out of Mistress May by Prince of Rosedale (24837)—Mistress Margaret by Pater Familias (18521)—Madrid by Valasco (15443)—Mistress Mary by Baron Warlaby (7813)—Waterwitch by Royal Buck (10750)—Hecate by Hopeful (10332)—by Hamlet (8126).

SHERIFF GWYNNE 24818 (35518)—Roan, calved Dec. 22, 1873, bred by Hugh Aylmer, West Dereham Abbey, got by High Sheriff (26392), out of Daffy Gwynne 6th by Ravenspur (20628)—Young Daffy Gwynne 5th by Young Duke of Cambridge 2d (17709)—Young Daffy Gwynne 3d by Lord Royston 2d (16449)—Daffy Gwynne by Sir Harry (10819)—Daphne Gwynne by Conservative (3472)—White Moll Gwynne by Wallace (5586)—Dorothy Gwynne by Marmion (406)—Daphne Gwynne by Merlin (430)—Nell Gwynne by Layton (366)—Nell Gwynne by Phenomenon (491)—Princess by Favorite (252)—by Favorite (252)—by Hubback (319)—by Snowdon's Bull (612)—by Waistell's Bull (669)—by Masterman's Bull (422)—by Studley Bull (626).

EDITH EMILY (vol. 20, p. 506, E., under dam)—Red, calved Feb. 22, 1872, bred by J. A. Mumford, Brill House, Oxfordshire, got by Caballer (28114), out of Etona by 3d Grand Duke (16182)—Lass of Dayrell by Earl of Dayrell (14474)—Kate 2d by Young Planet (11905)—Kate 1st by Admiral (8806)—by Ruler (10765)—bred by Sir H. R. Hoare, Bart.

7

LADY BARNES (vol. 20, p. 585, E., under dam)—Roan, calved March 31, 1872, bred by C. A. Barnes, Charleywood, Herts., got by Julius 4th (31451), out of Lady Beauford by Lord Wallace (24473)—British Queen by Sir Charles (16948)—Miss Amelia by Marquis of Bute (11788)—Miss Beauford by Red Roan Kirtling (10691)—Celia by 3d Duke of Northumberland (3647)—by Bashaw (1692)—by Helmsman (2109)—by Columella (904)—by Regent (544)—by Palatine (478)—by Palmflower (480)—by Patriot (486)—by Driffield (223)—by C. Holmes' Bull (314).

Sold to T. S. Smith, Afton, Ontario, Canada, for $575.

BULL CALF—Bred by C. A. Barnes, got by Prince Alfred (35083), out of Lady Barnes by Julius 4th (31451), &c., as in dam, above.

PRINCESS OF DENMARK 2D—Red, calved July 24, 1871, bred by R. H. Crabb, Baddow, Essex, got by Cambridge Duke 3d (23503), out of Princess of Denmark by Baron Roxwell (21240)—Eugenie by Danby (14366)—Princess Alice by Essex Hero (9096)—Duchess by Duke of Marlborough (3645)—by Captain (3273)—by Emperor (1974)—by Snowball (2647)—by son of St. Albans (2584)—by Col. Trotter's son of Lawnsleeves (365)—by Barnaby (1678).

Under dam, E. H. B., vol. 20, p. 704, and A. H. B., vol. 15, p. 835. Sold to S. Meredith & Son, Cambridge City, Ind., June 15, 1875, for $1,000.

MOUNTAIN ROSE 3D (vol. 15, p. 787)—Roan, calved Aug. 8, 1872, bred by T. Garne, Broadmoor, Gloucestershire, got by Buccaneer (25693), out of Mountain Rose 2d by Masterpiece (24561)—Mountain Rose by The Druid (20948)—Young Moss Rose by Bashaw (12449)—Moss Rose by Marchmont (9367)—Tortworth Rose by Fitz Hardinge (8073)—Moss Rose by Augustus (6751)—by son of Anthony (1640)—by a bull of Mr. Champion's of Blyth.

Under dam, E. H. B., vol. 20, p. 665. Sold to S. Meredith & Son, of Cambridge City, Ind., June 16, 1875, for $525.

HALSTEAD SURPRISE (vol. 14, p. 563)—Red and white, calved Sept. 30, 1870, bred by R. H. Crabb, got by Old Sam (32449), out of Princess Augusta by Baron Roxwell (21240)—Duchess of Cambridge by Orestes (15027)—Hethel Thorn by Tortworth (10966)—Myrtle by Essex Hero (9096)—Duchess by Duke of Marlborough (3645)—by Captain (3273)—by Emperor (1974)—by Snowball (2647)—by son of St. Albans (2584)—by son of Lawnsleeves (365)—by Barnaby (1678)—by Barber's Bull.

Under dam, E. H. B., vol. 20, p. 701.

NECTAR (vol. 14, p. 750)—Red, calved in Aug., 1872, bred by W. S. Marr, Upper Mill, Tarves, Aberdeen, Scotland, got by Heir of Englishman (24122), out of Nectar 9th by Romeo (27327)—Nectar 5th by Lord Lyons (22173)—Nectar 3d by Lord of Lorne (18258)—Nectar by Jemmy (11611)—Nectarine by Balco (9918)—Peach by Homer (2134)—Bloom by Sir Launcelot (5166)—by Viceroy (5561)—by Bulmer (1760)—by Appleton (1647)—by Newtonian (2368)—by a bull bred by Mr. Cattley.

Sold to Geo. Murray, Racine, Wis.

CHANDOS 22350—Red and white, calved April 14, 1875, bred by S. Beattie, got by Royal Butterfly 20th (25007), out of Chaplet by Grand Duke 15th (21852), &c., as in Chaplet, above.

Sold to Gen. Meredith, Indiana.

BARONESS CONYERS (vol. 15, p. 426)—Roan or red and white, calved Sept. 16, 1871, bred by Mr. J. Outhwaite, Bainesse, Yorks, got by Baron Killerby (27949), out of Sylvia by Champion (23529)—Sunflower by son of Apollo (9899)—Sally by Chieftain (10048)—Ruby by Postmaster (9487)—by Albert (7767)—by Noble (4579)—by Corelli (3485)—by Wauldby (2818)—by son of Wonderful (700).

Under dam, E. H. B., vol. 21, p. 876, and E. H. B., vol. 24, p. 382. Sold June 16, 1875, to S. Meredith & Son, Cambridge City, Ind., for $1,650; returned to England and sold Sept. 4, 1877, to C. H. Cock, for 60 guineas.

BUTTERFLY'S DUCHESS (vol. 14, p. 452)—Roan, calved July 16, 1870, bred by Mr. G. Garne, Churchill Heath, Oxon, got by Royal Butterfly 20th (25007), out of Delicacy by The Druid (20948)—Destiny by Progression (16770)—Damsel by Enterprise (11443)—Blonde by Patriot (10595)—by son of Elevator (6969)— by No Mistake (8357)—by Young Consul (6893)—by Fairfax (1023)—by Speculation (1472). '

Under dam, E. H. B., vol. 19, p. 473. Sold to Geo. Murray, Racine, Wis.

CHAPLET (vol. 15, p. 478)—Red, calved January 16, 1870, bred by Mr. F. Leney, Wateringbury, Kent, got by Grand Duke 15th (21852), out of Columbine by Lord of the Harem (16430)— Countess by Gloster's Grand Duke (12949)—Charmer by 4th Duke of York (10167)—Chaplet by Usurer (9763)—by Duke of Cornwall (5947)—by Morpeth (7254)—by Helicon (2107)—by Henwood (2114)—by Nestor (452)—by Harold (291)—by Meteor (432)—by Comet (155)—by Cupid (177).

Under dam, E. H. B., vol. 19, p. 450. Sold to S. Meredith & Son, Cambridge City, Ind., for $850.

DELIGHT (vol. 15, p. 503)—Roan, calved March 8, 1868, bred by Mr. W. Lambert, Elrington Hall, Northumberland, got by Pizarro (20497), out of Venus by Master Annandale (14916)— Sprightly by Young Earl (14467)—Splendor by Snowy Down (8607)—Angela by Prince Albert (4778)—by son of Cumberland (5256)—by Exmouth (3747)—by Prince (4765)—by Leopold (2199).

Under dam, vol. 18, p. 761, E. H. B. Sold to S. T. Spangler, Winthrop, Iowa, for $750.

TEA ROSE (vol. 15, p. 925)—Roan, calved Feb. 21, 1871, bred by Mr. T. Marshall, Howes Annam, Scotland, got by Lord Bacon (26607), out of Moss Rose by Knight Errant (18154)—Flora 2d by Earl of Derby (12810)—Moss Rose by Prince Edward (6334)— Flora by Traveler (6617)—by Charles (3343)—by Commodore (5874)—by Cupid (5900)—by Parson (1306)—by St. John (572)— by Pope (514)—by Chilton (136).

Sold to A. L. Stebbins, Port Huron, Mich., for $550.

VERBENA ROYAL (vol. 14, p. 894)—Roan, calved July 13, 1872, bred by Mr. J. Downing, Ashfield, County Cork, Ireland, got by Royal Duke (25014), out of Vestal Queen by Hero of Thorndale (18061)—Vidonia 4th by Western Wonder (17225)—Vespasia by Australian (12414)—Young Vidonia by Rolla 2d (13618)—by Sir Thomas Fairfax (5196)—by Norfolk (2377)—by Burley (1766)—by Pilot (496)—by Warlaby (672)—by Albion (14)—by Lame Bull (359)—by Shipton (587)—by son of Suwarrow (636)—by Son of Twin Brother to Ben (88)—by Twin Brother to Ben (660).

Under dam, vol. 20, p. 803, E. H. B. Sold to S. Meredith & Son, Cambridge City, Ind.

ROYAL OXFORD GWYNNE 21796 (35396)—Red and white, calved Aug. 5, 1873, bred by Major Webb, Elford House, got by Baron Oxford (23375), out of Gipsy Gwynne by Grand Duke of Lightburne (26290)—Goody Gwynne by 5th Grand Duke (19875)—Golden Gwynne by May Duke (13320)—Sylvia Gwynne by Duke of Cambridge (12747)—Silky Gwynne by Capt. Hardinge (10023)—Sophia Gwynne by St. Thomas (10077)—Sall Gwynne by Prime Minister (2456)—Cripple Gwynne by Marmion (406)—Daphne Gwynne by Merlin (430)—Nell Gwynne by Layton (366)—Nell Gwynne by Phenomenon (491)—Princess by Favorite (252)—by Favorite (252)—by Hubback (319)—by Snowdon's Bull (612)—by Waistell's Bull (669)—by Masterman's Bull (422)—by Studley Bull (626).

Sold to Geo. Murray, Racine, Wis.

PRINCESS OF WALES (vol. 20, p. 16087)—Roan, calved Feb. 8, 1873, bred by R. J. Maxwell, Gumbleton Glanatore, County Cork, Ireland, got by Red Cross (32247), out of Truelove (vol. 27, p. 514, E.) by Ducrow (19591)—Sweetheart by Sir Calidore (22890)—Miranda by Lord Raglan (13240)—Sally by Lord Anthony Stanley (10432)—Sylvia by Marquis—Sally by Monarch.

TEA ROSE 2D (vol. 17, p. 13186)—Roan, calved May 27, 1875, bred by T. M. Hows, got by Lord Derby (34524), out of Tea Rose by Lord Bacon (26607), &c., as in Tea Rose, above.

April 6, 1875. S. BEATTIE and W. M. MILLER.

By Ship Nova Scotia, from Liverpool to Whitevale, Ontario, Canada.

STATIRA 9TH (vol. 15, p. 914, and vol. 19, p. 739, E.)—Roan, calved in April, 1867, bred by F. Sartoris, Rushden Hall, Northamptonshire, got by 12th Duke of Oxford (19633), out of Statira 6th by Britannicus 2d (19349)—Statira by Duke of Gloster (11382)—Stately by Balco (9918)—Statice by Sir Launcelot (5166) —by Major (4345)—by Ganthorpe (2049)—by Don Juan (1923)— by Shylock (2622).

Sold to Meredith & Son, Cambridge, Ind.

PRINCESS MAUD (vol. 15, p. 835, and vol. 19, p. 682, E., under dam)—Roan, calved July 13, 1869, bred by E. Musgrove, West Tower, Aughton, Lancashire, got by 13th Duke of Oxford (21604), out of Princess Victoria 3d by Lord Oxford 2d (20215) —Pink of Fashion by Cherry Duke 4th (17552)—Princess Alice by Gen. Canrobert (12926)—Princess by Earl of Derby (10177)— Poppy by 2d Cleveland Lad (3408)—by 2d Earl of Darlington (1945)—by son of 2d Hubback (2683)—a cow, bought of Mr. Bates.

Sold to Avery & Murphy, Detroit, Mich.

KIRKLEVINGTON DUCHESS 8TH (vol. 20, p. 15773)—Red, calved Dec. 15, 1871, bred by R. P. Davies, Horton, Gloucestershire, got by Grand Duke of Clarence (28750), out of Kirklevington 15th by 4th Duke of Oxford (11387)—Kirklevington 10th by Delhi (15865)—Kirklevington 8th by Gen. Canrobert (12926)— Kirklevington 7th by Earl of Derby (10177)—Kirklevington 4th by Earl of Liverpool (9061)—Kirklevington by Duke of Northumberland (1940)—by Belvedere (1706)—by son of 2d Hubback (2683) —a cow of Mr. Bates', descended from the stock of Mr. Maynard of Eryholme.

Recorded under dam, E. H. B., vol. 20, p. 580. Sold to J. R. Craig, and he to W. S. Slater, Mass.

SURMISE DUCHESS 5TH (vol. 16, p. 12363)—Roan, calved Jan. 15, 1873, bred by Sir C. M. Lampson, Bart., Rowfant, Sussex, got by Grand Duke of Geneva (28756), out of Surmise Duchess by

Patrician (24728)—Hebe by Rowfant 1st (22767)—Surmise by Duke of Gloster (11382)—Silence by Earl of Derby (10177)—Secret by Duke of Sutherland (6945)—by Locomotive (4242)—by Short Tail (2621)—by Gambier (2046)—by Young Wynyard (2859)—by a bull of Mr. Colling's—by a bull of Mr. Colling's.

Recorded under dam, E. H. B., vol. 21, p. 802. Sold to M. H. Cochrane, Hillhurst, Canada.

STATICE (vol. 15, p. 914, under dam)—White, calved Dec. 3, 1875, bred by F. Sartoris, Rushden Hall, got by Royal Lancaster (29870), out of Statira 9th by 12th Duke of Oxford (19633), &c., as in dam, Statira 9th, above.

PRINCESS MAUD 2D (vol. 15, p. 835, imp. in dam)—White, calved Nov. 4, 1875, bred by E. Musgrove, got by Royal Lancaster (29870), out of Princess Maud by 13th Duke of Oxford (21604), &c., as in dam, Princess Maud, above.

YOUNG BRACELET—Roan, calved February 15, 1873, bred by Sir C. M. Lampson, Bart., Rowfant, Sussex, got by Grand Duke of Geneva (28756), out of Bracelet 5th by Knightley (22051)—Bracelet 3d by Touchstone (20986)—Bracelet 2d by 7th Duke of York (17754)—Bracelet by 2d Duke of Bolton (12739)—Bijou by Grand Duke (10284)—Buttercup by Homer (2134)—Butterfly by Sir Launcelot (5166)—Bolton by Bulmer (1760)—by Don Juan (1923).

CARELESS 8TH (vol. 17, p. 12815)—Red and white, calved May, 1873, bred by J. Winnall, The Hawthornes, Worcestershire, got by Duke of Carolina (33588), out of Careless 5th by Wellington (21090)—Careless 4th by Duke of Cambridge (15921)—Careless 2d by Squire Gwynne (12140)—Caroline by 2d Cleveland Lad (3408)—Careless by Short Tail (2621)—by son of 2d Hubback (2683)—Craggs, bought of Mr. Bates and descended from the stock of Mr. Maynard, of Eryholme.

Recorded under dam, vol. 21, p. 803, E. H. B. Sold to M. H. Cochrane.

DUCHESS OF RABY (vol. 16, p. 12043)—Red and white, calved Feb. 26, 1874, bred by W. Leigh Clare, Raby Hall, Cheshire, got by Royal Lancaster (29870), out of Fennel 6th by Lord Oxford 2d (20215)—Fennel 3d by Cherry Duke 2d (14265)—Fennel by 5th Duke of York (10168)—Filbert by 2d Cleveland Lad (3408)—Felicia by 4th Duke of Northumberland (3649)—by Short Tail (2621)—by Belvedere (1706)—by a son of Young Wynyard (2859)—descended from J. Brown's Red Bull (97).

Under dam, vol. 21, p. 629, E.   Sold to M. H. Cochrane.

SURMISE DUCHESS 10TH (vol. 15, p. 920)—Roan, calved April 22, 1874, bred by Sir C. M. Lampson, Bart., got by Grand Duke of Geneva (28756), out of Nancy by Rowfant 1st (22767)—Surmise by Duke of Gloster (11382)—Silence by Earl of Derby (10177)—Secret by Duke of Sutherland (6945)—by Locomotive (4242)—by Short Tail (2621)—by Gambier (2046)—by Young Wynyard (2859) —by a bull of Mr. Colling's.

Under dam, vol. 21, p. 802, E.   Sold to W. Major & Son, Canada.

PRINCESS OF RABY (vol. 17, p. 13112)—Red and white, calved June 4, 1874, bred by W. L. Clare, got by Royal Lancaster (29870), out of Princess Victoria 3d by Lord Oxford 2d (20215)—Pink of Fashion by Cherry Duke 4th (17552)—Princess Alice by Gen. Canrobert (12926)—Princess by Earl of Derby (10177)—Poppy by 2d Cleveland Lad (3408)—by 2d Earl of Darlington (1945) —by son of 2d Hubback (2683)—a cow of Mr. Bates'.

Recorded under dam, vol. 21, p. 629, E.

KIRKLEVINGTON PRINCESS 2D (vol. 15, p. 629)—Roan, calved Sept. 18, 1874, bred by J. W. Larking, Ashdown House, Sussex, got by Grand Duke of Geneva (28756), out of Kirklevington Duchess 9th by Grand Duke of Clarence (28750)—Duchess of Kent by Lord Liverpool (22168)—Kirklevington 14th by 4th Duke of Oxford (11387)—Kirklevington 7th by Earl of Derby (10177)—Kirklevington 4th by Earl of Liverpool (9061)—Kirklevington 1st by Duke of Northumberland (1940)—by Belvedere (1706)—by a son of 2d Hubback (2683)—a cow of Mr. Bates', descended from the stock of Mr. Maynard, of Eryholme.

Recorded under dam, vol. 21, p. 804, E.   Sold to Avery & Murphy, Mich.

DUKE OF EDINBURGH—Red and white, calved April 3, 1874, bred by R. Lodge, Southport, Lancashire, got by 5th Duke of Wetherby (31033), out of Princess Victoria by 4th Duke of Oxford (11387), &c., as in vol. 18, p. 670, E.

Died at Liverpool during shipment.

DUKE JOHN (see under dam, E. H. B., vol. 21, p. 801)—Red calved Dec. 18, 1874, bred by Sir C. M. Lampson, Bart., Rowfant, got by Grand Duke of Geneva (28756), out of Fancy Duchess by 9th Duke of Geneva (28391)—Fancy 2d by Rowfant 1st (22767)—Fancy by 4th Duke of Thorndale (17750)—Surmise by Duke of Gloster (11382)—Silence by Earl of Derby (10177)— Secret 3d by Duke of Sutherland (6945)—by Locomotive (4242)— by Short Tail (2621)—by Gambier (2046)—by Young Wynyard (2859)—by bulls of Messrs. C. & R. Colling.

1850 to 1856. N. J. BECAR.
New York.

ZOE (vol. 11, p. 494, E., under dam)—Roan, calved May 13, 1853, bred by J. S. Tanqueray, Hendon, got by 5th Duke of York (10168), out of Jannetta by Lycurgus (7180)—Jocasta by Friar Tuck (3848)—Junta by Warden (5595)—Joyance by Javelin (4093) —Joy by Blyth (797)—Janet by Wellington (684)—by Phenomenon (491)—by Favorite (252)—by Favorite (252)—by Hubback (319)— by Snowdon's Bull (612)—by Waistell's Bull (669)—by Masterman's Bull (422)—by Studley Bull (626).

MISS BELLEVILLE (vol. 11, p. 359, E., under dam)—Roan, calved Feb., 1853, bred by J. M. Hopper, Newham Grange, got by Belleville (6778), out of Carnation by Goldsmith (10277)—Crocus by Petrarch (7329)—Violet by Forester (3825)—Mary by a son of Fleetham (2028)—from the herd of Mr. Whitaker.

IRIS (vol. 12, p. 423, E.)—Roan, calved June 8, 1850, bred by J. C. Grant Duff, Eden, got by Louis D'or (9336)—Lady Love by Lord Warden (7167)—Belinda by Ranunculus (2479)—Sylph by Sir

Walter (2637)—by Hotspur (1117)—by Coxcomb (928)—by Midas (435)—by Comet (155)—by R. Colling's son of Favorite (252)—by a son of Favorite (252)—by Hubback (319).

OLIVIA JORDAN (vol. 4, p. 502)—Red and white, calved July 16, 1855, bred by J. S. Tanqueray, Hendon, got by Duke of Cambridge (12742), out of Iris by Louis D'or (9336), &c., as in Iris, above.

LADY BARRINGTON 12TH (vol. 11, p. 517, E.)—Roan, calved Sept. 15, 1849, bred by R. Bell, Kirklevington, Eng., got by 4th Duke of York (10167), out of Lady Barrington 8th by 2d Duke of Oxford (9046)—Lady Barrington 5th by 4th Duke of Northumberland (3649)—Lady Barrington 3d by Cleveland Lad (3407)—Lady Barrington 2d by Belvedere (1706)—by a son of Herdsman (304)—by Wonderful (700)—by Alfred (23)—by Young Favorite (6994).

LADY BOOTH (vol. 11, p. 518, E.)—Roan, calved Dec., 1850, bred by John Emmerson, Eryholme, Eng., got by Chilton (10054), out of Rosalba by Buckingham (3239)—Rosabella by Highflyer (2122)—Rachel by Frederick (1060)—by Planet (502).

DELIA (vol. 11, p. 404, E., under dam)—Roan, calved Nov. 24, 1854, bred by J. S. Tanqueray, Hendon, got by Duke of Gloster (11382), out of Delia Gwynne by Conservative (3472)—Red Nell Gwynne by Chorister (3378)—Poll Gwynne by Wellington (2824)—Dorothy Gwynne by Marmion (406)—Nell Gwynne by Layton (366)—by Phenomenon (491)—Princess by Favorite (252)—by Favorite (252)—by Hubback (319)—by Snowdon's Bull (612)—by Waistell's Bull (669)—by Masterman's Bull (422)—by Studley Bull (626).

OXFORD 6TH—Red, calved Nov. 6, 1846, bred by Thos. Bates, got by 2d Duke of Northumberland (3646), out of Oxford 2d by Short Tail (2621)—Matchem Cow by Matchem (2281)—by Young Wynyard (2859).

E. H. B., vol. 9, p. 509, and A. H. B., vol. 4, p. 503.

VICTORIA 26TH—Red, calved March 25, 1853, bred by R. Holmes, Moycashel, Ireland, got by Baron Warlaby (7813), out of

Victoria 4th by Prince Albert (11933)—Victoria 2d by Belzoni (783)—Victoria by Satellite (1420)—No. 1 Chilton Sale by Cato (119)—Pope Cow by Pope (514)—Flora by Favorite (252)—Nymph by White Bull (421)—Lily by Favorite (252)—Miss Lax by Dalton Duke (188)—Lady Maynard by R. Alcock's Bull (19)—by J. Smith's Bull (608)—by Jolly's Bull (337).

E. H. B., vol. 12, p. 646.

JACINTHA—Red, calved Feb. 15, 1846, bred by J. Beasley, Overstone, got by Fawsley (6004), out of Junta by Warden (5595) —Joyance by Javelin (4093)—Joy by Blyth (797)—Janet by Wellington (684)—by Phenomenon (491)—by Favorite (252)—by Favorite (252)—by Favorite (252)—by Hubback (319)—by Snowdon's Bull (612)—by Waistell's Bull (669)—by Masterman's Bull (422)—by Studley Bull (626).

E. H. B., vol. 10, p. 406.

ACTRESS (vol. 9, p. 570, E.)—Roan, calved May 11, 1849, bred by Mr. R. C. Lowndes, Clubmoor, Eng., sold to Hon. J. Wentworth, Ill., got by Harkaway (9184), out of Smut by Colonel (8667)—Moss Rose by Locomotive (4242)—Adelaide by Cleveland (3403)—by Young Eryholme (1981)—by Wonderful (700)—by Merlin (429)—by Alfred (23)—by Cupid (177)—by Suwarrow (636).

APRICOT (vol. 10, p. 256, E.)—Roan, calved Dec. 27, 1849, bred by T. Bell, Kirklevington, Eng., got by 3d Duke of York (10166), out of Annie by 4th Duke of Northumberland (3649)— Anna by Short Tail (2621)—Acomb by Belvedere (1706).

GARLAND 2D (vol. 11, p. 467, E.)—Roan, calved Nov. 18, 1849, bred by H. N. Hill, Berrington, Eng., got by Pestalozzi (10603), out of Garland by Hector (4000)—Moss Rose by Emperor (1974) —Rosebud by Margrave (2243)—by Leopold (2199)—by Hector (2103)—by Traveler (655)—by Surly (2715)—by Colonel (152).

GAZETTE—Got by Monk (11824), out of Garland 2d by Pestalozzi (10603), &c., as in Garland 2d, above.

SURPRISE (vol. 4, p. 572, and vol. 12, p. 603, E., under **dam**) —Roan, calved Jan. 23, 1854, bred by J. S. Tanqueray, got by Gilliver (11529), out of Silence by Earl of Derby (10177)—Secret by 3d Duke of Sutherland (6945)—Secret 2d by Locomotive (4242) —Secret by Short Tail (2621)—White Rose by Gambier (2046)— White Rose by Young Wynyard (2859)—by R. Colling's Bull—**by** C. Colling's Bull.

SONGSTRESS—White, calved April 30, 1849, bred by W. R. Baker, Bayfordbury, got by Snowball (10846), out of Melody by Sir Thomas Fairfax (5196)—Magic by Wallace (5586)—by Wellington (2824)—by Marmion (406)—Daphne by Merlin (430)—Nell Gwynne by Layton (366)—Nell Gwynne by Phenomenon (491)— by Favorite (252)—by Favorite (252)—by Hubback (319)—by Snowdon's Bull (612)—by Waistell's Bull (669)—by Masterman's Bull (422)—by Studley Bull (626).

E. H. B., vol. 10, p. 583.

AUG. 19, 1871. E. G. BEDFORD.

Imported by E. G. Bedford, Paris, Bourbon Co., Ky. By Ship Erin, from London, Aug. 19, 1871. Landed in New York, Sept., 1871. Selected in England by John Thornton, Esq.

CANNONDALE (vol. 11, p. 541, and under dam, vol. 18, p. 618, E.)—Roan, calved Aug. 15, 1867, bred by Lord Walsingham, Merton Hall, Norfolk, got by Royal Wharfdale (22805), out of Minnie by Lord Cobham (20164)—Cannon Ball by Robinson Crusoe (13610)—Cannonade by Bridegroom (11203)—Cherryripe by Sir Walter (2639)—Young Cherry by Young Waterloo (8757)— Cherry by Waterloo (2816)—Old Cherry by Waterloo (2816)—by Kitt (2179)—by Kitt (2179)—by Page's Bull (6269)—by Middleton's Bull (438).

In calf to Lord Blithe (22126). Sold Oct. 28, 1874, at Paris Fair Grounds, Paris, Ky., by public auction to H. P. Thompson, Winchester, Ky., for $1,500.

CANNONDALE 2D (vol. 11, p. 541)—Roan, calved Oct 15, 1871, bred by Lord Walsingham, Merton Hall, Norfolk, got by Lord

Blithe (22126), out of Cannondale by Royal Wharfdale (22805)—Minnie by Lord Cobham (20164)—Cannon Ball by Robinson Crusoe (13610)—Cannonade by Bridegroom (11203)—Cherryripe by Sir Walter (2639)—Young Cherry by Young Waterloo (8757)—Cherry by Waterloo (2816)—Old Cherry by Waterloo (2816)—by Kitt (2179)—by Kitt (2179)—by Page's Bull (6269)—by Middleton's Bull (438).

This heifer was gotten in England and calved at Mr. Bedford's, after the arrival of her dam at his place. Sold at Mr. E. G. Bedford's sale at Fair Grounds, Paris, Ky., Oct. 28, 1874, to E. L. Davison, Springfield, Ky., for $2,025. Again sold by Mr. Davison at his sale at the farm of Wm. Warfield, near Lexington, Ky., October 10, 1876, to J. C. Breckenridge and W. Warfield, for $560.

GAZELLE 16TH (vol. 11, p. 670, and under dam, vol. 19, p. 527, E.)—Roan, calved Nov. 24, 1869, bred by E. Bowly, Siddington House, Cirencester, Gloucestershire, got by 6th Earl of Walton (26078), out of Gazelle 8th by 7th Duke of York (17754)—Gazelle 2d by Earl of Walton (17787)—Selina by 4th Duke of Oxford (11387)—Buttercup by Snowstorm (12119)—Helena by Hampden (8129)—by Leo (4208)—by Henwood (2114)—by Sir Stephen (1456)—by Prince of Waterloo (528)—by May Flower (425)—by a bull of Mr. Nicholson's, descended from the stock of J. Brown.

In calf to 2d Duke of Tregunter (26022).

GAZELLE OF WOODLAND VILLA (vol. 12, p. 811)—Roan, calved April 10, 1872, bred by Mr. Bowly, Siddington House, Cirencester, Gloucestershire, got by 2d Duke of Tregunter (26022), out of Gazelle 16th by 6th Earl of Walton (26078)—Gazelle 8th by 7th Duke of York (17754)—Gazelle 2d by Earl of Walton (17787)—Selina by 4th Duke of Oxford (11387)—Buttercup by Snowstorm (12119)—Helena by Hampden (8129)—by Leo (4208)—by Henwood (2114)—by Sir Stephen (1456)—by Prince of Waterloo (528)—by May Flower (425)—by a bull of Mr. Nicholson's, descended from the stock of J. Brown, of Aldborough, Yorkshire.

Sold Oct. 28, 1874, to J. Niccolls, of Bloomington, Ill., for $925.

LADY ADELA (vol. 11, p. 754, and under dam, vol. 20, p. 582, E.)—Red and white, calved July 18, 1869, bred by C. A. Barnes, Charleywood, Rickmansworth, Herts., got by Lord Wallace (24473), out of Lady Agnes by The Florist (20952)—Lady Antoinette by Sailor Boy (20770)—Lady Anna Burdett by Sir Charles (16948)—Lady Elizabeth 2d by Marquis of Bute (11778)—Countess of Hardwicke by Percy (6283)—Celia by 3d Duke of Northumberland (3647)—Cornflower by Bashaw (1692)—Columbine by Helmsman (2109)—Columbine by Columella (904)—Charlottina by Regent (544)—Charlotte Palatine by Palatine (478)—Charlotte by Palmflower (480)—Crimson by Patriot (486)—Young Milbank by Driffield (223)—Milbank by C. Holmes' Bull (314).

Sold by public auction Oct. 28, 1874, to J. Niccolls, of Bloomington, Ill., for $1,075.

FRANCISCA (vol. 11, p. 666)—Red, calved March 29, 1869, bred by Mr. J. Gamble, Shouldhamthorpe, Downham, Norfolk, got by Zealot (25480), out of Flora by Prince Leopold (20557)—Bland by Red Duke (18676)—Blanche by Squire Blanche (12139)—Flora by Crusade (7938)—Flora by Cotherstone (6903)—Flora 2d by Yellow Boy (5694)—Flora by Red Robin (2492)—Flora by Burley (1766)—by a son of Young Albion (15)—by Sir Marton (1453)—by Palmsun (7311)—by Parrington's Bull.

Barren. Sold to Joseph Scott. See dam, vol. 18, p. 495, E.

May 17, 1870. W. A. BELL.
Shipped by the Daniel Webster, from London, for Colorado.

AJAX (Under dam, E. H. B., vol. 19, p. 639)—Roan, calved Jan. 12, 1869, bred by Captain Aveling, Needham House, Wisbeach, got by The Yeoman (25305), out of Moreen by Tom Sayers (19010)—Mohair by Old Buck (15017)—Merino by Frantic (12897)—

Cashmere by Boccaccio (7838)—by Duke of Rothsay (6943)—by Belshazzar (1703)—by Noble Henry (2374)—by Abraham (2905)—by Mustachios (4527)—by Simon (5134)—by Young George (3885) —by George (276).

BLUCHER (28049)—Roan, calved March 1, 1869, bred by T. E. Pawlett, Beeston, Sandy, Bedfordshire, got by Baron Warlaby (23381), out of Fathom by Gen. Havelock (16110)—Faith by Sir Charles (12075)—Fanchette by Petrarch (7329)—Fame by Raspberry (4875)—by Young Matchem (4422)—by Isaac (1129)—by Young Pilot (4702)—by Pilot (496)—by Julius Cæsar (1143).

GERTRUDE (vol. 19, p. 540, E., under dam)—Roan, calved March 1, 1868, bred by G. W. Roberts, King's Walden, Herts., got by Baron Torr (23380), out of Grizza by King Tom (20071)—Grilla by Euxine (12845)—Grizzy by The Stuart (7623)—Grizelda by Daniel (3555)—by Isaac (1129)—by Warlaby (672)—by a son of Sir Dimple (594)—by Layton (2190).

PRELUDE (vol. 18, p. 671, E., under dam)—Red heifer, calved March 21, 1868, bred by J. Gamble, Shouldhamthorpe, Norfolk, got by Zealot (25480), out of Prize by Forester (19767)—Pansy by Plato (18552)—Precious by Baron Albany (11151)—Prudence by Orontes (4623)—Zeal by Roman (2561)—by Mercury (2301)—by Monarch (2324)—by St. Albans (2584)—by Jupiter (342)—by Sir Oliver (605)—by Trunnell (659)—by Favorite (252)—by Favorite (252)—by Dalton Duke (188)—by R. Alcock's Bull (19)—by J. Smith's Bull (608)—by Jolly's Bull (337).

1857. THOMAS BETTS.

SUSAN 3D—Red roan, calved March 20, 1854, bred by R. Stratton, got by Waterloo (11025), out of Susan 2d by Hero of the West (8150)—Susan by Radical (7390)—Spider by Lottery.

Brought to Kentucky, June 3, 1859, by Hon. B. J. Clay, of Bourbon Co., Ky.

Feb., 1870. MR. BIHLER.
From Liverpool to San Francisco, Cal.

LADY HUDSON BAXTER (vol. 18, p. 567, E., under dam)—
Red and white, calved May 13, 1867, bred by B. Baxter, Elslack
Hall, Skipton, got by Prince Edgar (22585), out of Lady Hudson
4th by Lord Ravensworth (20222)—Hudson 3d by Gen. Canrobert
(12927)—Hudson 2d by Horrox (11591)—Hudson by The Duke
(8676).

CHERRY PRINCE 2D (vol. 18, p. 734, E., under dam)—Roan,
calved April 21, 1868, bred by T. Atherton, Chapel House, Spike,
got by Baron Wild Eyes (25604), out of Southwick Cherry by 3d
Grand Duke (16182)—Cherry Bloom by Heir-at-Law (13005)—
Cherry Blossom by Roland (2556)—Old Cherry by Pirate (2430)—
by Houghton (318)—by Marshal Blucher (416)—from the stock of
Messrs. Wright & Charge.

Aug., 1874. BIRRELL & JOHNSTON,
Of Greenwood, Ontario, Can. By Ship Manitobian, from Glasgow to Quebec.

ALEXANDRINA 6TH (vol 16, p. 11951, with dam omitted)—
Dark roan, calved Feb. 28, 1873, bred by W. S. Marr, Upper Mill,
Aberdeenshire, got by Gladstone (31253), out of Alexandrina 5th
by Macduff (26773)—Alexandrina 4th by Lord Lyons (22173)—
Alexandrina by Alaric (21155)—Alice by Somerset (10858)—Anna
by Hawthorn (7071)—Kilmeny by The Peer (5455)—Premium by
George (2054)—by Togston (5487)—bred by Mr. Laing.

PRISCILLA 7TH (vol. 3, p. 707, C. H. B.)—Red, calved
March 7, 1873, bred by James Bruce, Fochabers, Scotland, got by
Lord St. Leonards (29202), out of Priscilla 3d by Prince Arthur
(16723)—Priscilla by Water King (13980)—Peggy by Dannecker
(7949)—Ruby by The Pacha (7612)—Daffodil, by 2d Duke of
Northumberland (3646)—Myrtle by Sir Henry (1446)—Palmflower
by Juniper (1145)—Peeress by Lancaster (360)—Jessamine by
Neswick (1266)—Old Yellow Cow.

BARON BRUCE [4272]—Roan, calved March 16, 1875, bred by J. Bruce, got by Bridesman (30587), out of Priscilla 7th by Lord St. Leonards (29202), &c., as in dam. above.

Oct. 19, 1872. MR. BLANCHARD,
Of Appleby, Ontario, Can. By Ship Germany, from Liverpool.

COUNT GRINDELWALD (30808) [2919]—Roan, calved Jan. 6, 1872, bred by Mr. T. Stamper, Highfield House, Yorks., got by Grindelwald (26323), out of Countess by Wathstone Hero (25417)—Cygnet by Captain (14229)—Citron by Captain (14229)—Caroline by Chevy Chase (7897)—by Monarch (9407)—by Gainford (2044)—by Samson (5080)—by Young Sovereign (5286)—by White Walton (5652).

FR. BLOODGOOD.
Albany, N. Y.

BLOODGOOD COW—See 908, A. H. B.

April, 1876. THOMAS BOAK.
Milton, Ontario, Can.

FAREWELL (vol. 4, p. 151, C. H. B.)—Roan cow, calved Oct. 6, 1872, bred by R. Thompson, Inglewood Bank, Penrith, got by Royal Westmoreland (35416), out of General's Daughter by General Haynaw (11520)—Red Rose 6th by Gainford 2d (10255)—Red Rose 5th by Emperor (9084)—Red Rose by Sir Thomas Newton (6495)—Hackthorpe by General (3867).

PRINCE INGLEWOOD [5936]—Roan, calved March 13, 1876, bred by R. Thompson, got by Master Dragonfly (34807), out of Farewell by Royal Westmoreland (35416)—General's Daughter by Gen. Haynaw (11520), &c., as in dam, above.

DUKE OF CUMBERLAND 58590 (33610)—Roan, calved Feb. 20, 1874, bred by John Lamb, Burrell Green, Penrith, Cumberland, got by Hubback Junior (31395), out of Belle of Burrell Green by Ignoramus (28887)—Bellissima by Borderer (15676)—Bella by Omer Pasha (13429)—Belinda by Silk Velvet (12070)—Belloria by Buchan Hero (3238) — Urania by Dan O'Connell (3557)— Thimble by Brutus (1752).

Sept. 19, 1878. BOARD OF AGRICULTURE, HALIFAX, NOVA SCOTIA.

By Ship Canadian, from Liverpool.

ROSE OF DELHI—Roan heifer, calved Sept. 5, 1876, bred by I. & J. Gaitskill, Hall Santon, Cumberland, got by Sultan (30088), out of White Lady by King James (28972)—White Lily by Croupier (23656)—White Rose by First Fruits (16048)—Moss Rose by Gavazzi (11508)—Moss Rosebud by Wilberforce (9830)—Rosebud by Swintonian (9702)—Swinton Rose by Young Symmetry (7577)— by Shipton (2620)—by Champion (7007)—by Snowball (7521).

ELIZA STEWART—Red heifer, calved March, 1877, bred by James Beattie, Newby House, Annan, Scotland, got by Titan (35805), out of Eliza by Mac Turk (14872)—Edith by Pride (10631) —Brilliant 3d by Baron of Ravensworth (7811)—Brilliant by Sir Thomas (5194)—by Noble Henry (2374)—by Abraham (2905)—by' Mustachios (4527)—by Simon (5134)—by Young George (3885)— by George (276).

FAVORITE—Red bull, calved May 5, 1877, bred by Mr. Graham, Yauwath Hall, Cumberland, got by Hubback Jr. (31395), out of Fanny by Waterloo Commander (32811)—Fanny by British Prince (19354).

LORD WINDSOR—Roan bull, calved June 17, 1877, bred by W. Burneycat, Grenaby, Isle of Man, got by King Victor (28986), out of Windsor's Flower by Sir Windsor Broughton (27507)—Spring Flower by Frederick Colling (26194)—Regina by Heirloom (18045) —Sunflower by Lablache (11656)—Miss Unthank by Duke of Richmond (8000)—by His Royal Highness (4039)—by Gainford

(2044)—by Sampson (5080)—by Morpeth (4511)—by Lonsdale (379) —by Irishman (329).

MORISCO (40366)—Roan, calved July 10, 1877, bred by I. & J. Gaitskill, got by Sultan's Fame (35699), out of Princess of the Realm by Peer of the Realm (27057)—Rosina by Juryman (20043) —Rowena by Duke of Moscow (14447)—Ruby by Red Rover (11982) —Sprightly by Sir Isaac (9645)—Superb by Victory (5566)— Splendid by Matchem 3d (4420)—by Young Eryholme (1981)—by Wellington (2825)—by Alfred (23)—by Windsor (698)—by Cupid (177)—by Barker's Bull of Layton (53).

1860. THOMAS BOLSLER.
Islington, Ontario, Can.

YOUNG PRINCE JOHN [567]—Roan, calved April 24, 1858, bred by Joseph Crust, Garton Field, Yorkshire, got by Prince John (13511), out of Fidelity by Vatican (12261)—Fidelle by 2d Cleveland Lad (3408)—Fidget 2d by Duke of Northumberland (1940)— Fidget by 2d Earl of Darlington (1945)—Fletcher by a son of Wynyard (2859)—descended from J. Brown's Red Bull (97).

NOTE.—By reference to E. H. B., vol. 13, p. 458, Fidelity by Vatican (12261) produced, April 9, 1858, 3d Cleveland Lad by Easby Fairfax (12820); in 1859, March 6th, she produced 4th Cleveland Lad (17574).

Aug. 10, 1872. JOHN BOWLY.
By Ship Rhine, from London, for Virginia.

GRAND DUKE OF GENEVA (31287)—Red, calved Sept. 7, 1871, bred by T. G. Curtler, Revere House, Worcester, got by 8th Duke of Geneva (28390), out of Lady Ketura by 3d Duke of Claro (23729)—Ketura 3d by 7th Duke of York (17754)—Ketura 2d by Archduke (17316)—Ketura by Duke of Ulster (12774)—Kathleen by Havannah (10308)—Kitty by Sweet William (8646)—by Cedric (3311)—by Nimrod (4571)—by Cœlebs (897)—by Neswick (1266) —by Fisher's Old Bull (3799).

Nov., 1877. J. BOWMAN,

Of Victoria, Kansas. By Ship Texas, to Philadelphia.

PROUD BUTTERFLY (vol 21, p. 789, E. H. B., and vol. 17, p. 13113, A. H. B.)—Red and white, calved July 4, 1874, bred by R. Jefferson, Preston Hows, got by Gay Cavalier (31223), out of Phœbe Butterfly by Duke of Wharfdale (19648)—Double Butterfly by Royal Butterfly (16862)—Alice Butterfly by Master Butterfly (13311)—Alice 2d by Duke of Atholl (10150)—Madeline by Marcus (2262)—Landlady by Matchem (2281)—by Pilot (496)—by Young Albion (15)—Gaudy by Albion (14)—Old Gaudy by Suwarrow (636)—by Son of Twin Brother to Ben (88)—by Twin Brother to Ben (660).

PEARL BUTTERFLY (vol. 18, p. 13860)—Red and white, calved June 11, 1878, bred by R. Jefferson, got by British Boy (30597), out of Proud Butterfly by Gay Cavalier (31223), &c., as in dam, above.

1857. R. R. BOWN.

Brantford, Canada.

MASTER GRAHAME 4159 [444]—Roan, calved Oct. 23, 1855, bred by T. W. Brampston, The Skreens, got by Orestes (15027), out of Lady Mary by Tortworth (10996)—Mary Grahame by The Hon. C. Fairfax (17605)—Lady Grace by Blast (1724)—by Speeton—by Magnum Bonum (2882)—by Favorite (257)—by Palmflower (480)—by a grandson of Neswick (1266)—by Charge's Gray Bull (872)—from the stock of Geo. Coates.

ROAN DUCHESS—Got by Lord Ducie (13181), out of Duchess by Red Duke (8694)—Jemima by Tenantry (13829)—Lady by King Lear (8196)—Ora by Orestes (4623)—Minerva by Mercury (2301)—Empress by Monarch (2324)—Duchess by St. Albans (2584)—by Jupiter (342)—by Sir Oliver (605)—by Trunnell (659)—by Favorite (252)—by Favorite (252)—by Dalton Duke (188)—by the old Studley Bull.

BESSIE—Got by Bankfield (12434), out of Lady Jane by Capt. Shaftoe (6823)—Misfortune by Duke of Cumberland (3641)—by Chevalier (3365)—by Solomon—by Mr. Moore's White Bull—by a son of General (272). See [650].

1836. F. BOYD.
Near Toronto, Canada.

JESPA—From the herd of the Marquis of Exeter, got by Emperor (1014), out of Juno by Alexander (1624).

1882. R. BRAND and I. McCORD.
To Quebec, by Steamship Ontario, Sept. 14, 1882, by Rudolph Brand, Chicago, Ill., and Ira McCord, Orland, Ill.

BLOOMING HEATHER 2D (vol. 25, p. 983)—Red and white, calved Feb. 4, 1881, bred by Thomas Marshall Howes, Annan, Scotland, got by Roan Duke (45471), out of Blooming Heather by Red Cross Knight (35219)—Blooming Daisy by Blood Royal (28047)—Eliza by Mack Turk (14872)—Edith by Pride (10631)—Brilliant 3d by Baron of Ravensworth (7811)—Brilliant by Sir Thomas (5194)—by Noble Henry (2374)—by Abraham (2905)—by Mustachios (4527)—by Simon (5134)—by Young George (3885)—by George (276).

RED GRIZZY 2D (vol. 25, p. 983)—Red, calved Feb. 8, 1881, bred by Thomas Marshall, Howes, Annan, Scotland, got by Roan Duke (45471), out of Red Grizzy by Bright Hope (28081)—Young Grizzy by Gipsey Chief (15385)—by Viscount (15471)—by Kossuth (11646)—by Sir William (8598)—by Roger (13615)—by Studley (628)—by Young Rockingham (2547)—by Wonder (2853)—by Denton (198)—by Ladrone (353)—by Henry (301).

LADY NITHSDALE (vol. 25, p. 550)—Red and white, calved March 27, 1880, bred by Andrew Stobo, Porterstown, Thornhill, Eng., got by Capt. Hardy (39549), out of Hoddam Princess (vol. 25, p. 679, E.) by Prince Arthur Patrick (29600)—Jenny Lind by The Fiddler (27629)—Jessie Helen by Sir John (22905)—Blanche

by Mac Turk (14872)—Hoddam Heiress by Heir-at-Law (13005)—
by Hudibras (10339)—by Baronet (1686)—by Spectator (2688)—by
Fitz Remus (2025)—by Whitworth (695).

LADY NITHSDALE 2D (vol. 25, p. 550)—Light roan, calved
Feb. 6, 1883, bred by Andrew Stobo, got by King Edward (44990),
out of Lady Nithsdale by Capt. Hardy (39549), &c., as in dam,
above.

MYRTLE BEAUTY (vol. 25, p. 550, and vol. 23, p. 559, E.,
under dam)—Roan, calved Feb. 29, 1876, bred by T. Marshall,
Howes, Annan, Scotland, got by Lord Derby (34524), out of White
Beauty (vol. 23, p. 559, E.) by King Charles (26500)—Young
Beauty by Duke of Richmond (19640)—Beauty by Kossuth (11646)
—Rose by Sir William (8598)—by Roger (13615)—by Studley (628)
—by Young Rockingham (2547)—by Wonder (2853)—by Denton
(198)—by Ladrone (353)—by Henry (301).

EMIGRANT 51531—Red and white, calved Jan. 30, 1883, bred
by T. Marshall, got by King Edward (44990), out of Myrtle Beauty
by Lord Derby (34524), &c., as in dam, above.

ROAN GAINFORD (vol. 25, p. 550)—Roan, calved March 30,
1882, bred by T. Marshall, Howes, Annan, Scotland, got by Roan
Duke (45471), out of Red Gainford (vol. 25, p. 577, E.) by Lord
Derby (34524)—Lady Gainford 2d by Bright Hope (28081)—Lady
Gainford by Earl of Derby (12810)—Young Grizzy by The Gipsey
Chief (15385)—Grizzy by Viscount (15471)—Beauty by Kossuth
(11646)—Rose by Sir William (8598)—by Roger (13615)—by Stud-
ley (628)—by Young Rockingham (2547)—by Wonder (2853)—by
Denton (198)—by Ladrone (353)—by Henry (301).

CLARA BOOTH—Red and white, calved June 5, 1883, bred by
Thos. Marshall, got by King Edward (44990), out of Blooming
Heather 2d by Roan Duke (45471), &c., as in dam, above.

MINNIE BOOTH—Roan, calved June 5, 1884, bred by Thos.
Marshall, got by King Edward (44990), out of Red Grizzy 2d by
Roan Duke (45471), &c., as in dam, above.

CAPENOCH 51331 (46038)—Red and white, calved March 30, 1881, bred by Andrew Stobo, Porterstown, England, got by Capt. Hardy (39549), out of Royal Duchess 3d by Prince Arthur Patrick (29600)—Royal Duchess by Royal George (25023)—Cherry Duchess by Mr. Banting (22366)—Red Duchess 4th by Bumper (10005)—Red Duchess 3d by Young Peer (13462)—Caroline 3d by Duke of St. Albans (8001)—Caroline by Tyne (7653)—by Wolsington (2852)—by Brandon (3206)—by Young Sovereign—by Londsdale (379)—by Irishman (329).

1837. MICHAEL BOYNE.
Ohio.

GROSVENOR (3946)—Red and white, calved October 13, 1836, bred by Mr. Paley, got by Talleyrand (2728), out of Clarinda by Buckingham (1755)—Clara by Election (1961)—Young Charlotte by Pilot (1319)—Charlotte by Clarence (888)—by George (275)—by Ben (70)—by The Lame Bull (358)—by Punch (531).

This bull was brought over with the cattle of Ohio Importing Company.

JOHN BRENTNAL.

DURHAM 1487—Bred by Mr. Smith, Dishley, got by Paragon (1303), out of Tuberose by Lancaster (360)—Cherry by Ladrone (353)—by Danby (190)—by a son of Favorite (252).

DISHLEY—Bull, bred by Mr. Smith, Dishley.

BRENTNAL—Cow, bred by Mr. Smith, Dishley. See 1628, A. H. B.

OLD RED JACKET. See vol. 6, p. 343, C. H. B.

BRODIE & HUNGERFORD.
Jefferson County, N. Y.

ST. NICHOLAS 4347—Roan, calved Feb., 1853, bred by J. M. Hopper, got by Master Belleville (11795), out of Zenobia 2d by Belleville (6778)—Zenobia by Ernest (3735)—Dido by Marton

(4408)—by Uptaker (2784)—by Wonderful (700)—by Mars (413)—by Trunnell (659)—by George (273)—by Dash (191).

EMPRESS JOSEPHINE (vol. 5, p. 268)—Calved Aug., 1850, got by Isaac (9239), dam by Cotherstone (6903)—by Yellow Boy (5694)—by Mowthorpe (2343)—by Fleatham (2028)—by Easby (232).

FLOURISH (vol. 5, p. 288, and vol. 11, p. 458, E., under dam, but not named)—Red and white, calved Jan. 10, 1854, bred by John Emmerson, Yorkshire, got by Young Chilton (11278), out of Flounce by Chilton (10054)—Fortuna by Magician (7185)—Whitworth by Miracle (2320)—by Spectator (2688)—by Fitz Remus (2025)—by Jupiter (342).

LADY NEWHAM ALIAS MATILDA 2D (vol. 12, p. 490, E. H. B., as Matilda 2d)—Roan, calved in March, 1853, bred by J. M. Hopper, Yorkshire, got by Belleville (6778), out of Matilda by Belleville (6778)—Madeline by Newham (4563)—Ganymede by Uptaker (5534)—Garland by Matchem (2281)—by Fitz Remus (2025)—by Cato (119)—by Whitworth (695).

ROSAMOND—Roan, calved July, 1852, bred by J. M. Hopper, got by Quarrington (10671), out of May Rose 3d by Belleville (6778)—May Rose 2d by Belleville (6778)—Sylvester by Ernest (3735)—Silk Velvet by Pedestrian (4676)—May Rose by Miracle (2320)—Georgiana by Fitz Remus (2025)—by Whitworth (695)—by Charles (127).

MR. BROOKS.
New York City.

MONARCH 4TH (20370)—Roan, calved Jan. 23, 1862, bred by J. Brett, Burton Joyce, got by Monarch 3d (20369), out of Ruby 9th by Monarch 2d (14958)—Ruby 6th by Improver (20003)—Ruby 5th by Lord Arthur Fairfax (7152)—Ruby 4th by Homer (8153)—by Snap (2646)—by Shakespeare (1429)—by Rubens (568).

Oct. 24, 1840. GEO. BROWN.

By Ship Alexander, to Baltimore.

SON OF MAGNUM BONUM (2243)—See 23858, and vol. 2, p. 385, A. H. B.

SUPERIOR—Got by Charley, out of Princess by Cupid.   See A. H. B., 23858.

FAN—Got by Magnum Bonum Jr. [son of (2243)], out of Superior by Charley—by Cupid.

GEO. BROWN.

Bow Park, Canada.

LORD BARRINGTON 17550 (31616)—Roan, calved June 1, 1870, bred by H. J. Sheldon, Brailes House, got by Duke of Brailes (23724), out of Grand Duchess of Barrington by 7th Grand Duke (19877)—Countess of Barrington 2d by 9th Duke of Oxford (17738) —Countess of Barrington by Grand Duke 3d (16182)—Laurel by Grand Turk (12969)—Lally by Earl of Derby (10177)—Olive Leaf 3d by Earl of Liverpool (9061)—Olive Leaf 2d by 2d Duke of Cambridge (3638)—Olive Leaf by Belvedere (1706)—Lady Barrington by son of Herdsman (304)—Young Alicia by Wonderful (700)—Old . Alicia by Alfred (23)—by Young Favorite (6994).

Sept. 2, 1873. GEO. BROWN.

Toronto, Canada. By Ship Phœnician, from Glasgow.

DUKE OF BARRINGTON 4TH 16708 (30924) — Red and white, calved May 2, 1872, bred by H. J. Sheldon, Brailes House, Shipston-on-Stour, Warwickshire, got by 9th Duke of Geneva (28391), out of Lady Louisa Barrington by Duke of Brailes (23724) —Countess of Barrington 2d by 9th Duke of Oxford (17738)— Countess of Barrington by 3d Grand Duke (16182)—Laurel by Grand Turk (12969)—by Earl of Derby (10177)—by Earl of Liverpool (9061)—by 2d Duke of Cambridge (3638)—by Belvedere (1706) —by Son of Herdsman (304)—by Wonderful (700)—by Alfred (23) —by Young Favorite (6994).

ROYAL TUDOR 18272 (35411)—Red and white, calved April 21, 1872, bred by Hugh Aylmer, West Dereham Abbey, got by Royal Broughton (27352), out of Christina by Prince Christian (22581)—Calendula by Majestic (13279)—Calomel by Hamlet (8126) —Chalk by Leonard (4210)—by Buckingham (3239)—from the stock of Sir M. W. Ridley, Bart.

1856. B. B. BROWNING.
Madison Co., Ohio.

NELSON GWYNNE 3191—White, calved April 9, 1855, bred by Mr. Troutbeck, Carlisle, England, got by Benedict (7828), out of Pricky Gwynne by Sir Thomas (5194)—Old Pricky Gwynne by Marmion (406)—Daphne by Merlin (430)—Nell Gwynne by Layton (366)—Nell Gwynne by Phenomenon (491)—by Favorite (252)—by Favorite (252)—by Hubback (319)—by Snowdon's Bull (612)—by Waistell's Bull (669)—by Masterman's Bull (422)—by Studley Bull (626).

1865. R. W. CAMERON.
Staten Island, N. Y.

PANSY—Roan, calved Jan. 14, 1865, bred by The Queen, Windsor Castle, got by Prince Alfred (13494), out of Young Primrose by Goldsmith (10227)—Primrose by Lord Foppington (10437) —Sunflower by Sam Slick (8534).

A. H. B., vol. 7, p. 500.

VICTORIA—Roan, calved Nov. 2, 1864, bred by The Queen, Windsor Castle, got by Royal Prince (20743), out of Carolina by Prince Alfred (13494)—Coldcream by Earl of Dublin (10178)— Pansy by Gray Friar (9172)—Furbelow by Little John (4238)— Erato by Marcellus (2260)—Beatrice by Caliph (1774)—Quickly by Swing (2721)—Alamode by Argus (759)—Valuable by Defender (194)—Violet by Petrarch (488)—by own brother to Colling's white heifer—by Butterfly (104)—by Globe (278).

A. H. B., vol. 7, p. 577

PRINCESS—Red and white, calved Nov. 2, 1865, bred by The Queen, Windsor Castle, got by Prince of Saxe Coburg (20576), out of Pansy by Prince Alfred (13494), &c., as in Pansy, above.

4TH KENT OF OXFORD 5648—Calved Oct. 19, 1863, bred by John Kilk, Bentley Priory, got by 3d Kent of Oxford (20049), out of Violet by Chevalier (14250)—Venus by Bletsoe (9970)—Vetch by Viscount Winkle (9788)—Victoria by Sultan (7566)—Vetch by Norfolk (2377)—Verbena by Burley (1766)—Young Alexina by Pilot (496)—Alexina by Warlaby (672)—Agnes by Albion (14)—Bright Eyes by Lame Bull (359)—by Shipton (587)—by a son of Suwarrow (636)—by a son of Twin Brother to Ben (88)—by Twin Brother to Ben (660).

PRINCE LEOPOLD 6057—Roan, calved Jan. 13, 1865, bred by The Queen, Windsor Castle, got by Enterprise (19704), out of Annette by Prince Alfred (13494)—Alix by Earl of Dublin (10178)—Latakia by Gray Friar (9172)—Annette by Little John (4232)—Anna by Young Norman (4584)—Walnut by White Boy (1580)—Pauline by Wyvill's Bull—by a bull bred by Mr. Charge.

MARCHIONESS—Roan, calved Feb. 11, 1864, bred by John Kilk, Bentley Priory, Stonemore, got by Grand Duke of Sussex (17995), out of Fabiola by Emperor Napoleon (14509)—Frenzy by Fair Eclipse (11456)—Off-She-Goes by Emperor (6973)—Sally O'Moore 3d by Sir Walter (2639)—Sally O'Moore 2d by Young Remus (2523)—Sally O'Moore by Remus (550)—Grizzle by Hollings (2131)—Lady by His Honor (2126)—Young Ridley by Partner (2409)—Old Ridley by Hutton's Bull (2145)—Lofty, descended from the stock of R. Alcock.

A. H. B., vol. 7, p. 451.

Aug., 1871. HUGH CAMPBELL.
Kincardine, Canada.

NONPAREIL 31ST—Red, calved March 16, 1870, bred by S. Campbell, Kinellar, Blackburn, Scotland, got by Sir Christopher (22895), out of Nonpareil 26th by Scarlet Velvet (16916)—

Nonpareil 24th by Lord Sackville (13249)—Nonpareil 23d by The Baron (13833)—Nonpareil 17th by Matadore (11800)—Nonpareil 10th by Prince Edward Fairfax (9506)—Countess Lincoln by Diamond (5918)—Nonpareil 3d by Young Frederick (3836)—by Commodore (1858)—by Tathwell Studley (5401)—by Blyth Comet (85).

A. H. B., vol. 23, p. 18001.

June 1, 1876. CANADA WEST FARM STOCK ASSOCIATION.

Bow Park, Brantford, Ont. By Ship Polynesian, from Liverpool.

WATERLOO 34TH—Roan cow, calved May 10, 1867, bred by C. R. Saunders, Nunwick Hall, Penrith, got by Wallace (23166), out of Waterloo 18th by Bosquet (14183)—Waterloo 15th by The Hero (10934)—Waterloo 13th by 3d Duke of Oxford (9047)—Waterloo 9th by 2d Cleveland Lad (3408)—Waterloo 6th by Duke of Northumberland (1940)—by Norfolk (2377)—by Waterloo (2816)—by Waterloo (2816).

E. H. B., vol. 18, p. 775, under dam, and A. H. B., vol. 16, p. 12382.

ROYAL CHARMER 3D—Red cow, calved Feb. 27, 1868, bred by W. W. Slye, Beaumont Grange, near Lancaster, got by Grand Duke of Lancaster (19883), out of Royal Charmer by 2d Duke of Cambridge (12743)—Sunrise by Mameluke (13289)—Sweetheart 3d by Daybreak (11338)—by Accordion (5708)—by Little John (4232)—by Caliph (1774)—by Sir Walter (2637)—by Hotspur (1117)—by Coxcomb (928)—by Midas (435)—by Comet (155)—by son of Favorite (252)—by same son of Favorite (252)—by Hubback (319).

E. H. B., vol. 19, p. 716, and A. H. B., vol. 16, p. 12340.

ROYAL CHARMER 4TH—Red heifer, calved Aug. 24, 1875, bred by T. Lister Groby, Leicester, got by Duke of Rutland (33726), out of Royal Charmer 3d by Grand Duke of Lancaster (19883), &c., as in dam, above.

A. H. B., vol. 16, p. 12340 ; E. H. B., vol. 23, p. 372, under dam, and is given as bred by W. Fowler, Cottesmore.

WILD EYES—Roan cow, calved March 3, 1868, bred by P. Stephenson, Rainton Thirsk, Yorks., got by Lord Lally 3d (24408), out of Wild Eyes 22d by Wild Duke (19148)—Wild Eyes 20th by Lord Barrington (13170)—Wild Eyes 16th by 2d Duke of Oxford (9046)—Wild Eyes 15th by 4th Duke of Northumberland (3649)— Wild Eyes 8th by Duke of Northumberland (1940)—Wild Eyes 2d by Belvedere (1706)—Wild Eyes by Emperor (1975)—by Wonderful (700) -by Cleveland (145)—by Butterfly (104)—by Hollon's Bull (313)—by Mowbray's Bull (2342)—by Masterman's Bull (422)— from the stock of Mr. Dobison.

A. H. B., vol. 16, p. 12388, and E. H. B., vol. 18, p. 785, under dam.

BARONESS BATES—Roan cow, calved April 18, 1872, bred by the Earl of Bective, Underley Hall, got by Baron Oxford 5th (27958), out of Lady Bates 7th by 3d Duke of Geneva (23753)— Lady Bates 5th by Duke of Geneva (19614)—Lady Bates 3d by 4th Duke of Oxford (11387)—Lady Bates 2d by The Buck (13836)— Lady Bates by Duke of Gloster (11382)—Lady Blanche by 4th Duke of York (10167)—Lady Barrington 8th by 2d Duke of Oxford (9046)—Lady Barrington 5th by 4th Duke of Northumberland (3649)—Lady Barrington 3d by Cleveland Lad (3407)—Lady Barrington 2d by Belvedere (1706)—Lady Barrington by Son of Herdsman (304)—Young Alicia by Wonderful (700)—Old Alicia by Alfred (23)—by Young Favorite (6994).

A. H. B., vol. 16, p. 11969, and E. H. B., vol. 20, p. 584, under dam.

KIRKLEVINGTON DUCHESS 10TH—Roan cow, calved Oct. 1, 1872, bred by R. P. Davies, Horton, Gloucestershire, got by 2d Duke of Tregunter (26022), out of Kirklevington 18th by 3d Lord Oxford (22200)—Kirklevington 10th by Delhi (15865)—Kirklevington 8th by Gen. Canrobert (12926)—Kirklevington 7th by Earl of Derby (10177)—Kirklevington 4th by Earl of Liverpool (9061)—Kirklevington 1st by Duke of Northumberland (1940)—Nell Gwynne by Belvedere (1706)—Northallerton by Son of 2d Hubback (2683)—cow of Mr. Bates', descended from the stock of Mr. Maynard of Eryholme.

A. H. B., vol. 16, p. 12128, and E. H. B., vol. 20, p. 580, under dam.

KIRKLEVINGTON DUCHESS 17TH—Roan, calved Dec. 10, 1873, bred by R. P. Davies, got by 2d Duke of Gloster (28392), out of Kirklevington Rose by Earl of Gloster (21644)—Kirklevington 14th by 4th Duke of Oxford (11387)—Kirklevington 7th by Earl of Derby (10177), &c., as in Kirklevington Duchess 10th, above.

A. H. B., vol. 16, p. 12128, and E. H. B., vol. 21, p. 665, under dam.

KIRKLEVINGTON DUCHESS 20TH—Red, calved Oct. 6, 1874, bred by R. P. Davies, got by 2d Duke of Gloster (28392), out of Kirklevington Duchess 4th by 13th Duke of Oxford (21604) —Kirklevington 14th by 4th Duke of Oxford (11387)—Kirklevington 7th by Earl of Derby (10177), &c., as in Kirklevington Duchess 10th, above.

A. H. B., vol. 16, p. 12128, and E. H. B., vol. 21, p. 665, under dam.

PRIESTESS—Roan cow, calved Nov. 9, 1870, bred by A. J. Robarts, Lillingstone, Dayrell, Buckingham, got by Wild Duke (27808), out of Grand Princess by 7th Grand Duke (19877)—Princess by Royal Butterfly 5th (18756)—Diadem by Marmaduke (14897) —Darlington 5th by 4th Duke of Oxford (11387)—Darlington 2d by Percy (9472)—Darlington by Thomas (5471)—Pretty Maid by Eryholme (3736)—by Reformer (4914)—by Young Favorite (3770) —by Wellington (2825).

A. H. B., vol. 16, p. 12289, and E. H. B., vol. 19, p. 539, under dam.

BLANCHE 12TH—Roan cow, calved Feb. 16, 1871, bred by Duke of Devonshire, got by Baron Oxford 4th (25580), out of

Blanche 3d by 10th Duke of Oxford (17739)—Blanche by Dundas (17763)—Sylph by Gloucester (14619)—by Childers (10052)—by Selim (6454)—by Rex (6385)—by Norfolk (2377)—by Belvedere (1706)—by Belvedere (1706)—by Lancaster (360)—by Petrarch (488)—by Major (397)—by Chapman's son of Punch (122)—by Dickson's Grandson of Punch (213)—by Checks (132)—by R. Grimston's Bull (282)—by J. Coates' Bull (148).

A. H. B., vol. 16, p. 11990, and E. H. B., vol. 21, p. 675.

BLANCHE 14TH—Roan heifer, calved Dec. 26, 1873, bred by Duke of Devonshire, got by 24th Duke of Oxford (31002), out of Blanche 12th by 4th Baron of Oxford (25580), &c., as above.

A. H. B., vol. 16, p. 11990, and E. H. B., vol. 21, p. 675, under dam.

HAIDEE 2D—Red and white cow, calved March 7, 1872, bred by H. J. Sheldon, Brailes House, Warwickshire, got by 18th Duke of Oxford (25995), out of Virginia by Duke of Brailes (23724)—Miss Knightley by Bull's Run (19368)—Gionetta by Sarawak (15238)—Smock Frock by Earl of Dublin (10178)—London Pride by Janizary (8175)—Aline by Snowball (8602)—Lila by Caliph (1774)—Amy by Norman (2379)—Walnut by White Boy (1580)—Pauline by Wyvill's Bull—by a bull of Mr. Charge's.

A. H. B., vol. 16, p. 12097, and E. H. B., vol. 20, p. 810, under dam.

DARLINGTON PRINCESS—Red and white, calved Sept. 20, 1875, bred by Geo. Ashburner, got by Duke of Oxford (31004), out of Priestess by Wild Duke (27808), &c., as in Priestess, above.

E. H. B., vol. 23, p. 314, under dam.

OXFORD'S WATERLOO 6TH—Red and white heifer, calved March 7, 1875, bred by R. Lodge, The Rookery, Bishopdale, Yorks., got by 18th Duke of Oxford (25995), out of Oxford's Waterloo 4th by 13th Duke of Oxford (21604), &c., as in Oxford's Waterloo 4th, above.

A. H. B., vol. 16, p. 12273, and E. H. B., vol. 23, p. 537, under dam.

LADY FLORENCE 5TH—Roan heifer, calved March 31, 1875, bred by H. J. Sheldon, got by 2d Duke of Collingham (23730), out of Lady Florence 2d by 18th Duke of Oxford (25995)—Lady Florence by Duke of Brailes (23724)—Countess by Duke of Cambridge (12742)—Chrysalis by Earl of Dublin (10178)—Garland by Gray Friar (9172)—Fillet by Fawsley (6004)—Marguerite by Marcellus (2260)—Pearl by Rufus (2576)—Ruby by Wellington (683)—by Windsor (698)—by Windsor (698)—by an own brother to North Star (459).

A. H. B., vol. 16, p. 12146, and E. H. B., vol. 22, p. 559, under dam.

LADY USK—Roan heifer, calved May 12, 1875, bred by Lord Fitz Hardinge, Berkeley Castle, Gloucestershire, got by Oxford's Tony (35000), out of Lady Ursula by Marquis of Bickerstaffe (29292)—Ursula 21st by 7th Duke of York (17754)—Ursula 12th by Gen. Canrobert (12927)—Duchess of Gloster by Duke of Gloster (11382)—Ursula by Usurer (9763)—Crystal by Prince Ernest (4818)—Coquette by Alamode (725)—Young Cowslip by Ratify (2481)—Cowslip by Wellington (681)—by Favorite (252)—by Punch (531).

A. H. B., vol. 16, p. 12166, and E. H. B., vol. 23, p. 438, under dam.

PRINCESS GWYNNE—Roan heifer, calved Aug. 11, 1875, bred by W. H. Salt, Maplewell, Loughborough, got by 5th Lord Oxford (31738), out of Duchess Gwynne by Duke of Wetherby (17753)—Polly Gwynne by Flying Dutchman (10235)—by St. Thomas (10777)—by Prime Minister (2456)—by Wallace (5586)—by Marmion (406)—by Merlin (430)—by Layton (366)—by Phenomenon (491)—by Favorite (252)—by Favorite (252)—by Hubback (319)—by Snowdon's Bull (612)—by Waistell's Bull (669)—by Masterman's Bull (422)—by Studley Bull (626)—bred by Mr. Stephenson, of Ketton, in 1839.

A. H. B., vol. 16, p. 12291, and E. H. B., vol. 22, p. 550, under dam.

OXFORD'S WATERLOO 4TH—Red cow, calved Jan. 22, 1869, bred by T. Atherton, Speke, near Liverpool, got by 13th Duke of

Oxford (21604), out of Oxford's Waterloo by Lord Oxford 2d (20215)—Waterloo 19th by 2d Grand Duke (12961)—Waterloo 15th by Matadore (11800)—Waterloo 12th by 3d Duke of York (10166) —Waterloo 4th by Cleveland Lad (3407)—by Norfolk (2377)—by Waterloo (2816)—by Waterloo (2816).

E. H. B., vol. 21, p. 818, and A. H. B., vol. 16, p. 12272.

BLANCHE 8TH—Red and white cow, calved Jan. 31, 1869, bred by the Duke of Devonshire, Holker Hall, Lancashire, got by Grand Duke 10th (21848), out of Blanche 3d by 10th Duke of Oxford (17739)—Blanche by Dundas (17763)—Sylph by Gloucester (14619) —by Childers (10052)—by Selim (6454)—by Rex (6385)—by Norfolk (2377)—by Belvedere (1706)—by Belvedere (1706)—by Lancaster (360)—by Petrarch (488)—by Major (397)—by Chapman's son of Punch (122)—by Dickson's Grandson of Punch (213)—by Check's (132)—by R. Grimston's Bull (282)—by J. Coates' Bull (148).

E. H. B., vol. 21, p. 674, and A. H. B., vol. 16, p. 11990.

LADY THORNDALE BATES 2D—Roan cow, calved Oct. 27, 1869, bred by W. W. Slye, got by 4th Duke of Thorndale (17750), out of Lady Bates 3d by 4th Duke of Oxford (11387)—Lady Bates 2d by The Buck (13836)—Lady Bates by Duke of Gloster (11382)— by 4th Duke of York (10167)—by 2d Duke of Oxford (9046)—by 4th Duke of Northumberland (3649)—by Cleveland Lad (3407)— by Belvedere (1706)—by son of Herdsman (304)—by Wonderful (700)—by Alfred (23)—by Young Favorite (6994).

E. H. B., vol. 20, p. 616, and A. H. B., vol. 16, p. 12166.

AMERICA'S DUCHESS—Red heifer, calved July 17, 1873, bred by W. W. Slye, got by Grand Duke of Thorndale (31297), out of America by Marmaduke (14897)—Asia by 2d Grand Duke (12961)—Apricot by Fusileer (11499)—Augusta by 3d Duke of York (10166)—Annie by 2d Cleveland Lad (3408)—Annabella by Duke of Cleveland (1937)—Acomb by Belvedere (1706)—a cow bought of Mr. Bates.

A. H. B., vol. 16, p. 11959, and E. H. B., vol. 23, p. 370, under dam.

GENEVA'S MINSTREL 2D—Red and white heifer, calved May 29, 1874, bred by E. H. Cheney, Gaddesby Hall, Leicester, got by 9th Duke of Geneva (28391), out of Minstrel 3d by 10th Duke of Oxford (17739)—Minstrel 2d by Prince of Gloster (13517) —Minstrel by Count Conrad (3510)—Magic by Wallace (5586)— by Wellington (2824)—by Marmion (406)—by Merlin (430)—by Layton (366)—by Phenomenon (491)—by Favorite (252)—by Favorite (252)—by Hubback (319)—by Snowdon's Bull (612)— Beauty by Waistell's Bull (669)—by Masterman's Bull (422)— by Studley Bull (626)—bred by Mr. Stephenson, of Ketton, 1839.

A. H. B., vol. 16, p. 12085, and E. H. B., vol. 23, p. 371.

DUCHESS OF BARRINGTON 2D—Roan heifer, calved July 25, 1874, bred by W. W. Slye, got by Grand Duke of Thorndale 2d (31298), out of Lady Bates 3d by 4th Duke of Oxford (11387)— Lady Bates 2d by The Buck (13836)—Lady Bates by Duke of Gloster (11382), &c., as in Lady Thorndale Bates 2d, above.

A. H. B., vol. 16, p. 12038, and E. H. B., vol. 23, p. 371, under dam.

GRAND DUCHESS OF OXFORD 29TH—Roan heifer, calved July 28, 1874, bred by Duke of Devonshire, Holker Hall, got by Baron of Oxford 4th (25580), out of Grand Duchess of Oxford 19th by Grand Duke 10th (21848)—Grand Duchess of Oxford 6th by Imperial Oxford (18084)—Grand Duchess of Oxford 4th by Grand Duke of Wetherby (17997)—Oxford 15th by 4th Duke of York (10167)—Oxford 6th by 2d Duke of Northumberland (3646)—Oxford 2d by Short Tail (2621)—Matchem Cow by Matchem (2281)— by Young Wynyard (2859).

A. H. B., vol. 16, p. 12095, and E. H. B., vol. 21, p. 675, under dam.

DUKE OF OXFORD 45TH 29333—Roan, calved Jan. 13, 1877, bred by the Duke of Devonshire, got by 7th Baron of Oxford (36199), out of Grand Duchess of Oxford 29th by 4th Baron of Oxford (25580), &c., as in dam, above.

GRAND DUKE OF THORNDALE 2D (31298)—Red bull, calved May 9, 1872, bred by W. W. Slye, got by 9th Duke of Geneva (28391), out of Grand Duchess 20th by 4th Duke of Thorndale (17750)—Grand Duchess 8th by Prince Imperial (15095)—Grand Duchess 2d by Grand Duke (10284)—Duchess 51st by Cleveland Lad (3407)—by Belvedere (1706)—by 2d Hubback (1423)—by 2d Hubback (1423)—by The Earl (646)—by Ketton 2d (710)—by Comet (155)—by Favorite (252)—by Daisy Bull (186)—by Favorite (252)—by Hubback (319)—by J. Brown's Red Bull (97).

DUKE OF OXFORD 30TH 26349 (33712)—Red, calved Feb. 1, 1874, bred by the Duke of Devonshire, got by 5th Duke of Wetherby (31033), out of Grand Duchess of Oxford 21st by Baron Oxford 4th (25580)—Grand Duchess of Oxford 11th by Grand Duke 10th (21848)—Grand Duchess of Oxford 5th by Priam (18567)—Countess of Oxford by Earl of Warwick (11412)—by 4th Duke of York (10167)—by 2d Duke of Northumberland (3646)—by Short Tail (2621)—by Matchem (2281)—by Young Wynyard (2859).

BARON PAULINE 25636 [4595]—Rich roan, calved Nov. 8, 1876, bred by Jas. How, Broughton, Huntingdon, got by Pretender (35068), out of Pauline 11th by Prince of the Realm (22627), &c., as in Pauline 11th, above.

6TH DUKE OF BARRINGTON 26165 [4968]—Red and white, calved Jan. 8, 1877, bred by W. W. Slye, Beaumont Grange, got by Grand Duke of Thorndale 2d (31298), out of Duchess of Barrington 2d by Grand Duke of Thorndale 2d (31298), &c., as in dam, above.

5TH DUKE OF BARRINGTON 26164 [4967]—Roan, calved July 28, 1876, bred by Earl of Bective, got by 2d Duke of Tregunter (26022), out of Lady Thorndale Bates 2d by 4th Duke of Thorndale (17750), &c., as in dam, above.

DUKE OF KIRKLEVINGTON 26275—White, calved Nov. 25, 1876, bred by R. Pavin Davies, Horton, got by Oxford's King (34997), out of Kirklevington Duchess 10th by 2d Duke of Tregunter (26022), &c., as in dam, above.

DUKE OF KIRKLEVINGTON 2D 26276—Red roan, calved Dec. 30, 1876, bred by R. Pavin Davies, got by Oxford's King (35997), out of Kirklevington Duchess 17th by 2d Duke of Gloster (28392), &c., as in dam, above.

AZALIA—Roan, calved June 29, 1875, bred by Geo. Ashburner, Low Hall, got by Duke of Oxford (31004), out of Azelea by Baron Fennel (27937)—Anemone 2d by Gen. Napier (24023)—Anemone by Duke of Kent (19619)—Acacia by Count De Gourcy (17632)—Asia by 2d Grand Duke (12961)—Apricot by Fusileer (11499)—Augusta by 3d Duke of York (10166)—Annie by 2d Cleveland Lad (3408)—Annabella by Duke of Cleveland (1937)—Acomb 2d by Belvedere (1706)—Acomb by Son of 2d Hubback (2683)—cow, bought by Mr. Bates of Mr. Maynard.

A. H. B., vol. 16, p. 11969.

MINSTREL 4TH—Red and white, calved Nov. 24, 1876, bred by Wm. H. Salt, Maplewell, got by 5th Duke of Gloster (36494), out of Geneva's Minstrel 2d by 9th Duke of Geneva (28391), &c., as in dam, above.

SERAPHINA'S DUCHESS—Roan, calved Jan. 17, 1875, bred by J. P. Foster, got by 22d Duke of Oxford (31000), out of Seraphina 19th by Imperial Oxford (18084)—Seraphina 11th by May Duke (13320)—Seraphina 7th by Duke of Sussex (12772)—Seraphina 2d by Sweet William (7571)—Seraphina by Earl of Essex (6955)—Sapphire by Stratton (5336)—Ruby by Fanatic (1996)—Rufe by Red Rover (4902)—by Rufus (2576)—by Emperor (1014).

E. H. B., vol. 22, p. 418, under dam, and vol. 16, p. 12349, A. H. B.

POLLY GWYNNE 9TH—Red and white, calved June 15, 1873, bred by Mr. Hetherington, Middle Farm, got by 2d Grand Duke of Lightburne (26291), out of Polly Gwynne 3d by Wild Duke 5th (27807)—Polly Gwynne 2d by Wild Duke 4th (21107)—Polly Gwynne by Flying Dutchman (10235)—Young Dowager Gwynne by St. Thomas (10777)—Dowager Gwynne by Prime Minister (2456)—White Moll Gwynne by Wallace (5586)—Dorothy Gwynne by Marmion (406)—Daphne Gwynne by Merlin (430)—Nell

Gwynne by Layton (366)—Nell Gwynne by Phenomenon (491)—Princess by Favorite (252)—Bright Eyes by Favorite (252)—Bright Eyes by Hubback (319)—Bright Eyes by Snowdon's Bull (612)—Beauty by Waistell's Bull (669)—Duchess of Athol by Masterman's Bull (422)—Tripes by Studley Bull (626).

A. H. B., vol. 16, p. 12283.

POLLY GWYNNE 11TH—Roan, calved March, 1876, bred by Geo. Fox, Elmhurst, got by Grand Duke of Weston 3d (34079), out of Polly Gwynne 9th by 2d Grand Duke of Lightburne (26291), &c., as in dam, above.

A. H. B., vol. 16, p. 12283.

DARLINGTON 26TH — White, calved Aug. 6, 1876, bred by Geo. Ashburner, got by Cherry Duke of Lightburne (36349), out of Priestess by Wild Duke (27808), &c., as in dam, above.

A. H. B., vol. 17, p. 12843.

July 13, 1876. CANADA WEST FARM STOCK ASSO-CIATION.

By Ship Polynesian, from Liverpool.

JANIE STUART—Red, calved Nov. 29, 1869, bred by S. Canning, Snitterfield, Stratford-on-Avon, got by Gen. Bragg (26232), out of Lady Stuart by John O'Groat (18115)—Grand Duchess by Grand Sultan (16189)—Duchess of Sussex by Duke of Sussex (12772)—Countess of Beverly by Lord Foppington (10437)—Cowslip Belle by 2d Cleveland Lad (3408)—Cicely by Duke of Northumberland (1940)—Craggs by Son of 2d Hubback (2683)—Craggs, bought by Mr. Bates, descended from the stock of Mr. Maynard.

A. H. B., vol. 16, p. 12111, and E. H. B., vol. 21, p. 909.

PAULINE 8TH—Red and white cow, calved Dec. 29, 1870, bred by J. How, Broughton, Huntingdon, got by Lord Blithe (22126), out of Pauline 6th by Heir of Windsor (26364)—Pauline 3d by Ravenspur (20628)—Pauline by British Boy (11206)—Hilda by Hopewell (10332)—Hester by Hamlet (8126)—a cow, bred at Killerby by Mr. Booth.

A. H. B., vol. 16, p. 12275, and E. H. B., vol. 21, p. 781.

PAULINE 11TH—Roan cow, calved Oct. 10, 1872, bred by J. How, got by Prince of the Realm (22627), out of Pauline 5th by British Hope (21324)—Pauline 3d by Ravenspur (20628), &c., as above.

A. H. B., vol. 16, p. 12275, and E. H. B., vol. 20, p. 682, under dam.

PAULINE 15TH—Roan heifer, calved Sept. 8, 1875, bred by J. How, got by Great Hope (24082), out of Pauline 7th by King Charles (24240)—Pauline 5th by British Hope (21324)—Pauline 3d by Ravenspur (20628)—Pauline by British Boy (11206), &c., as above.

A. H. B., vol. 16, p. 12275, and E. H. B., vol. 22, p. 464, under dam.

PERILLA—Red heifer, calved March 28, 1875, bred by Jas. How, got by Great Hope (24082), out of Persicaria, by King Victor (28986)—Pauline 4th by Hopewell (19973)—Pauline by British Boy (11206), &c., as above.

A. H. B., vol. 16, p. 12278, and E. H. B., vol. 22, p. 464, under dam.

LADY HUDSON'S DUCHESS 4TH—Red heifer, calved April 22, 1874, bred by Messrs. F. Leney & Son, Waterbury, Kent, got by 6th Duke of Oneida (30997), out of Lady Hudson's Duchess 2d by Grand Duke 15th (21852)—Lady Hudson by 4th Duke of Oxford (11387)—Hudson 3d by Gen. Canrobert (12927)—Hudson 2d by Horrox (11591)—Hudson by the Duke (8676).

A. H. B., vol. 16, p. 12151, and E. H. B., vol. 21, p. 909, under dam.

MARCHIONESS 9TH ALIAS KIRKLEVINGTON 19TH—Red and white heifer, calved June 1, 1875, bred by the Earl of Bective, Underley Hall, got by 3d Duke of Gloster (33653), out of Siddington 6th by 7th Duke of York (17754)—Siddington 2d by 4th Duke of Oxford (11387)—Kirklevington 7th by Earl of Derby (10177)—Kirklevington 4th by Earl of Liverpool (9061)—Kirklevington 1st by Duke of Northumberland (1940)—by Belvedere (1706)—by Son of 2d Hubback (2683)—a cow of Mr. Bates', descended from the stock of Mr. Maynard, of Eryholme.

E. H. B., vol. 22, p. 319, under dam, and vol. 16, p. 12128.

AURORA—Red heifer, calved June 21, 1875, bred by the Earl of Bective, got by Oxford Beau 3d (32013), out of Ariel Marchioness by 3d Duke of Clarence (23727)—Ariel Duchess by Duke of Wharfdale (19648)—Ariel 5th by Czarovitz (17954)—Ariel 3d by Gen. Canrobert (12926)—Ariel 2d by Cherry Duke 2d (14265)—by 2d Cleveland Lad (3408)—by 4th Duke of Northumberland (3649) —by Duke of Cleveland (1937)—by Belvedere (1706)—a cow of Mr. Bates'.

E. H. B., vol. 22, p. 317, under dam, and A. H. B., vol. 16, p. 11968.

OXFORD BELLE 3D—Roan heifer, calved July 10, 1875, bred by Col. Kingscote, got by Duke of Hillhurst (28401), out of Countess of Oxford by 7th Duke of Airdrie (23718)—Gem of Oxford by 2d Grand Duke (12961)—Romeo's Oxford by Romeo (13619)— Oxford 5th by Duke of Northumberland (1940)—Oxford 2d by Short Tail (2621)—Matchem Cow by Matchem (2281)—by Young Wynyard (2859).

A. H. B., vol. 16, p. 12271, and E. H. B., vol. 23, p. 520, under dam.

ROYAL KENT CHARMER—Roan, calved May 20, 1872, bred by W. W. Slye, Lancaster, got by Grand Duke of Kent 2d (28729), out of Royal Charmer by 2d Duke of Cambridge (12743)—Sunrise by Mameluke (13289)—Sweetheart 3d by Daybreak (11338)—Sweetheart by Accordion (5708)—Charmer by Little John (4234)—Graceful by Caliph (1774)—Sylph by Sir Walter (2637)—by Hotspur (1117) —by Coxcomb (928)—Rachel by Comet (155)—by a son of Favorite (252)—by son of Favorite (252)—by Hubback (319).

A. H. B., vol. 17, p. 13162.

PAULINE 20TH—Red and white, calved April 23, 1877, bred by James How, Broughton, got by King Victor (28986), out of Pauline 8th by Lord Blithe (22126)—Pauline 6th by Heir of Windsor (26364)—Pauline 3d by Ravenspur (20628)—Pauline by British Boy (11206)—Hilda by Hopewell (10332)—Hester by Hamlet (8126) —a cow, bred at Killerby by Mr. Booth.

A. H. B., vol. 18, p. 13859.

WILD EYES DUKE [6503]—Red and white, calved March 19, 1877, bred by E. H. Cheney, Gaddesby Hall, got by 5th Duke of Gloster (36494), out of Wild Eyes 31st by 3d Duke of Claro (23729), &c., as in Wild Eyes 31st, above.

LADY ANNIE BATES—Roan heifer, calved Sept. 30, 1875, bred by the Earl of Bective, got by 2d Duke of Tregunter (26022), out of Lady Bates 7th by 3d Duke of Geneva (23753)— Lady Bates 5th by Duke of Geneva (19614)—Lady Bates 3d by 4th Duke of Oxford (11387)—Lady Bates 2d by The Buck (13836) —Lady Bates by Duke of Gloster (11382)—Lady Blanche by 4th Duke of York (10167)—Lady Barrington 8th by 2d Duke of Oxford (9046)—by 4th Duke of Northumberland (3649)—by Cleveland Lad (3407)—by Belvedere (1706)—by Son of Herdsman (304)—by Wonderful (700)—by Alfred (23)—by Young Favorite (6994).

E. H. B., vol. 22, p. 318, under dam, and A. H. B., vol. 16, p. 12038, as Duchess of Barrington 4th.

July 27, 1876. CANADA WEST FARM STOCK ASSOCIATION.

By Ship Circassian, from Liverpool.

KNIGHTLEY GRAND DUCHESS—Roan cow, calved Nov. 22, 1867, bred by Messrs. F. Leney & Son, Wateringbury, Kent, got by 4th Grand Duke (19874), out of Nymphalin by Bull's Run (19368)—Sylphide by Sarawak (15238)—Pintail by Janizary (8175) —Catilina by Caliph (1774)—by Scipio (1421)—by Billy (787)—by Western Comet (689)—by Favorite (252)—from Studley White Bull (627).

E. H. B., vol. 21, p. 804, and A. H. B., vol. 16, p. 12130.

LADY FAWSLEY 2D—Roan cow, calved Feb. 22, 1870, bred by H. J. Sheldon, Brailes House, got by Duke of Brailes (23724),

out of Lady Fawsley by Duke of Darlington (21586)—Hyampea by
Æsop (19197)—Polytint by Earl of Dublin (10178)—Cornbind by
Janizary (8175)—Golden Rod by Snowball (8602)—Florence by
Little John (4232)—Cathleen by Caliph (1774)—Rosy by Rob Boy
(557)—by Satellite (1420)—By Sir Dimple (594)—by Styford
(629).

A. H. B., vol. 16, p. 12146, and E. H. B., vol. 19, p. 578, under
dam.

ROSE O'LEE—Roan heifer, calved Oct. 26, 1875, bred by J. W.
Larking, got by Grand Duke of Geneva (28756), out of Rosy by
Grand Duke of Kent (26289)—Lactea Oxoniensis by Imperial Oxford (18084)—Lactea by Sarawak (15238)—Cornbind by Janizary
(1875)—Golden Rod by Snowball (8602)—Florence by Little John
(4232)—Cathleen by Caliph (1774)—Rosy by Rob Roy (557)—by
Satellite (1420)—by Sir Dimple (594)—by Styford (629).

A. H. B., vol. 16, p. 12334, and E. H. B., vol. 22, p. 477, under
dam.

4TH DUKE OF CLARENCE 26188 (33597)—Roan bull, calved
Oct. 28, 1874, bred by Col. Gunter, Wetherby Grange, Yorkshire,
got by 18th Duke of Oxford (25995), out of Duchess 109th by 2d
Duke of Claro (21576)—Duchess 100th by 3d Duke of Wharfdale
(21619)—Duchess 87th by 7th Duke of York (17754)—Duchess
80th by Grand Duke of Oxford (16184)—Duchess 72d by 4th Duke
of Oxford (11387)—Duchess 67th by Usurer (9763)—Duchess 59th
by 2d Duke of Oxford (9046)—Duchess 56th by 2d Duke of Northumberland (3646)—Duchess 51st by Cleveland Lad (3407)—Duchess
41st by Belvedere (1706)—Duchess 32d by 2d Hubback (1423)—
Duchess 19th by 2d Hubback (1423)—Duchess 12th by The Earl
(646)—Duchess 4th by Ketton 2d (710)—Duchess 1st by Comet
(155)—by Favorite (252)—by Daisy Bull (186)—by Favorite (252)
—by Hubback (319)—by James Brown's Red Bull (97).

DUKE OF OXFORD 38TH 26351 (38172)—Red and white,
calved Jan. 13, 1876, bred by the Duke of Devonshire, Holkar
Hall, got by 5th Duke of Wetherby (31033), out of Grand Duchess

of Oxford 14th by Grand Duke 10th (21848)—Grand Duchess of Oxford 7th by Lord Oxford (20214)—Grand Duchess of Oxford by 3d Grand Duke (16182)—Countess of Oxford by Earl of Warwick (11412)—Oxford 15th by 4th Duke of York (10167)—Oxford 6th by 2d Duke of Northumberland (3646)—Oxford 2d by Short Tail (2621)—Matchem Cow by Matchem (2281)—by Young Wynyard (2859).

PRINCESS ONEIDA—Roan heifer, calved Aug. 22, 1874, bred by Sir T. C. Constable, Bart., got by Duke of Oneida 1st (30996), out of Princess Victoria 7th by 13th Duke of Oxford (21604)— Princess Victoria by 4th Duke of Oxford (11387)—Princess Louisa by Delhi (15866)—Princess Alice by Gen. Canrobert (12927)—by Earl of Derby (10177)—by 2d Cleveland Lad (3408)—by 2d Earl of Darlington (1945)—by Son of 2d Hubback (2683)—a cow of Mr. Bates'.

E. H. B., vol. 21, p. 639, under dam, and A. H. B., vol. 16, p. 12293.

PRINCESS VICTORIA ONEIDA—Roan heifer, calved June 21, 1875, bred by Sir T. C. Constable, Bart., Burton Constable, Hull, got by Oneida Prince (34948), out of Princess Victoria 9th by 9th Duke of Geneva (28391), Princess Victoria 5th by Lord Oxford 2d (20215)—Princess Alice by Gen. Canrobert (12927)— Princess by Earl of Derby (10177), &c., as above.

E. H. B., vol. 23, p. 372, under dam, and A. H. B., vol. 16, p. 12294.

DUCHESS OF BARRINGTON 3D—Red and white heifer, calved Sept. 4, 1875, bred by W. W. Slye, got by Grand Duke of Thorndale 2d (31298), out of Lady Bates 3d by 4th Duke of Oxford (11387)—Lady Bates 2d by The Buck (13836)—Lady Bates by Duke of Gloster (11382)—Lady Blanche by 4th Duke of York (10167)— Lady Barrington 8th by 2d Duke of Oxford (9046)—Lady Barrington 5th by 4th Duke of Northumberland (3649)—Lady Barrington 3d by Cleveland Lad (3407)—Lady Barrington 2d by Belvedere (1706) — Lady Barrington by Son of Herdsman (304) — Young

Alicia by Wonderful (700)—Old Alicia by Alfred (23)—by Young Favorite (6994).

A. H. B., vol. 16, p. 12038.

LADY EMILY 5TH—Red, calved July 15, 1872, bred by H. J. Sheldon, got by 18th Duke of Oxford (25995), out of Lady Emily Darlington by Duke of Darlington (21586)—Lady Emily 2d by 7th Duke of York (17754)—Lady Emily by Duke of Bolton (12738)—Eugenie by Gray Friar (9172)—Fidelia by Fawsley (6004) —Flourish by Little John (4232)—Lilla by Caliph (1774)—Amy by Norman (2379)—Walnut by White Boy (1580)—Pauline by Wyville's Bull—by a bull of Mr. Charge's.

A. H. B., vol. 16, p. 12145, and E. H. B., vol. 20, p. 593, under dam.

BARON KNIGHTLEY 25610—Red and white, calved May, 1876, bred by the Marquis of Blandford, got by Duke of Rosedale (33721), out of Lady Emily 5th by 18th Duke of Oxford (25995), &c., as in dam, above.

WILD EYES 31ST—Roan heifer, calved Oct. 24, 1873, bred by W. Angerstein, Weeting Hall, Brandon, Norfolk, got by 3d Duke of Claro (23729), out of Wild Eyes 30th by 7th Duke of York (17754)— Wild Eyes 24th by 4th Duke of Oxford (11387)—Wild Eyes 22d by Wild Duke (19148)—Wild Eyes 20th by Lord Barrington 1st (13170) —by 2d Duke of Oxford (9046)—by 4th Duke of Northumberland (3649)—by Duke of Northumberland (1940)—by Belvedere (1706) —by Emperor (1975)—by Wonderful (700)—by Cleveland (145)— by Butterfly (104)—by Hollon's Bull (313)—by Mowbray's Bull (2342)—by Masterman's Bull (422)—descended from M. Dobison's stock.

E. H. B., vol. 22, p. 297, under dam, and A. H. B., vol. 16, p. 12388.

WILD EYES 33D—Red roan heifer, calved July 18, 1874, bred by Wm. Ashburner, Nether House, Ulverstone, got by 2d Grand Duke of Kent (28759), out of Wild Eyes 31st by Grand Duke of Cambridge 3d (31285)—Wild Eyes 29th by Knight of the Harem

(24278)—Wild Eyes 27th by Gainford 5th (12913)—Wild Eyes 26th by 2d Cleveland Lad (3408)—Wild Eyes 5th by Short Tail (2621)—Wild Eyes by Emperor (1975), &c., as above.

A. H. B., vol. 16, p. 12389, and E. H. B., vol. 23, p. 316, under dam.

LADY FAWSLEY 8TH—White, calved May 5, 1876, bred by George Fox, Elmhurst Hall, got by 24th Duke of Airdrie (36460), out of Lady Fawsley 2d by Duke of Brailes (23724), &c., as in Lady Fawsley 2d, above.

A. H. B., vol. 16, p. 12146, and E. H. B., vol. 23, p. 371, under dam.

LADY FAWSLEY 6TH—Roan heifer, calved Dec. 9, 1873, bred by George Fox, got by 18th Duke of Oxford (25995), out of Lady Lady Fawsley 2d by Duke of Brailes (23724), &c., as above.

A. H. B., vol. 16, p. 12146, and E. H. B., vol. 21, p. 720, under dam.

LADY FAWSLEY 7TH—Roan heifer, calved in March, 1875, bred by George Fox, Elmhurst, got by 9th Duke of Geneva (28391), out of Lady Fawsley 2d by Duke of Brailes (23724), &c., as in dam, above.

A. H. B., vol. 16, p. 12146.

AMERICA'S OXFORD—Red heifer, calved April 1, 1872, bred by W. W. Slye, Beaumont Grange, got by 18th Duke of Oxford (25995), out of America by Marmaduke (14897)—Asia by 2d Grand Duke (12961)—Apricot by Fusileer (11499)—Augusta by 3d Duke of York (10166)—Annie by 2d Cleveland Lad (3408)—Annabella by Duke of Cleveland (1937)—Acomb by Belvedere (1706)—a cow of Mr. Bates'.

A. H. B., vol. 16, p. 11959, and vol. 20, p. 391, E. H. B., under dam.

BARON ACOMB 25580—Red bull, calved in May, 1876, bred by W. W. Slye, got by 2d Grand Duke of Thorndale (31298), out of America's Oxford by 18th Duke of Oxford (25995), &c., as above.

BARON ACOMB 2D 25581—Red bull, calved Sept. 3, 1876, bred by W. W. Slye, got by 2d Grand Duke of Thorndale (31298), out of America's Duchess by Grand Duke of Thorndale (31297), &c., as above.

2D BARON KNIGHTLEY 25611—White bull, calved May 17, 1876, bred by J. W. Larking, got by Grand Duke of Geneva (28756), out of Knightley Grand Duchess by 4th Grand Duke (19874), &c., as in dam, above.

1874. MR. CANTRELL,
Of Mexico. By Ship The Nile, from Southampton.

NOBLEMAN (34916)—Red bull, calved March 22, 1873, bred by G. Garne, Churchill Heath, got by St. Swithin (22833), out of Nettle by Monk (24616)—Nemophila by Cynric (19542)—Nectarine by Gen. Pelissier (14605)—Necklace by Uncle Tom (13912)—by Fitz Hardinge (8073)—by Raffler (7391)—by Consul (1868)—by son of Speculation (1472).

PRIDE OF THE HEATH 2D—Roan heifer, calved Jan. 31, 1872, bred by G. Garne, Churchill Heath, Chipping Norton, Oxon, got by Royal Butterfly 20th (25007), out of Pride of the Heath by Cynric (19542)—Peach by Havelock (14676)—Peace by Valiant (7662)—Young Portrait by Fitz Hardinge (8073)—Portrait by Lord John (4259)—by Young Consul (6893)—by Newnham (2365)—by Satellite (1420)—by Jupiter (342)—by Sir Oliver (605)—by Trunnell (659)—by Favorite (252)—by Favorite (252)—by Dalton Duke (188)—by R. Alcock's Bull (19)—by J. Smith's Bull (608)—by Jolly's Bull (337).

E. H. B., vol. 20, p. 696, under dam.

Aug., 1868. W. CARR.

PRINCE OF BUCKINGHAM 8857—Red roan, calved April 6, 1867, bred by W. Carr, Stackhouse, Yorkshire, got by Imperial Windsor (18086), out of Princess Louise by King Arthur (13110)—

Bustle by Valiant (10989)—Bonnet by Buckingham (3239)—Bliss
by Leonard (4210)—Young Broughton by Young Matchem (2282)
—Broughton by Jerry (4097)—by Young Pilot (4702)—by Pilot
(496)—by son of Apollo (36).

Dec. 26, 1872. W. CARR,
Of Compton, Can. By Ship Manitobian, from Liverpool.

LADY CHRISTABEL—Roan, calved May 15, 1868, bred by
W. Carr, Stackhouse, Yorks., got by Windsor Fitz-Windsor (25458),
out of Lady Clare by Prince of the Realm (22627)—Claribel by
Valasco (15443)—Farewell by Royal Buck (10750)—Little Fanny
by Exquisite (8048)—by Hamlet (8126)—by Marcus (2262)—by
Matchem (2281)—by Alderman (1622)—by Pilot (496)—by Remus
(550)—by Sir Charles (592)—by R. Colling's son of Favorite (252)
—by son of Favorite (252)—Strawberry.

E. H. B., vol. 18, p. 560, under dam.

Oct. 20, 1875. W. S. CHAPMAN,
Of San Francisco, Cal. By Ship Erin, from Liverpool.

WILD EYES 31st—Red and white cow, calved April 24, 1871,
bred by Mr. J. Robinson, Bootle, Cumberland, got by Grand Duke
of Cambridge 3d (31285), out of Wild Eyes 29th by Knight of the
Harem (24278)—Wild Eyes 27th by Gainford 5th (12913)—Wild
Eyes 26th by 2d Cleveland Lad (3408)—Wild Eyes 5th by Short
Tail (2621)—Wild Eyes by Emperor (1975)—by Wonderful (700)—
by Cleveland (145)—by Butterfly (104)—by Hollon's Bull (313)—
by Mowbray's Bull (2342)—by Masterman's Bull (422)—descended
from the stock of M. Dobison.

E. H. B., vol. 22, p. 547, under dam.

AMETHYST (32946)—Red bull, calved April 21, 1874, bred by
G. Ashburner, Low Hall, got by Grand Duke of Lightburne 3d
(28761), out of Oxford's Gem by Oxford 4th (24706)—Double Oxford
by Oxford (20449)—Lady Oxford by 10th Duke of Oxford (17739)
—Lily by Hope (13042)—by Duke of Richmond (8000)—by Bachelor

(5770)—by Romulus (6405)—by Sillery (5131)—by Young Western Comet (1575)—by Western Comet (689)—by son of Favorite (252).

FAMOUS KNIGHT (33878)—Roan bull, calved Jan. 13, 1874, bred by W. Torr, Aylesby Manor, Lincolnshire, got by Knight of the Shire (26552), out of Fair Jute by Breastplate (19337)—Fair Dane by Fitz Clarence (14552)—Flower Nymph by Vanguard (10994)—Flower Girl by Londesboro (6141)—by Rinaldo (4949)—by Sir Thomas (2636)—by Sir Alexander (591)—by Marske (418)—by North Star (459)—by Wellington (680)—by Favorite (252)—by Favorite (252)—by Ben (70)—by Hubback (319)—by Snowdon's Bull (612)—by Sir James Pennyman's Bull (601).

Aug., 1864. DAVID CHRISTIE,
Of Paris, Ont. By Ship Sardinian to Quebec.

QUEEN OF ATHELSTANE—Red, calved April 29, 1860, bred by James Douglas, got by Sir James the Rose (15290), out of Playful by 4th Duke of York (10167)—Place 3d by 4th Duke of Northumberland (3649)—Place 2d by Duke of Northumberland (1940)—Place by 2d Earl of Darlington (1945)—Place by son of 2d Hubback (2683)—a cow of Mr. Bates'.

C. H. B., vol. 1, p. 440, and E. H. B., vol. 15, p. 659, under dam.

CROWN PRINCESS OF ATHELSTANE—Roan, calved May 3, 1865, bred by James Douglas, got by Next of Kin (20405), out of Queen of Athelstane by Sir James the Rose (15290), &c., as above.

C. H. B., vol. 1, p. 252.

PRINCESS OF ATHELSTANE—Red, calved July 6, 1863, bred by James Douglas, got by Watchman (17216), out of Queen of Athelstane by Sir James the Rose (15290), &c., as above.

C. H. B., vol. 1, p. 437.

PLACIDA—Red cow, calved July 8, 1858, bred by T. B. Sydserff, Ruchlaw, got by Master of Athelstane (14933), out of Julia

Cruise by Crusade (7938)—Butterfly by Raspberry (4875)—Daisy by Wellington (5625)—by Fleetham (2028)—by Admiral (5)—by Young Denton (963)—by Young Albion (15).

C. H. B., vol. 2, p. 428, and E. H. B., vol. 15, p. 659.

DOUGLAS OF ATHELSTANE [173]—Red and white, calved May 16, 1864, bred by James Douglas, got by Knight of Athelstane (20075), out of Placida by Master of Athelstane (14933), &c., as in dam, above.

CROWN PRINCE OF ATHELSTANE (21512) [157]—Roan bull, calved June 1, 1864, bred by James Douglas, Athelstaneford, Scotland, got by Next of Kin (20405), out of Queen of Athelstane by Sir James the Rose (15290)—Playful by 4th Duke of York (10167)—Place 3d by 4th Duke of Northumberland (3649)—Place 2d by Duke of Northumberland (1940)—Place 1st by 2d Earl of Darlington (1945)—Place by son of 2d Hubback (2683)—a cow of Mr. Bates', of Kirklevington.

PRIDE OF ATHELSTANE—Red, calved July 6, 1861, bred by James Douglas, got by Sir James the Rose (15290), out of Lady of Athelstane by Hymen (13058)—Playful by 4th Duke of York (10167)—Place 3d by 4th Duke of Northumberland (3649)—Place 2d by Duke of Northumberland (1940)—Place 1st by 2d Earl of Darlington (1945)—Place by 2d Hubback (2683)—a cow of Mr. Bates'.

C. H. B., vol. 1, p. 432, and E. H. B., vol. 16, p. 530, under dam.

Sept., 1868. HON. DAVID CHRISTIE,
Of Paris, Ont. To Quebec.

KNIGHT OF ST. GEORGE 8472 (26544) [1630]—Red and white, calved March 12, 1867, bred by W. Carr, Stackhouse, Yorkshire, got by Prince of the Realm (22627), out of Windsor's Queen by Windsor (14013)—Wide Awake by Royal Buck (10750)—Bonnet by Buckinghan (3239)—Bliss by Leonard (4210)—by Young Matchem (2282)—by Jerry (4097)—by Young Pilot (4702)—by Pilot (496)—by son of Apollo (36).

1854. CLARK COUNTY (O.) IMPORTING COMPANY.

(Dr. Arthur Watts and Alexander Waddle, Agents.)

By Ship F. W. Bally. Landed at Philadelphia, June 30, 1854.

LORD OF THE ISLES 3090—White, calved Sept. 5, 1853, bred by F. H. Fawkes, Farnley Hall, got by Bridegroom (11203), out of Leda by Lord Marquis (10459)—Laura by Petrarch (7329)—Fair Spots by Sir Thomas Fairfax (5196)—Spots by Garton (2052)—Latona by Harold (291)—by Count (170)—by Badsworth (47)—by Coates' son of Twin Brother to Ben (660).

Recorded under dam, E. H. B., vol. 11, p. 544. Sold to Alexander Waddle, Clark Co., Ohio, for $575.

YOUNG AMERICA 1123—Red roan, calved Aug. 2, 1854, bred by Mr. Wilkinson, got by Prince Royal (7371), out of Zealous by St. Albans (7462), &c., as in dam, below.

BUCKINGHAM 2D 297 (12509)—Roan, calved June 10, 1852, bred by Mr. Mitchell, Cleasby, got by Oxygen (9464), out of Queen Bess by Hamlet (8126)—White Rose by Pam (4643)—Strawberry by Young Matchem (2282)—Rosamond by Jack Tar (1133)—by Pilot (496)—by Young Albion (15).

Sold to W. D. Pierce, Clark Co., Ohio, for $1,000.

THE DUKE 1029 (13847)—Roan, calved Feb. 19, 1852, bred by John Clark, Aldborough, got by Adam (12338), dam by Whittington (12299)—by Mehemet Ali (7227)—by Guardian (3947)—by Paganini (2405)—by Paul Jones (8333)—by Pirate (2430)—by Sedbury (1424)—by Charge's Gray Bull (872).

Sold to W. C. Davis, Montgomery Co., Ohio, for $625.

NEW-YEAR'S DAY (13383)—Roan, calved Jan. 1, 1853, bred by Lee Norman, Corbollis, got by Magnet (11765), out of Moss Rose by Killerby (7122)—Maradan by Orator (2390)—Martha by Darlington (3561)—Mary Anne by Favorite (1028)—by Crispin (174)—by Rose's Red Bull (5009)—by Turnell's Red Bull (1536)—by a bull of Mr. Cornforth's.

Sold to C. M. Clark, Clark Co., Ohio, for $3,500.

CZAR 395—Roan, calved Dec. 31, 1852. bred by J. Farrell, got by Baron Warlaby (7813), out of Maid of Athens by Druid (10140) —Maid of Aln by Regent (2517)—Edith by Borderer (3191)—by Eclipse (1949)—by Togston (5487)—by Bolingbroke (3184)—by son of Midas (435)—by Twin Brother to Ben (660).

Sold to A. J. Paige, Clark Co., Ohio, for $1,900.

MEDALIST 697 (13324)—White, calved Jan. 18, 1853, bred by Wm. Torr, Aylesby Manor, got by Crown Prince (10087), out of Magnet by Baron Warlaby (7813)—Mosaic by Leonard (4210)— Moonbeam by Prince Comet (1342)—by Constellation (163)—by Prince of Waterloo (528)—by Young Favorite (255).

Sold to Arthur Watts, Chillicothe, Ohio, for $2,100.

LORD STANWICK (13253)—White, calved Jan. 24, 1853, bred by Mr. Wood, Stanwick Park. Darlington, got by Whittington (12299), out of Lady Alice by Noble (4578)—Lady Agnes by Newton (2367)—Lady Mary by Emperor (3716)—Lady Sarah by Satellite (1420) — Portia by Cato (119)—by Jupiter (342) — by George (273)—by Chilton (136)—by Irishman (329)—by B (45).

Sold to Alex. Waddle, Clark Co., Ohio, for $500.

RODOLPH 923—Roan, calved May 31, 1853, bred by Lord Feversham, Duncombe Park, got by Financier (9122), out of Blue Bell by Cleveland Lad (3407)—Blanche by Triumph (5518)—Betsy by Grazier (10085)—Favorite by Parrington (4653)—Light Roan Twin by Baron (58)—Dairymaid by Windsor (698)—out of a grand-daughter of Washington (674).

Sold to W. C. Davis, Montgomery Co., Ohio, for $200.

SHYLOCK 965 (13698)—Roan, calved July 22, 1853, bred by Wm. Torr, Aylesby Manor, got by Crown Prince (10087), out of Solar Ray by Leonard (4210)—Sunshine by Remus (4932)—Sun-beam by Prince Comet (1342)—by Count (170)—by Charles 2d (879)—by Constellation (163)—by Young Favorite (255).

Sold to John Hadley, Clinton Co., Ohio, for $300.

COWS.

LANCASTER 17TH—Roan, calved Feb. 21, 1852, bred by Mr. Wilkinson, Lenton, got by Prince Royal (7371), out of Lancaster 10th by George 3d (7038)—Lancaster 9th by Spectator (2688)—by Albion (1619)—by Lancaster (360)—by a son of Windsor (698)—by Comet (155).

A. H. B., vol. 3, p. 491, and vol. 11, p. 537, E., under dam. Sold to Wm. D. Pierce, Clark Co., Ohio, for $900.

ROAN LADY—Roan, calved April 18, 1852, bred by Mr. Wilkinson, Lenton, got by St. Albans (7462), out of Wiseton Lady by Humber (7102)—Roguery by Mercury (2301)—Pageant by Monarch (2324)—No. 13 by St. Albans (2584)—by Jupiter (342)—by Oliver (605)—Raspberry by Trunnell (659)—Lily by Favorite (252).

Vol. 3, p. 627. Sold to Wm. D. Pierce, Clark Co., Ohio, for $1,000.

LANCASTER 19TH—Red, calved Sept. 13, 1852, bred by Mr. Wilkinson, Lenton, got by St. Albans (7462), out of Lancaster 13th by Queen's Roan (7389)—Laurel by Will Honeycomb (5660)—Lancaster 9th by Spectator (2688)—by Albion (1619)—by Lancaster (360)—by a son of Windsor (698)—by Comet (155).

Vol. 2, p. 434, and vol. 11, p. 537, E., under dam. Sold to L. B. Sprague, Clark Co., Ohio.

VENUS—Roan, calved Dec. 29, 1852, bred by H. Ambler, Watkinson Hall, Halifax, got by Lord Byron (11710), out of Psyche by Abraham Parker (9876)—Octavia by 2d Duke of Northumberland (3646)—Junia by Robertson (2538)—by Prince Edward (2462)—by Sir Francis B (1443)—by Sultan (1485)—by Wellington (683)—by North Star (458).

A. H. B., vol. 3, p. 682, and vol. 11, p. 650, E., under dam. Sold to Wm. D. Pierce, Clark Co., Ohio, for $1,075.

ZENOBIA—Roan, calved Dec. 22, 1852, bred by H. Ambler, Watkinson Hall, Halifax, got by Crusade (7938), out of Zalia by Major (4344)—Zalia by Zadig (8794)—Tuberose by Constitution

(3476)—by Scrip (2604)—by Burley (1766)—by Isaac (1129)—by Pilot (496)—by Albion (14)—by Lame Bull (359)—by Shipton (587)—by a son of Suwarrow (636)—by Son of Twin Brother to Ben (88)—by Twin Brother to Ben (660).

A. H. B., vol. 2, p. 607, and vol. 11, p. 762, E., under dam. Sold to Alex. Waddle, Clark Co., Ohio, for $625.

NELL 2D—Roan, calved March 22, 1853, bred by Mr. Wilkinson, got by Monarch (13347), out of Young Nell by Queen's Roan (7389)—Nell by Will Honeycomb (5660)—by Spectator (2688)—by Albion (1619)—by Sir Peter (606)—by Punch (531)—by Washington (675)—by Washington (674).

Sold to A. Toland, for $350.

BUTTERFLY 13TH—Red roan, calved Sept. 28, 1853, bred by Mr. Wilkinson, got by Monarch (13347), out of Butterfly 12th by Queen's Roan (7389)—by George 3d (7038)—by Will Honeycomb (5660)—by Spectator (2688)—by Sir Roger de Coverley (5187)—by Brother to Albion (1619)—by Alexander (1624)—by son of Butterfly (104).

A. H. B., vol. 6, p. 208, and vol. 11, p. 354, E., under dam, without name.　Sold to H. Stickney, for $290.

AYLESBY LADY—Roan, calved May 13, 1848, bred by Wm. Torr, Aylesby Manor, got by Baron Warlaby (7813), out of Adelaide by Ganthorpe (2049)—Mary by Cossack (1880)—by Sir Henry (1446)—by Young Albion (730)—by Marske (418).

A. H. B., vol. 2, p. 291, and vol. 10, p. 261, E. H. B.　Sold to A. J. Paige, Clark Co., Ohio, for $1,425.

ROMAN 13TH—Roan, calved Sept. 24, 1849, bred by Mr. Wilkinson, Nottingham, Eng., got by Will Honeycomb (5660), out of Roman 3d by Prince (7615)—by Lenton (4205)—by Spectator (2688)—by Sir Roger de Coverley (5187)—by a son of Alexander (1624)—by Favorite (6996).

Vol. 2, p. 537.　Sold to Jacob Pierce, Clark Co., Ohio, for $1,300.

DAHLIA—Red, calved March 6, 1850, bred by Sir T. Cartwright, Aynhoe, got by Upstart (9760), out of Daisy by Sweet William (8646)—Daffodil by Harold (8131)—Jenny Dennison by Cedric (3311)—by Son of Favorite (1028)—by Cœlebs (897)—by Rose's Bull (5009)—by Fisher's Bull (3799).

A. H. B., vol. 2, p. 341, and vol. 11, p. 394, E. Sold to A. J. Paige, Clark Co., Ohio, for $1,100.

ZEALOUS—Roan, calved March 18, 1850, bred by Mr. Wilkinson, Nottingham, got by St. Albans (7462), out of Zeal by Roman (2561)—Roguery by Mercury (2301)—Pageant by Monarch (2324)—No. 13 by St. Albans (2584)—by Jupiter (342)—by Sir Oliver (605)—Raspberry by Trunnel (659)—Strawberry by Favorite (252)—Lily by Favorite (252)—Miss Lax by Dalton Duke (188)—by R. Alcock's Bull (19)—by J. Smith's Bull (608)—by Jolly's Bull (337).

A. H. B., vol. 2, p. 606, and vol. 10, p. 638, E., under dam. Sold to Alexander Waddle, for $1,000.

NECTAR—Roan, calved Jan. 20, 1851, bred by T. Birchall, Ribbleton Hall, got by North Star (9447), out of Beeswing by Lord Adolphus Fairfax (4249)—Bessy by Thick Hock (6601)—Barmpton Rose by Expectation (1988)—by Belzoni (1709)—by Comus (1861)—by Denton (198).

Vol. 11, p. 608, E. Sold to J. Davis, for $600.

SIR ARTHUR 2208—Roan, calved Sept. 7, 1854, bred by Sir T. Cartwright, got by Gilliver (11529), out of Dahlia by Upstart (7960), &c., as in Dahlia, above.

LAVENDER 3D—Red and white, calved Aug. 5, 1851, bred by Mr. Wilkinson, got by St. Albans (7462), out of Lavender 2d by Queen's Roan (7389)—by Will Honeycomb (5660)—by Spectator (2688)—by Albion (1619)—by Lancaster (360)—by a son of Windsor (698)—by Comet (155).

Vol. 2, p. 437, A., and vol. 10, p. 446, E., under dam, without name. Sold to Arthur Watts, Chillicothe, Ohio, for $500.

BLUSHING BEAUTY—Roan, calved June 21, 1853, bred by Wm. Torr, Aylesby, got by Crown Prince (10087), out of Blushing

Maid by Helmsman (8141)—Brunette Beauty by Lord Adolphus Fairfax (4249)—Brownie by Ormsby (4621)—Bonadea by Count Comet (3509)—Cherub by Soldier (2656)—Cherry by Gray Robin (1090)—Ruby by Barmpton (54)—by Aylesby (44)—by brother to R. Colling's White Heifer.

Vol. 2, p. 311, A., and vol. 11, p. 344, E., under dam. Sold to Alexander Waddle, Clark Co., Ohio, for $425.

ROMAN 14TH—Red roan, calved Jan. 18, 1855, bred by Mr. Wilkinson, got by St. Albans (7462), out of Roman 13th by Will Honeycomb (5660)—Roman 3d by The Prince (7615)—by Lenton (4205)—by Spectator (2688)—by Sir Roger de Coverley (5187)—by a son of Alexander (1624)—by Favorite (6996).

A. H. B., vol. 4, p. 539.

ZEPHYR—Roan, calved in Jan., 1853, bred by Mr. Fawkes, Farnley Hall, got by Beaufort (9943), out of Lady Zariffa by Laudable (9282)—Zuleika by Norfolk (2377)—Medora by Ambo (1636) —Blossom by Memnon (2295)—Sister to Isabella by Pilot (496)—by Agamemnon (9)—by Burrell's Bull (1768).

Vol. 2, p. 607, A. H. B. Sold to L. B. Sprague, Clark Co., Ohio, for $400.

EASTER-DAY—Roan, calved March 27, 1853, bred by F. H. Fawkes, Farnley Hall, got by Lord Marquis (10459), out of Loyal by Triumph (8717)—Lydia by Matchless (4428)—Laura by Boughton (2868)—Lily by Roman (2559)—by Columella (904)—by Albion (14)—by Cinnamon (139)—by Neswick (1266).

Vol. 2, p. 360, A., and vol. 11, p. 554, E., under dam. Sold to C. M. Clark, Clark Co., Ohio, for $1,125.

BLUSH 17TH—Red, calved April 5, 1853, bred by Thos. Barnes, Westland, Moynalty, Meath, Ireland, got by Baron Warlaby (7813), out of Blush 14th by Hamlet (8126)—Blush 12th by Albion (7771) —Blush 9th by 2d Comet (5107)—Blush 7th by Lucifer (4293)—Blush 4th by Prince George (2464)—Blush 2d by Volunteer (——) —by Kearney's Bull (4144).

Vol. 3, p. 321, A. H. B. Sold to C. Green, Bloomington, Ill., for $470.

ROSY—Red and white, calved Sept. 16, 1853, bred by T. Barnes, Westland, Moynalty, County Meath, Ireland, got by Royal Buck (10750), out of Rosebud by Ury (5536)—by 2d Comet (5101)—by Lucifer (4293)—by Prince George (2464)—by Frederick (7023)—by Kearney's Bull (4144).

Vol. 3, p. 641, A. H. B. Sold to C. Green, Bloomington, Ill., for $400.

SILK—Red and white, calved July 25, 1853, bred by T. Barnes, Westland, Moynalty, County Meath, Ireland, got by Hopewell (10332), out of Phœnix by 2d Comet (5701)—Pauline by Prince of Northumberland (4826)—Paulina by Mason's son of Matchem (2676)—by Falstaff (1993)—by Richard (1376)—by Jupiter (342)—by Rufus (570)—by Pope (514).

Vol. 2, p. 556, A. H. B. Sold to Charles Phellis, Madison Co., Ohio, for $205.

ROSE OF PANTON—Roan, calved in Sept., 1853, bred by Mr. Dudding, got by Leonidas (10414), out of Rhoda by Gen. Washington (6036)—Red Rose by Plenipo (4724)—by Alamode (725)—by Childers (1824)—by Young Wyham Favorite (7734)—by Quarternion (1351)—by Rocket (1390)—by Eclipse (——)—by Waddingworth (668).

Sold to Dr. Toland for $375.

1871. CLARK COUNTY (Ky.) IMPORTING COMPANY.

(LEWIS HAMPTON AND W. C. VANMETER, Agents.)

By Ship Hudson, April 11, 1871, to New York.

COWS.

RED PRINCESS—Red, calved Aug. 23, 1865, bred by D. Nesham, Gainford Hall, Darlington, got by King of Trumps (22047), out of Viscountess by 8th Duke of York (21621)—Victress by Victory (19073)—Dewdrop by Bacchus (17338)—Lady Arabella by Duke of Cleveland (3640)—by Bellerophon (3119)—by Warrior (9806).

Recorded under dam, E. H. B., vol. 18, p. 771, and A. H. B., vol. 11, p. 1025. Bought at Company's sale Aug. 26, 1871, by Asa Bean, Clark Co., Kentucky, for $800.

PATCHOULI 4TH—Roan, calved June 7, 1866, bred by J. Christy, Boynton Hall, Chelmsford, got by Duke of Grafton (21594), out of Patchouli 2d by Tragedian (20989)—Patchouli by Commedian (15789)—Primula by Victor (15458)—Polyanthus by Orestes (15027)—Oxslip by Progress (11950)—Cowslip by Roan Robin (10721)—by Mars (7212)—by Earl of Liverpool (3677)—by Arthur (3099)—by Imperial (4068)—by Juniper (1145)—by Columella (904)—by Shakespeare (1429)—by Blyth Comet (85)—by Neswick (1266)—by Favorite (1033).

E. H. B., vol. 17, p. 662, under dam, A. H. B., vol. 12, p. 1117. Sold to Jas. Houlton, Warren Co., Ill., for $870.

MIRANDA—Red heifer, calved Feb. 6, 1867, bred by T. E. Pawlett, Beeston, Sandy Beds, got by Baron Killerby (23364), out of Miracle by Prince James (20554)—Heather Belle by Hero (18055)—Fanny by Rubens (5027)—Farewell by Young Matchem (4422)—by Isaac (1129)—by Young Pilot (4702)—by Pilot (496)—by Julius Cæsar (1143).

Recorded E. H. B., vol. 18, p. 620, under dam, and A. H. B., vol. 11, p. 915. Bought at sale of Company, Aug. 26, 1871, by Thos. G. Sudduth, Clark Co., Ky., for $975—for cow and calf.

DULCIMER—Red and white heifer, calved Feb. 13, 1867, bred by Lord Penrhyn, Penryhn Castle, Bangor, Wales, got by 11th Grand Duke (21849), out of Dulcinea by Duke of Geneva (19614) —Duchess 1st by Master Rembrandt (16545)—Duchess Nanny by Jasper (11609)—Duchess Nancy by 2d Duke of Oxford (9046)— Nettle by 2d Duke of Northumberland (3646)—by Belvedere (1706) —by Son of 2d Hubback (2683)—a cow of Mr. Bates', of Kirklevington.

Recorded under dam, E. H. B., vol. 18, p. 468. Sold to Dr. Wash. Miller, for $570. Died without produce.

WELCOME—Red and white heifer, calved Feb. 27, 1867, bred by D. Nesham, Gainford Hall, Darlington, got by Windsor 2d (23226), out of Susan by 3d Duke of Athol (12734)—Mignonette by Bates (12451)—Henriette by Ingram (9236)—by Liberator (7140)—by Prince Albert (4791)—by Young Matchem (4422)—by Young Red Rover (4904) or Rockingham (2551)—by Whisker (1579)—by Pilot (496).

Recorded E. H. B., vol. 18, p. 743, under dam. Sold to T. C. Vanmeter, for $700.

LADY SPENCER 2D—Red, calved March 1, 1870, bred by R. Eastwood, Thorney Holme, Whitewell, Clitheroe, got by Baron Oxford (23375), out of Lady Spencer by May Duke (13320)—Joke by John Ford (9253)—Festival by Lord Warden (7167)—Farce by Zenith (5702)—Melodrama by Orontes (4623)—Sacontola by William (2840)—Clarion by Childers (1824)—No. 25 Chilton Sale by Richard (1376)—by Jupiter (342)—Charles (127)—by Windsor (698)—by Chilton (136)—by Colonel (152).

Recorded under dam, E. H. B., vol. 19, p. 592. Sold to W. H. Nelson, for $1,220.

DAPHNE—Roan heifer, calved March 11, 1870, bred by Lord Penrhyn, Penrhyn Castle, Bangor, Wales, got by 3d Duke of Wharfdale (21619), out of Dulcinea by Duke of Geneva (19614)—Duchess 1st by Master Rembrandt (16545)—Duchess Nanny by Jasper (11609)—Duchess Nancy by 2d Duke of Oxford (9046)—Nettle by 2d Duke of Northumberland (3646)—Nell Gwynne by Belvedere (1706)—North Allerton by Son of 2d Hubback (2683)—a cow of Mr. Bates'.

Recorded under dam, E. H. B., vol. 19, p. 490. Sold to B. F. Vanmeter, for $710. Sold at Company's sale, Aug. 26, 1871, under name of Lady Penrhyn. I think she never produced.

PRINCESS MAUD—Roan heifer, calved March 20, 1870, bred by R. Searson, Cranmore Lodge, Market Deeping, got by Duke of Devonshire (21588), out of Primitive by Birthday (19313)—Pristine by Uxbridge (13930)—Priscilla by Earl Stanhope (5966)—

14

Perpetua by Master Charley (7215)—Daffodil by Plenipo (4724)—Duchess by Alamode (725)—Lady Sarah by Childers (1824)—by Wellington (5619)—by Panton Major (6274)—by Waddington (668).

Sold to Abram Vanmeter, for \$330; was out of health and died soon after sale.

RARITY—Roan cow, calved June 17, 1866, bred by J. Clayden, Littlebury, Saffron Waldon, Essex, got by Costa (21487), out of Ruby by Lord of the Harem (16430)—Cornelian by Mameluke (13289)—Coral by Cardinal (11246)—Charmer 3d by Earl of Dublin (10178)—Charmer by Little John (4232)—by Caliph (1774)—by Sir Walter (2637)—by Hotspur (1117)—by Coxcomb (928)—by Midas (435)—by Comet (155)—by R. Colling's son of Favorite (252)—by Hubback (319).

Recorded under dam, E. H. B., vol 17, p. 731, and A. H. B., vol. 11, p. 1015. Bought at sale of Company, Aug. 26, 1871, by Asa Bean, Clark Co., Kentucky, for \$1,080.

MIRANDA 2D—Red cow, calved Aug. 12, 1871, bred by T. E. Bawlett, got by Majestic (29255), out of Miranda by Baron Killerby (23364), &c., as in Miranda, above.

A. H. B., vol. 12, p, 1056. Sold with dam to T. G. Sudduth.

PRIDE OF THE WEST—Red heifer, calved Sept. 29, 1867, bred by G. Garne, Churchill Heath, Chipping Norton, Oxon., got by 6th Duke of Airdrie (19602), out of Peach by Havelock (14676)—Peace by Valiant (7662)—Young Portrait by Fitz Hardinge (8073)—Portrait by Lord John (4259)—by Young Consul (6893)—by Newnham (2365)—by Satellite (1420)—by Jupiter (342)—by Sir Oliver (605)—by Trunnell (659)—by Favorite (252)—by Favorite (252)—by Dalton Duke (188)—by R. Alcock's Bull (19)—by J. Smith's Bull (608)—by Jolly's Bull (337).

Recorded under dam, E. H. B., vol. 18, p. 651. Sold to W. Nelson, for \$1,250.

ROSE OF WICKEN—Red and white heifer, calved Oct. 3, 1867, bred by Lord Penrhyn, Penrhyn Castle, Bangor, Wales, got by 2d Duke of Geneva (21591), out of Red Rosette by 2d Duke of Thorndale (17748)—Red Bonnett by Game Boy (14586)—Aurora by Walter (6658)—Graceful by Marmion (4383)—Victoria by Young Rubens (5026)—Splendid by Matchem 3d (4420)—by Young Eryholme (1981)—by Belzoni (1709)—by Comus (1861)—by Denton (198)—by Henry (301).

Recorded, E. H. B., vol. 18, p. 686, under dam, and A. H. B., vol. 12, p. 1212. Bought at sale by Dr. W. Cunningham, Clark Co., Ky., for $850.

FATIKO—Red heifer, calved March 9, 1868, bred by D. Nesham, Gainford Hall, Darlington, got by Fitz Arthur (26161), out of Alexandra by Problem (16767)—Miss Rothschild by Lord Mayor (14828)—Carnation by Omer Pacha (13417)—Young Strawberry by Colonel (5428)—Strawberry by Guardian (3947)—Sally of the Tees by Magnum Bonum (2243)—Old Sally by Young Rockingham (2547)—by North Star (460)—by Denton (198)—by Ladrone (353)—by Henry (301).

Recorded, E. H. B., vol. 18, p. 369, under dam, and A. H. B., vol. 11, p. 647. Sold to A. H. Hampton, Winchester, Ky., for $850. Produced a bull calf; died in 1873.

ROSETTE 5TH—Red and white heifer, calved May 29, 1868, bred by Mr. Botcherby, Middle-one-row, Darlington, got by Vain Hope (23102), out Rosette 3d by Earl of Derby 2d (15960)—Rosette by Young Magistrate (14877)—by Hero (12203)—by Liverpool (4235)—by Colonel (5428).

Recorded, E. H. B., vol. 19, p. 714, and A. H. B., vol. 11, p. 1066. Bought by Asa Bean, of Clark Co., Ky., at Company's sale, Aug. 26, 1871, for $900.

DAPHNE—Roan heifer, calved Jan. 10, 1869, bred by T. E. Pawlett, Beeston, Sandy Beds, got by Fitz Killerby (26166), out of Florence by Prince Hopewell (22592)—Faithful by Sir James (16980)—Faith by Sir Charles (12075)—Fanchette by Petrarch (7329)—by Raspberry (4875)—by Young Matchem (4422)—by Isaac

(1129)—by Young Pilot (4702)—by Pilot (496)—by Julius Cæsar (1143).

Recorded, A. H. B., vol. 11, p. 786 as Lady Pawlett, and so sold at Company's sale, Aug. 20, 1871, to Louis Hampton, Winchester, Ky., for $900. Produced no females.

COWSLIP 2D—Red and white heifer, calved Feb. 14, 1869, bred by Lord Penrhyn, Wicken Park, got by 3d Duke of Wharfdale (21619), out of Cherry Blossom by Duke of Geneva (19614)—Cherry Lips by Cherry Duke 2d (14265)—Cowslip 5th by Chieftain (10048)—Cowslip 2d by Duke of Norfolk (5952)—Cowslip by Waterloo (2816)—by Kitt (7127)—by Kitt (7127)—by Page's Bull (6269)—by Middleton's Bull (438).

Recorded under dam, E. H. B., vol. 19, p. 440. Sold to Lewis Hampton for $1,300. Produced no females and died early.

TINY—Red and white heifer, calved May 9, 1869, bred by R. Searson, Cranmore Lodge, Market Deeping, got by Falstaff (21720), out of Devon Rose by Duke of Devonshire (21588)—Moss Rose by Sir Simon (18867)—Sweetbriar by Uxbridge (13930)—Duchess 3d by Sugar Plum (10894)—Dorcas by Earl of Stanhope (5966)—Duchess 2d by Master Charley (7215)—Dowager by Plenipo (4724)—Red Duchess by Blaize (76)—by Chieftain (886)—by Young Neswick (1268)—by Marfleet's Red Bull (1192).

Recorded, A. H. B., vol. 11, p. 1113. Bought at Company's sale, Aug. 26, 1871, by S. R. Redmon, Winchester, Ky., for $600.

HARTFORTH STRAWBERRY—Roan heifer, calved April 14, 1867, bred by R. F. Housman, Lune Bank, Lancaster, got by Baron Hartforth (23361), out of White Strawberry by Duke of Cambridge (12747)—Strawberry Lass by St. Thomas (10777)—by Chorister (3378)—by Tom Gwynne (5498)—by Wellington (2824)—by Marmion (406)—by Ossian (476).

Recorded under dam, E. H. B., vol. 19, p. 785. Sold to J. G. Clinkenbeard, for $900.

SWEET ROSE—Red heifer, calved April 24, 1867, bred by R. Searson, Cranmore Lodge, Market Deeping, got by Chieftain

(21421), out of Sweetbrier by Uxbridge (13930)—Duchess 3d by
Sugar Plum (10894)—Dorcas by Earl Stanhope (5966)—by Master
Charley (7215)—by Plenipo (4724)—by Blaize (76)—by Chieftain
(886)—by Neswick (9268)—by Marfleet's Red Bull (1192).

Recorded, A. H. B., vol. 11, p. 1107. Bought by T. G. Sudduth,
Clark Co., Ky., at Company's sale, Aug. 26, 1871, for $900.

GERTY—Roan heifer, calved July 31, 1867, bred by J. W.
Botcherby, Middleton-one-row, Darlington, got by Vain Hope
(23102), out of Garland by Grand Master (24078)—Bridget by
Highthorn (13028)—Cambridge by Cavaignac (10033)—Clytem-
nestra by Sir Walter (2639)—Curioso by Cordilleras (3484)—Lily
or White Rose by Young Remus (2523)—White Cow by Remus
(550)—by Hollings (2131)—by His Honor (2126)—by Partner (2409)
—by Hutton's Bull (2145)—Lady, descended from R. Alcock's stock.

Recorded under dam, E. H. B., vol. 18, p. 507, and A. H. B.,
vol. 11, p. 676. Bought at Company's sale, Aug. 26, 1871, by
Lewis Hampton, Winchester, Ky., for $895.

TINY 2D—Red, calved Oct. 30, 1871, bred by R. Scarson, got
by Lord Chatham (26625), out of Tiny by Falstaff (21720), &c., as
in Tiny, above.
See A. H. B., vol. 12, p. 1257.

CLOCHETTE—Red heifer, calved June 25, 1870, bred by J.
Christy, Boynton Hall, Chelmsford, got by Duke of Grafton (21594),
out of German Aster by Royal Arch (18749)—China Aster by Archduke 2d (15588)—Cream by Young Weathercock (15495)—Roan
Crocus by General Elliott (10266)—Red Crocus by Young Locksley
(4240)—Crocus by Sherborne (10805)—Crocus by Prince (4772)—
Hawthorn by Stanhope (5315)—Beauty by a son of Sir Kenneth
(1450)—by Pilot (1319)—by Thorpe (1515)—by Waistell's Bull
(1567).

Recorded, A. H. B., vol. 11, p. 560. Bought at Company's sale,
Aug. 26, 1871, by Gen. Lucius Desha, Cynthiana, Ky., for $850.

ROSELEAF—Died in England before shipment.

SWEET ROSE 2D—Red, calved June 25, 1872, bred by R. Searson, got by Lord Chatham (26625), out of Sweet Rose (vol. 11) by Chieftain (21421), &c., as in dam, above.

A. H. B., vol. 20, p. 16220.

BULLS.

DUKE OF BABRAHAM (25934)—Red bull, calved Sept. 5, 1867, bred by J. Christy, Boynton Hall, Chelmsford, got by Duke of Grafton (21594), out of Babraham Duchess by Guelder Rose (19910)—Young Celia 6th by Young Duke of Cambridge (14433)—Young Celia by Lord of the North (11743)—Celia by 3d Duke of Northumberland (3647)—Cornflower by Bashaw (1692)—Columbine by Helmsman (2109)—Columbia by Columella (904)—Charlottina by Regent (544)—Charlotte Palatine by Palatine (478)—Charlotte by Palmflower (480)—Crimson by Patriot (486)—Young Milbank by Driffield (223)—by C. Holmes' Bull (314).

Sold to W. S. Sudduth, for $790.

WELCOME'S SORCERER 15690—Red and white, calved Aug. 10, 1871, bred by D. Nesham, Gainford Hall, Darlington, got by Sorcerer (30031), out of Welcome by Windsor 2d (23226), &c., as in Welcome, above.

Sold with dam, to T. C. Vanmeter.

PEABODY (29535)—Roan bull, calved June 18, 1869, bred by T. Lawson, Stapleton Grange, got by Sir Christopher (22895), out of Miss Tod by Fitzroy (21760)—Princess Alice by Arthur Gwynne (19244)—Red Princess by Leader (11674)—Princess by Duke of Richmond (14453)—by Young Hector (2104)—by Sir Charles (1440)—by Sir Roland (1455)—by son of Phenomenon (491)—by Irishman (329).

Sold to W. C. Vanmeter, for $900.

PIONEER 12593 (29553)—Red bull, calved Dec. 18, 1870, bred by Mr. P. Pawlett, Buston, got by Baron Killerby (23364), out of Rose of Warlaby by British Flag (19351)—Rose of Hope by Prince

Alfred (13494)—Rose of Promise by Heir at Law (13005)—Rose of Autumn by Sir Henry (10824)—Pelerine by Buckingham (3239)—Mantalini by Marcus (2262)—Maiden by Matchem (2281)—Lady by Alderman (1622)—Lady Mowbray by Pilot (496)—Sylph by Remus (550)—Matilda by Sir Charles (592)—Alpine by R. Colling's son of Favorite (252)—Young Strawberry by son of Favorite (252)—Old Strawberry.

Sold to W. S. Sudduth, for $400.

WHARFDALE PRINCE 15996—Roan, calved Sept. 19, 1871, bred by Lord Penrhyn, got by 3d Duke of Grafton (28395), out Cowslip 2d by 3d Duke of Wharfdale (21619), &c., as in dam, above.

DUKE OF WICKEN 14130—Red, calved Sept. 8, 1871, bred by Lord Penrhyn, got by Cherry Duke (25752), out of Rose of Wicken by 3d Duke of Geneva (21591), &c., as in dam, above.

ROYAL BOOTH 15392—Roan, calved Nov. 19, 1871, bred by Mr. Huchison, Yorkshire, got by Merry Monarch (23349), out of Gerty by Vain Hope (23102), &c., as in dam, above.

1836. H. CLAY, JR.

LORD ALTHORP 658 (3005)—Roan, calved in 1835, bred by Mr. Thompson, got by Chance (1807), out of Luck's-all by Rufus—Bellona by Young Lancaster (1162)—Ruby by a son of Blaize—Buttercup by a son of Favorite (252)—Trotter's Georgiana by a son of Favorite (252)—by Punch (531).

CROCUS (cow)—Red and white, calved in 1829, got by Imperial (2151), out of Judith by Walker's Emperor—Primrose by Prince—Snowball by ——————, —a cow of Earl De Grey's.

IO (cow)—Roan, bred by J. Woodhouse, calved in 1836, in New York, got by Milton (2315), out of Crocus by Imperial (2151)—Judith by Mr. Walker's Emperor—Primrose by Prince—Snowball —a cow of Earl De Grey's.

1837. H. CLAY, JR.

MARY ANN—Got by Young Phœnix, out of Mary by Saladin (1417)—Red Rose by a bull of Mr. Maynard's—by Twin, a bull of Mr. Booth's.

CORA—Sire unknown, out of Mary Ann—1837.

PRINCESS—Roan, calved in April, 1832, bred by Mr. Smith, got by Edmund (1954), out of Selina by Grazier (1085)—Favorite by Northampton (2380)—Vanity by Reform (1361)—Old Cora.
Recorded, E. H. B., vol. 3, p. 624.

BRITANNIA—Roan, bred by Charles Ellerton, Smeaton, England.   No pedigree sent.

VENUS—Roan, got by Thorp (2757), dam by Mars.   Full pedigree not sent.

BEAUTY—Roan, calved at sea, got in England by Temperance, out of Venus by Thorp (2757)—by Mars.

VICTORIA—White, calved in 1837, gotten in England by Osgodly, out of Princess by Edmund (1954)—Selina by Grazier (1085)—Favorite by Northampton (2380)—Vanity by Reform (1361)—Old Cora.

NEPTUNE 743—Red and white, got by Roscius (2565), dam by Vaunter (1543)—by Blucher (84)—by Windsor, &c.

1839. H. CLAY, JR., and GEN. JAMES SHELBY.
Fayette County, Ky.

COWS.

JANE—Roan, calved in 1833 or 1834, got by Neptune (4554), out of Strawberry by Emperor (1013)—by Snowball (611)—by Baron (58)—Wright, bred by Lord Feversham.

DORCAS—Red and white, calved in 1833 or 1834, got by Sheridan (2616), out of Duchess by Ivanhoe (1331)—by Enchanter.

CHARITY—Roan, calved in 1834, got by Lakin's son of Jupiter (2169). dam by Roman (2559)—by Admiral (5)—by a son of Blyth Comet (85)—by a bull of Mr. Foljambe's.

NERISSA—Roan, calved in 1835 or 1836, bred by Rev. Thomas Harrison, got by The Chief, out of Netherby by Cupid (938)—Moss Rose by Barmpton (54)—by Western Comet (689)—by Son of Favorite (253).

See vol. 5, p. 731, E. H. B.

NOTE.—Mr. Clay refers us for The Chief to "Omission to Coates' Herd Book," probably (6575).

MOSS ROSE—Red roan, calved in 1835, bred by Mr. Topham, got by Eclipse (1949), out of Miss Points Jr. by Northern Light (1280)—Miss Points by Aid-de-Camp (722)—by Charles (127)—by Prince (521)—by Neswick (1266).

See vol. 3, p. 513, E. H. B., under dam.

COLUMBINE—Red and white, calved in 1832, got by Rockingham (2449), dam by Young Major (2254)—by George—by Cossack (925).

PET—White, calved in 1835, got by Mameluke (2257), dam by Tarrare (2735)—by R. Colling's Major (398)—Brown's Bull of Aldborough (820).

VIXEN—Roan, calved May 1, 1835, bred by Mr. Smith, West Rasen, got by Statesman (2700), dam by Vendor [a son of Reform (1361)]—by Reform (1361).

See vol. 3, p. 684, E. H. B.

PROTECTRESS—Red and white, calved in 1833, bred by Mr. Clark, Hellaby Hall, got by Topper (2708), out of Clara by Young Rectifier—Lucy by Protector (1346)—Magdalene by Marshall's Bull—by Style's Bull.

PRINCESS—Roan, calved in 1836, got by Cedric (3311), dam by Mr. Roger's son of Blaize (76)—by Cœlebs (797)—by Ross' Red Bull (2568)—by Neswick (1266)—by Fisher's Red Bull (2022).

DON JOHN 426—Roan, bred by Mr. Clark, Usworth, England, got by Farmer (2001), out of Martha by Sir Charles—by son

15

of Mason's Sir Charles (421)—by Mr. Powell's Bull, descended from the Styford stock.

COSSACK ALIAS JULIUS CÆSAR (3503)—Roan, calved Oct. 12, 1839, bred by Mr. Topham, got by Cossack (1880), out of Moss Rose by Eclipse (1949)—Miss Points Jr. by Northern Light (1280) —Miss Points by Aid-de-Camp (722)—by Charles (127)—by Prince (521)—by Neswick (1266).

Sold to Benjamin Warfield, Sr.

1854. CLINTON COUNTY (OHIO) IMPORTING COMPAY.

(MESSRS. H. H. HANKINS, J. G. COULTER AND A. R. SEYMOUR, Agents.)

COWS.

DAISY (vol. 2, p. 344)—Roan, calved May 26, 1850, bred by H. J. Spearman, got by Zadig (8796), out of Alice Grey by Nero 3196—a cow, bred by Wm. Grey, Hilfield Hall.

DUCHESS—Roan, calved in May, 1849, bred by Wm. Harrison, Greta-Bridge, got by Norfolk (9442), out of Nancy by Guardian (3947)—by Red Highflyer (2488)—by Rob Roy (557)—by Marshal Blucher (416)—by Haughton (318).

Vol. 2, p. 356. Sold to Mr. B. Wright and W. Palmer, Fayette Co., Ohio, for $1,677.

EMMA—Roan, calved Jan. 22, 1850, bred by Henry Smith. Drax Abbey, Yorkshire, got by Promoter (10658), dam by Liberty (7141)—by Belshazzar (3123)—by Norfolk (2377)—by Baronet (774) —by Marshal Beresford (415)—by Windsor (698)—by son of Patriot (486).

Vol. 2, p. 366. Sold to Thos. Kirk, Fayette Co., Ohio, for $750.

HOPE—White, calved March 14, 1848 or 1849, bred by Henry Smith, Drax Abbey, got by Duke of York (6947), out of Faith by Shakespeare (2614)—Famous by Soldier (2656)—by Young Favorite

(254)—Diana by Young Barmpton (55)—by Wellington (680)—by Phenomenon (491)—by Favorite (252).

Vol. 10, p. 401, E. H. B., and vol. 2, p. 402, A. H. B. Sold to Wm. Palmer, Fayette Co., Ohio, for $1,000.

MISS SHAFTOE—Red, calved in Dec., 1849, bred by W. Smith, got by Capt. Shaftoe (6833), out of Madcap by William (2840)—Reform by Mercury (8305)—Liberty by Ivanhoe (1131)—Tiffany by Regent (544)—Cambric by Lawnsleeves (365)—by Lawnsleeves (365).

Vol. 2, p. 484. Sold to Jesse Starbuck, Clinton Co., Ohio, for $650.

FAMILIAR—Roan, calved June 12, 1846, bred by H. Mitchell, Cleasby, got by Fitz-Leonard (7010), out of Isabella by Velocipede (5552)—Strawberry by Young Matchem (2282)—Rosamond by Jack Tar (1133)—by Pilot (496)—by Young Albion (15).

Vol. 10, p. 360, E. H. B., and vol. 3, p. 399, A. H. B. Sold to Jesse Pancake, Ross Co., Ohio, for $500.

SUNBEAM—Roan, calved in July, 1849, bred by Mr. Wetherell, near Darlington, got by Twilight (9758), out of Susan by Augustus (6752)—Sally by a son of Grazier (1085)—Sarah by Scarlet (5094)—Esther by Aid-de-Camp (722).

Vol. 2, p. 570. Sold to J. G. Coulter, Clinton Co., Ohio, for $450.

YOUNG EMMA—Light roan, calved in April, 1849, bred by R. Thornton, got by Sailor (9592), out of Emma by Paley (7310)—by Bulmer (1760)—by a son of Fairfax (1023)—by Shylock (2622)—by Whitworth (1584)—by Candour (107).

Vol. 2, p. 601, A. H. B., and Vol. 9, p. 350, E. H. B., under dam. Sold to Hankins & Palmer, Clinton Co., Ohio, for $450.

MISS WALTON 2D—Roan, calved Oct. 15, 1851, bred by R. Emmerson, Stapleton, got by Chilton (10054), out of Miss Walton by Belvedere 3d (3128)—by Sir Walter (2639)—by Marquis (2270)—by Edmond (1954)—by Barmpton (54).

Vol. 2, p. 485. Sold to John Hadley, Clinton Co., Ohio, for $325.

PRINCESS—Roan, calved 1850, bred by R. Thornton, Stapleton, Eng., got by Lord Newton,* out of Kate by Isaac (9239)—White Cow by Nelson (4549)—by Saladin (1417)—by Pilot (1319).

Vol. 2, p. 518. Sold to Hadley & Hankins, Clinton Co., Ohio, for $1,060.

　　* Lord Newton by Rebuke—dam by Newton (2367)—by Goldfinder (2066).

MOONBEAM—Red and white, calved Feb. 16, 1852, bred by Mr. Wetherell, Kirkbridge, got by Oxygen (9464), out of Sunbeam by Twilight (9758)—Susan by Augustus (6752)—Sally by a son of Grazier (1085)—Sarah by Scarlet (5094)—Esther by Aid-de-Camp (722).

Vol. 3, p. 571. Sold to Henry Kirk, Fayette Co., Ohio, for $500.

LADY JANE—Red, calved June 26, 1852, bred by Mr. Wetherell, of Kirkbridge, got by Whittington (12299), out of Lady Welbourn by Lord Lowther (7164)—Eliza by Panton Favorite (4646)—Miss Eliza by a son of Grazier (1085)—Elizabeth by a son of Grazier (1085)—Bessy by Vulcan (8746)—Bessy by Quaker (1349)—Betty by 1st Stonehill Bull (3798).

Vol. 3, p. 482, A. H. B., and vol. 11, p. 535, E. H. B., under dam. Sold to David Watson, Madison Co., Ohio, for $500.

LADY WHITTINGTON—Roan, calved June 27, 1853, bred by Mr. Wetherell, Kirkbridge, got by Whittington (12299), out of Lady Welbourn by Lord Lowther (7164)—Eliza by Panton Favorite (4646)—Miss Eliza by a son of Grazier (1085)—Elizabeth by a son of Grazier (1085)—Bessy by Vulcan (8764)—Bessy by Quaker (1349)—Betty by 1st Stonehill Bull (3798).

Vol. 5, p. 345, A. H. B., and vol. 11, p. 535, E. H. B., under dam. Sold to Wm. Reed, Clinton Co., Ohio, for $300.

STRAWBERRY—Roan, calved May 26, 1852, bred by John Foster, Newton Hall, near Redale, got by Wiseman (12317), out of Dairymaid by Hautboy (10305)—Kitchenmaid by Irishman (5446)—Maiden by Oliver (4909)—Maid by William (5661)—The Maid by Pilot (496).

Vol. 2, p. 568. Sold to James Fullington, Union Co., Ohio, for $675.

LOUISA—Roan, calved Dec. 25, 1853, bred by J. Robinson, Top Cliff, near Thirk, got by Crusader (10088), out of Miss Shaftoe by Capt. Shaftoe (6833)—Madcap by William (2840)—Reform by Mercury (8305)—Liberty by Ivanhoe (1131)—Tiffany by Regent (544)—Cambric by Lawnsleeves (365)—by Lawnsleeves (365).

Vol. 2, p. 449. Sold to J. R. Mills, Clinton Co., Ohio, for $300.

JESSAMINE—Roan, calved in Jan., 1853, bred by R. Emmerson, Eryholme, got by Young Chilton (11278), out of Jessamine (vol 10, p. 412, E.) by Guy Faux (7062)—Duchess of St. Albans by St. Albans 2d (5048)—from the stock of Mr. James, of Stamford.

Sold to J. O'B. Renick, Franklin Co., Ohio, for $475.

VICTORIA—Roan, calved 1850, bought of Mr. R. Emmerson, of Eryholme, without pedigree.

Vol. 2, p. 583, A. H. B. Sold to D. Peringer, for $1,000.

QUEEN—Red and white, calved April 19, 1854, bred by R. Emmerson, got by The Marquis 1031, out of Victoria.

Sold to Mr. H. S. Pavy, for $425.

DIANA—Roan, calved Oct. 22, 1854, bred by Henry Smith, calved the property of Wm. Palmer, Fayette Co., Ohio, got by Capt. Shaftoe (6833), out of Hope by Duke of York (6947), &c., as in Hope, above.

A. H. B., vol. 4, p. 318.

BULLS.

LORD RAINE 2D 665—Red and white, calved in Ohio, June 23, 1854, bred by R. Thornton, Stapleton, got by Lord Raine (13248), out of Young Emma by Sailor (9592)—Emma by Paley (7310)—by Bulmer (1760)—by a son of Fairfax (1023)—by Shylock (2622)—by Whitworth (1584)—by Candor (107).

Sold to Daniel Early, Clinton Co., Ohio, for $195.

YOUNG SIR ROBERT 1161—Roan, calved July 6, 1854, bred by R. Emmerson, Eryholme, got by The Marquis 1031, out of Daisy by Zadig (8796)—Alice Gray by New *—bred by Wm. Gray, Hill-field Hill.

Sold to Thomas McMillan, Clinton Co., Ohio, for $250.

* New by Speculation, a son of Guy Faux (7062)—by Young Barmpton (3088)—by Shortley (5124)—by Ajax (723), &c.

DUKE OF DARLINGTON 448—White, calved Jan. 14, 1855, bred by T. Wetherell, Darlington, got by Whittington (12299), out of Sunbeam by Twilight (9758), &c., as in Sunbeam, above.

PRINCE GEORGE 2076—Roan, calved Dec. 10, 1854, bred by H. Mitchell, got by Wellington 1087, out of imp. Familiar by Fitz Leonard (7010), &c., as in Familiar, above.

DUKE OF CORNWALL 2D—Roan, calved in April, 185-, bred by Wm. Harrison, Greta Bridge, got by Albert (8816), dam by Norfolk (9442)—Duchess of Cornwall by Albert (5729)—Lily by Young St. Ledger (2585)—Sally by North Star (459)—by Alexander (1623)—by a grandson of Favorite (252)—by Punch (531)—by Hubback (319).

Sold to David Quinn, Clinton Co., Ohio, for $700.

BILLY HARRISON 263—White, calved in July, 1852, bred by William Harrison, Greta Bridge, got by Master Belleville (11795), out of Duchess by Norfolk (9442)—Nancy by Guardian (3947)—by Red Highflyer (2488)—by Rob Roy (557)—by Marshall Blucher (416)—by Houghton (318).

Sold to Jesse Starbuck, Clinton Co., Ohio, for $1,500.

MOONRAKER 3175—Roan, calved Feb. 2, 1854, bred by Mr. Wetherell, Kirkbridge, got by Whittington (12299), out of Sunbeam by Twilight (9758)—Grandame Susan by Augustus (6752)—Sally by a son of Grazier (1085)—Sarah by Scarlet (5094)—Esther by Aid-de-Camp (722).

Sold to Thomas Connor, Fayette Co., Ohio, for $400.

WELLINGTON (13989)—Roan, calved May 21, 1852, bred by R. Lawson, Stapleton, near Darlington, got by Landsdown (9277), out of Cycle by King Lear (8196)—Zone by Orontes (4623)—Roguery by Mercury (2301)—Pageant by Monarch (2324)—by St. Albans

(2584)—by Jupiter (342)—by Sir Oliver (605)—Raspberry by Favorite (252)—Miss Lax by Dalton Duke (188)—Lady Maynard by R. Alcock's Bull (19)—by Jacob Smith's Bull (608)—by Jolly's Bull (337).

Sold to J. G. Coulter. H. H. Hankins and others, Clinton Co., Ohio, for $3,700.

ALFRED (12374)—Roan, calved April 10, 1852, bred by Mr. John Clarke, Aldborough, near Darlington, got by Adam (12338), dam by Old Red Bull *—by Paganini (2405)—by Paul Jones (8383) —by Pirate (2430)—by Sedbury (1424)—by Charge's Gray Bull (872).

Sold to D. S. King, Clinton Co., Ohio, for $900.

*Old Red Bull by Æolus (2938), dam by a son of Raine's Wellington (bred by Robt. Clark, of Barmpton)—grand dam by Col. Leather's Achmet.

WARRIOR (12287)—Roan, calved Sept. 26, 1850, bred by R. Booth, Warlaby, got by Water King (11024), out of Bagatelle by Buckingham (3239)—Jemima by Raspberry (4875)—Strawberry 3d by Young Matchem (4422)—Strawberry 2d by Young Alexander (2977)—Strawberry by Pilot (496)—Halnaby by Lame Bull (359)— by Easby (232)—by Suworrow (636).

Sold to B. Hinkson and H. H. Hankins, Clinton Co., Ohio, for $1,200.

WHITTINGTON 2D (14005)—Roan, calved May 17, 1851, bred by John Wood, Stanwick Park, got by Whittington (12299), out of Beeswax by Noble (4578)—Beeswing by St. Helena (5055)— Blossom by Reformer (2502)—Rosebud by Margrave (2263)—by Leopold (2199)—by Hector (2103)—by Surly (2715)—by Traveler (655)—by Colonel (152).

Sold to Sol. Brock, Fayette Co., Ohio, for $900.

THE MARQUIS 1031—Roan, calved in Oct., 1851, bred by Col. Henly, got by Landsdown (9277), out of Lady by Colonel (5428) —Young Strawberry by Guardian (3947)—Old Strawberry by a son of Barmpton (54)—by a brother to Brutus (100)—by Duke (127).

Sold to W. Bently, Clinton Co., Ohio, for $625.

1867. M. H. COCHRANE.

Hillhurst, Compton, Quebec. By Ship Austrian, from Glasgow to Montreal.

ROSEDALE—Roan cow, calved Feb. 13, 1861, bred by Lady Pigot, Branches Park, Newmarket, got by Valasco (15443), out of Rosy by Master Belleville (11795)—Red Rose by Vanguard (10994) —Dinah by Diamond (5018) — Strawberry by Young Matchem (2282)—Rosamond by Jack Tar (1133)—by Pilot (496)—by Young Albion (15).

Recorded, E. H. B., vol. 16, p. 668, and A. H. B., vol. 8, p. 538. Sold to W. S. King, Minn.

BARON BOOTH OF LANCASTER 7535—Red and white, calved Feb. 22, 1867, bred by Mr. Barclay, Keavil, Dumfermline. Fife, got by Baron Booth (21212), out of Mary of Lancaster by Lord Raglan (13244)—Lancaster 25th by Matadore (11800)—Lancaster 16th by The Marquis (10938) — Lancaster 12th by Will Honeycomb (5660)—Lancaster 10th by George 3d (7038)—Lancaster 9th by Spectator (2688)—by Albion (1619)—by Lancaster (360)—by son of Windsor (698)—by Comet (155).

Sold to J. H. Pickrell, Harristown, Ill.

CAPTAIN AITON 6512—Red and white bull, calved Sept. 8. 1867, bred by Lady Pigot, Branches Park, Newmarket, calved the property of M. H. Cochrane, got by Scottish Chief (22849), out of Rosedale by Valasco (15443)—Rosy by Master Bellville (11795)— Red Rose by Vanguard (10994)—Dinah by Diamond (5918)—Strawberry by Young Matchem (2282)—Rosamond by Jack Tar (1133)— by Pilot (496)—by Young Albion.

Aug. 15, 1868. M. H. COCHRANE.

Hillhurst, Compton, Prov. Quebec. By Ship Germany, from Liverpool.

DUCHESS 97TH—Red heifer, calved March 27, 1867, bred by Col. Gunter, Wetherby Grange, Yorks., got by 3d Duke of Wharfdale (21619), out of Duchess 92d by 4th Duke of Oxford (11387)— Duchess 84th by Archduke (14099)—Duchess 72d by 4th Duke of Oxford (11387)—Duchess 67th by Usurer (9763)—Duchess 59th by 2d Duke of Oxford (9046)—by 2d Duke of Northumberland

(3646)—by Cleveland Lad (3407)—by Belvedere (1706)—by 2d Hubback (1423)—by The Earl (646)—by Ketton 2d (710)—by Comet (155)—by Favorite (252)—by Daisy Bull (186)—by Favorite (252)—by Hubback (319)—by J. Brown's Red Bull (97).

Recorded, E. H. B., vol. 21, p. 685, and A. H. B., vol. 9, p. 563. Sold to W. S. King, Minn. Returned to England, by Ship Sarmatian, July, 1873, for Lord Dunmore.

WHARFDALE ROSE—Roan heifer, calved Sept. 27, 1867, bred by Col. Gunter, Wetherby Grange, got by 3d Duke of Wharfdale (21619), out of Oxford Rose by 6th Duke of Oxford (12765)— Moss Rose by Ravensworth (9532)—Graceful by Freebooter (7025) —Treasure by Ganthorpe (2049)—by Belshazzar (1704)—by Don Juan (1923)—by Shylock (2622)—by Muggeen's Bull.

Recorded under dam, E. H. B., vol. 18, p. 647, and A. H. B., vol. 9, p. 1002. Sold to W. R. Duncan, Towanda, Ill.

WILD EYES 26TH—Red and white heifer, calved June 24, 1865, bred by C. W. Harvey, Walton-on-the-Hill, Liverpool, got by Earl of Walton (17787), out of Wild Eyes 24th by 4th Duke of Oxford (11387)—Wild Eyes 22d by Wild Duke (19148)—Wild Eyes 20th by Lord Barrington 1st (13170)—Wild Eyes 16th by 2d Duke of Oxford (9046)—Wild Eyes 15th by 4th Duke of Northumberland (3649)—by Duke of Northumberland (1940)—by Belvedere (1706)—by Emperor (1975)—by Wonderful (700)—by Cleveland (145)—by Butterfly (104)—by Hollon's Bull (313)—by Mowbray's Bull (2342)—by Masterman's Bull (422)—descended from M. Dobison's stock.

Recorded under dam, E. H. B., vol. 17, p. 791, and A. H. B., vol. 9, p. 1008. Sold to W. S. King, Minneapolis, Minn.

GAY LADY—White heifer, calved Dec. 10, 1867, bred by Wm. Torr, Aylesby Manor, got by Lord Blithe (22126), out of Gay Nun by The Druid (18981)—Gay Bride by Brideman (12493)—Guiding Star by Crown Prince (10087)—Gleamy by Vanguard (10994)— Glitter by Londesbro' (6142)—Glowworm by Ranunculus (2479)—

16

Golden Locks by Remus (4932)—Golden Beam by Prince Comet (1342)—by Count (170)—by Constellation (163)—by Young Favorite (255).

Recorded, E. H. B., vol. 18, p. 508, under dam, and A. H. B., vol. 9, p. 636.

ROBERT NAPIER (27310)—Roan bull, calved April 18, 1868, bred by Wm. Torr, Aylesby Manor, got by Lord Blithe (22126), out of Riby Peeress by Breastplate (19337)—Riby Queen by Booth Royal (15673)—Riby Rose by Vanguard (10994)—Rennet by Fanatic (8054)—Rosebud by Auld Robin Gray (6753)—Red Rose by Scrip (2604)—Rose by Burley (1766)—Young Anna by Isaac (1129) —Anna by Pilot (496)—Ariadne by Albion (14)—Bright Eyes by Booth's Lame Bull (359)—by Shipton (587)—by son of Suwarrow (636)—by Son of Twin Brother to Ben (88)—by Twin Brother to Ben (660).

Sold to Wm. Warfield, Lexington, Ky.

1st LORD WILD EYES OF COMPTON alias COMPTON LORD WILD EYES 8540 (25819)—Roan, calved August 14, 1868, bred by Mr. Harvey, Walton-on-the-Hill, Liverpool, England, got by Lord Wild Eyes 5th (26762), out of Wild Eyes 26th by Earl of Walton (17787)—Wild Eyes 24th by 4th Duke of Oxford (11387)—Wild Eyes 22d by Wild Duke (19148)—Wild Eyes 20th by Lord Barrington 1st (13170)—Wild Eyes 16th by 2d Duke of Oxford (9046)—Wild Eyes 15th by 4th Duke of Northumberland (3649)—Wild Eyes 8th by Duke of Northumberland (1940)—Wild Eyes 2d by Belvedere (1706)—Wild Eyes by Emperor (1975)—by Wonderful (700)—by Cleveland (145)—by Butterfly (104)—by Hollon's Bull (313)—by Mowbray's Bull (2342)— by Masterman's Bull (422)—descended from M. Dobison's stock.

CAPTAIN GRAHAM 7656—Red and white, calved on board ship Germany, Aug. 27, 1868, bred by R. S. Bruere, Braithwaite Hall, got by Prince of the Realm (13510), out of Pink Thornleaf by Baron Booth (21212)—Windsor Lavender Leaf by Windsor (14013)—Lavender Leaf by Sylvan King (13819)—Lavender by

Silky Laddie (10947)—Myrtle by Rogue (5012)—Tulip by Chance (3329)—Leaf by Burton (3250)—by son of Comet (155).

STAR OF THE REALM 11021—Roan bull, calved Nov. 3, 1868, bred by R. S. Bruere, got by Prince of the Realm (22627), out of Star of Braithwaite by Baron Booth (21212)—Star of Windsor by Windsor (14013)—Vesper by King Arthur (13110)—Vesper by Morning Star (6225)—Primrose by Roland (2556)—by Priam (2452)—by Matchem (2281)—by son of Peter (487).

Sold to A. J. Alexander, Woodburn, Ky., and resold to M. H. Cochrane.

STAR OF BRAITHWAITE—Roan heifer, calved June 10, 1865, bred by R. S. Bruere, Braithwaite Hall, got by Baron Booth (21212), out of Star of Windsor by Windsor (14013)—Vesper by King Arthur (13110)—Vesper by Morning Star (6223)—Primrose by Roland (2556)—by Priam (2452)—by Matchem (2281)—by son of Peter (487).

Recorded, E. H. B., vol. 17, p. 748, under dam, and A. H. B., vol. 9, p. 963.

PINK THORNLEAF—Roan heifer, calved April 24, 1865, bred by R. S. Bruere, Braithwaite Hall, got by Baron Booth (21212), out of Windsor's Lavender Leaf by Windsor (14013)—Lavender Leaf by Sylvan King (13819)—Lavender by Silky Laddie (10947)—Myrtle by Rouge (5012)—Tulip by Chance (3329)—by son of Wynyard.

Recorded under dam, E. H. B., vol. 17, p. 794, and A. H. B., vol. 9, p. 875. Sold to W. S. King, Minn.

WARLABY FLOWER—Roan heifer, calved Aug. 28, 1866, bred by Wm. Torr, Aylesby Manor, got by Prince of Warlaby (15107), out of Clarence Flower by Fitz Clarence (14552)—British Flower by British Prince (14197)—The Flower by Baron Warlaby (7813)—Flower Girl by Londesbro' (6142)—Flora of Farnsfield by Rinaldo (4949)—by Sir Thomas (2636)—by Sir Alexander (591)—

by Marske (418)—by North Star (459)—by Wellington (680)—by Favorite (252)—by Ben (70).

Recorded under dam, E. H. B., vol. 17, p. 421, and A. H. B., vol. 9, p. 1001.

June 23, 1869. M. H. COCHRANE,

Of Hillhurst, Compton, Prov. Quebec. By Ship Gleniffer, from Glasgow.

BRITISH MAID—Roan, calved March 29, 1862, bred by R. Chaloner, King's Fort, Moynalty, Ireland, got by British Prince (14197), out of Village Rose by Blood Royal (14169)—Village Maid by Crown Prince (10087)—Wharfdale Maid by Sir Leonard (10827) —Village Belle by Pilgrim (4701)—by 2d Hubback (1423)—by Frederick (1060)—by Western Comet (689)—by Western Comet (689)—by Western Comet (689)—by son of Favorite (252).

Recorded under dam, E. H. B., vol. 15, p. 767, and A. H. B., vol. 9, p. 512. Sold to W. S. King, Minn.

FLORIBUNDA—Roan cow, calved July 4, 1862, bred by R. Chaloner, King's Fort, Moynalty, Ireland, got by Dr. McHale (15887), out of Flower Queen by Vanguard (10994)—Flower Girl by Londesbro' (6142)—Flora of Farnsfield by Rinaldo (4949)—by Sir Thomas (2636)—by Sir Alexander (591)—by Marske (418)— by North Star (459)—by Wellington (680)—by Favorite (252)— by Ben (70).

Recorded under dam, E. H. B., vol. 15, p. 497, and A. H. B., vol. 9, p. 628.

PRINCESS—Roan, calved April 17, 1867, bred by T. E. Pawlett, Beeston, Sandy Beds, got by Baron Killerby (23364), out of Fathom by General Havelock (16110)—Faith by Sir Charles (12075)—Fanchette by Petrarch (7329)—Fame by Raspberry (4875)—by Young Matchem (4422)—by Isaac (1129)—by Young Pilot (4702)—by Pilot (496)—by Julius Cæsar (1143).

Recorded under dam, E. H. B., vol. 18, p. 489, and A. H. B., vol. 9, p. 886.

ROSE OF JUNE—Roan heifer, calved June 10, 1867, bred by
T. E. Pawlett, Beeston, Sandy Beds, got by Baron Killerby (23364),
out of Rose of Promise by Heir-at-Law (13005)—Rose of Autumn by
Sir Henry (10824)—Pelerine by Buckingham (3239)—Mantalini by
Marcus (2262)—by Matchem (2281)—by Alderman (1622)—by
Pilot (496)—by Remus (550)—by Sir Charles (592)—by R. Colling's
son of Favorite (252)—by son of Favorite (252)—Strawberry.

Recorded, E. H. B., vol. 18, p. 710, and A. H. B., vol. 9, p. 936.

MAY FLOWER—Roan heifer, calved May 15, 1869, bred by
R. Chaloner, King's Fort, Ireland, got by King Richard (26523),
out of Floribunda by Dr. McHale (15887)—Flower Queen by Vanguard (10994)—Flower Girl by Londesbro' (6142)—Flora Farnsfield by Rinaldo (4949)—Formosa by Sir Thomas (2636)—by Sir
Alexander (591)—by Marske (418)—Sweetbrier by North Star (459)
—by Wellington (680)—by Favorite (252)—by Ben (70).

Recorded, A. H. B., vol. 11, p. 897. Sold to W. S. King, Minneapolis, Minn.

PRINCESS ROYAL—Roan, calved Aug. 27, 1869, bred by Mr.
Pawlett, Beeston, got by Prince Alfred (27107), out of Princess by
Baron Killerby (23364)—Fathom by Gen. Havelock (16110)—Faith
by Sir Charles (12075)—Fanchette by Patriarch (7329)—Fame by
Raspberry (4875)—Farewell by Young Matchem (4422)—Flora by
Isaac (1129)—by Young Pilot (4702)—by Pilot (496)—by Julius
Cæsar (1143).

Recorded, A. H. B., vol. 10, p. 789. Calved after her mother
was purchased by Mr. Cochrane. Sold to Hon. Geo. Brown, Bow
Park, Canada.

ROSE OF AUTUMN—Roan heifer, calved Sept. 2, 1869, bred
by Mr. Pawlett, Beeston, got by Prince Alfred (27107), out of Rose
of June by Baron Killerby (23364)—Rose of Promise by Heir-at-
Law (13005)—Rose of Autumn by Sir Henry (10824)—Pelerine
by Buckingham (3239)—Mantalini by Marcus (2262)—Maiden by
Matchem (2281)—Lady by Alderman (1622)—Lady Mowbray by
Pilot (496)—Sylph by Remus (550)—Matilda by Sir Charles (592)

—Alpine by R. Colling's son of Favorite (252)—Young Strawberry by a son of Favorite (252)—Old Strawberry.

Recorded, A. H. B., vol. 10, p. 829. Sold to Hon. Geo. Brown, Canada. Calved after her dam was bought by Mr. Cochrane.

Aug. 7, 1869. M. H. COCHRANE.
By Ship Germany, from Liverpool.

ISABELLA SOVEREIGN—Roan heifer, calved Feb. 2, 1867, bred by T. Barnes, Westland, Moynalty, Ireland, got by Royal Sovereign (22802), out of Isabella by British Prince (14197)—Sweetbrier by Nimrod (13388)—Charlotte by Selim (6454)—Rebecca by Rex (6385)—by Sir Thomas Fairfax (5196)—by Ambo (1636)—by Memnon (2295)—by Pilot (496)—by Agamemnon (9)—by Burrell's Bull of Burdon (1768).

Recorded under dam, E. H. B., vol. 18, p. 535, and A. H. B., vol. 9, p. 672. Sold to A. Vanmeter, Clark Co., Ky.

QUEEN OF DIAMONDS—Red and white heifer, calved Feb. 24, 1867, bred by J. Lynn, Stroxton, Grantham, Lincolnshire, got by Prizeman (24870), out of Queen of Hearts by May Duke (13320)—Queen Bess by 3d Duke of Oxford (9047)—Queen Anne by Freeman (10244)—by Hamlet (8128)—by Short Tail (2621)—by Emperor (1974)—by Young Lancaster (361)—by St. Albans (2584)—by Lawnsleeves (365).

Recorded under dam, E. H. B., vol. 18, p. 676, and A. H. B., vol. 9, p. 895. Sold to W. S. King, Minn.

FOREST QUEEN—Red heifer, calved Oct. 25, 1867, bred by Hugh Aylmer, West Dereham Abbey, Stakeferry, Norfolk, got by Prince Christian (22581), out of Flounce by Alderman 2d (17292)—Filigree by First Fruits (16048)—Flirt by Kirklevington (11639)—Flounce by Broughton Hero (6811)—Frill by Rockingham (2550)—Fancy by Remus (2524)—by Juniper (1144)—by White Comet (1582)—by Wright's grandson of Favorite (2073)—by Cattley's Gray Bull (1798).

Recorded under dam, E. H. B., vol. 18, p. 500, and A. H. B., vol. 9, p. 632. Sold to A. Vanmeter, Clark Co., Ky.

WEAL BLISS—Roan heifer, calved Nov. 19, 1867, bred by W.
Torr, Aylesby Manor, Lincolnshire, got by Lord Blithe (22126), out
of Weal Royal by Booth Royal (15673)—Weal Princess by British
Prince (14197)—Water Coryphee by Crown Prince (10087)—Water
Witch by 4th Duke of Northumberland (3649)—by Norfolk (2377)
—by Waterloo (2816)—by Waterloo (2816).

Recorded under dam, E. H. B., vol. 18, p. 778, and A. H. B.,
vol. 9, p. 1001.

BRIGHT LADY—Roan heifer, calved April 6, 1868, bred by
Wm. Torr, Aylesby Manor, Lincolnshire, got by Lord Blithe
(22126), out of Bright Countess by Breastplate (19337)—Bright
Princess by British Prince (14197)—Bright Dawn by Vanguard
(10994)—Bright Phœbus by Crown Prince (10087)—by Zadig
(8796)—by Auld Robin Gray (6753)—by J. Chrisp's Bull—by
Burley (1766)—by Isaac (1129)—by Pilot (496)—by Albion (14)
—by Booth's Lame Bull (359)—by Shipton (587)—by son of
Suworrow (636)—by Son of Twin Brother to Ben (88)—by Twin
Brother to Ben (660).

Recorded under dam, E. H. B., vol. 18, p. 404, and A. H. B.,
vol. 9, p. 511. Returned to England by ship Sardinia; landed at
Liverpool Aug. 4, 1877, and sold Sept. 4, 1877, by public auction,
to J. Torr, for 330 guineas.

GENERAL NAPIER 8199 (26239)—Roan bull, calved May 2,
1868, bred by Wm. Torr, Aylesby Manor, Lincolnshire, got by Lord
Blithe (22126), out of Glossary by Booth Royal (15673)—Guide
Book by Dr. McHale (15887)—Guiding Star by Crown Prince
(10087)—Gleamy by Vanguard (10994)—by Londesbro' (6142)—by
Ranunculus (2479)—by Remus (4932)—by Prince Comet (1342)—
by Count (170)—by Constellation (163)—by Young Favorite (255).
Sold to W. S. King, Minn.

SENATOR 3D 9077—Red and white bull, calved Jan. 19, 1869,
bred by T. Crisp, Butley Abbey, Suffolk, got by Earl of Westmoreland (21662), out of Silence 5th by Count de Vermandois (19507)
—Silence 2d by Sir Colin Campbell (13718)—Silence by Earl of

Derby (10177)—Secret 3d by Duke of Sutherland (6945)—Secret 2d by Locomotive (4242)—Secret by Short Tail (2621)—White Rose by Gambier (2046)—by Young Wynyard (2859)—by bulls of C. & R. Colling.

BRITISH QUEEN—Roan heifer, calved Dec. 1, 1869, bred by Mr. Chaloner, King's Fort, Ireland, calved in America, the property of M. H. Cochrane, Hillhurst, got by Sovereign (27538), out of British Maid by British Prince (14197)—Village Rose by Blood Royal (14169)—Village Maid by Crown Prince (10087)—Wharfdale Maid by Sir Leonard (10827)—Village Belle by Pilgrim (4701)—Wharfdale Lady by 2d Hubback (1423)—White Face by Frederick (1060)—Western Lady by Western Comet (689)—by Western Comet (689)—by Western Comet (689)—Haughton by a son of Favorite (252).

A. H. B., vol. 10, p. 426. Returned to England Aug. 4, 1877, by ship Sardinian ; sold Sept. 4, 1877, at public sale, to Rev. T. Standiforth, for 230 guineas.

Aug. 2, 1870. M. H. COCHRANE.
By Ship North America.

CHARLOTTE 4TH—Roan cow, calved March 29, 1863, bred by J. Logan, Maindee House, Newport, Monmouthshire, got by Duke of Knowlmere (19623), out of Charlotte by Noble Arthur (16621)—Celma by Highland Laddie (13026)—Crummy by Cotherstone (6902)—Crummy by Locksley (4240)—Crummy by Stanhope (5315)—Crummy by Prince (4772)—a cow bought from Tyneside.

Recorded, E. H. B., vol. 18, p. 418, and A. H. B., vol. 10, p. 441.

CANDIDATE'S DUCHESS 2D—Roan, calved Feb. 15, 1869, bred by D. R. Davies, Mere Old Hall, Knuttsford, Cheshire, got by Grand Duke of Essex 4th (24068), out of Candidate's Duchess by Duke of Wharfdale (19648)—Candidate by Jasper (11609)—Candido 3d by Voltiguer (12274)—Candido 2d by Lord March (10457)—Candido by Ribblesdale (7422)—Eliza Jane by Prince Albert (4801)—Eliza by Election (1961)—Mary Ada by Navarino

(2352)—by Top Knot (1521)—by Cobourg (1841)—by Young
Marske (2275)—Ossian (476)—Young Laura by Comet (155)—
Laura by Favorite (252)—Lady by Grandson of Bolingbroke (280)
—Phœnix by Foljambe (263)—Favorite by R. Alcock's Bull (19)
—by J. Smith's Bull (608)—Strawberry by Jolly's Bull (337).

Recorded, A. H. B., vol. 10, p. 433. Sold to W. T. Hughes,
Lexington, Ky.

MEADOW FLOWER 13TH—Roan cow, calved Sept. 27, 1867,
bred by T. T. Drake, Shardeloes, Amersham, Bucks., got by Wizard
(25468), out of Meadow Flower 8th by Honeycomb (16279)—
Meadow Flower 4th by Ellington (14491)—Meadow Flower 2d by
Scapulary (15244)—Meadow Flower by Vanguard (10994)—Mag-
dalen by Viceroy (7678)—Mulberry by Raspberry (4875)—Maid by
Isaac (1129)—Madam by Cecil (120).

Recorded under dam, E. H. B., vol. 18, p. 613, and A. H. B.,
vol. 10, p. 706.

SCOTSMAN 10951 (27435)—Roan, calved Feb. 27, 1868, bred
by the Duke of Buccleuch, Dalkeith Park, Edinburgh, got by Royal
Errant (22780), out of Comet by Lord Stanley (18275)—Fortunate
by Captain Balco (12564)—Violet by Kossuth (11646)—Rosie by
Roger (13615)—Fortunate by son of Thorp (2757)—Lady Derby 2d
by Thorp (2757)—Lady Derby by Albion (731)—by Wellington
(679)—by Sultan (631)—by Signior (588).

OLD SAM 10551 (32449)—Red, calved June 4, 1868, bred by
R. H. Crabb, Great Baddow, Chelmsford, Essex, got by Duke of
Grafton (21594), out of Roma by Baron Roxwell (21240)—Eugenie
by Dandy (14366)—Princess Alice by Essex Hero (9096)—Duchess
by Duke of Marlborough (3645)—Miss Chrisp by Captain (3273)—
Nanny by Emperor (1974)—Young Nanny by Snowball (2647)—
Nanny by son of St. Albans (2584)—by Col. Trotter's son of Lawnsleeves (365)—by Barnaby (1678)—by Barber's Bull—bred from the
stock of Messrs. James, of Stamford, Northumberland.

ROYAL PRINCE 12859—Roan bull, calved Dec. 23, 1870, bred
by H. Aylmer, got by Royal Broughton (27352), out of Princess

17

Christian by Prince Christian (22581)—Queen of the Herd by Royal Counsellor (20725)—Queen of the May by Lord of the Harem (16430)—Sally by Gloster's Grand Duke (12949)—Silky by Bristol (11205)—Champion by Vagabond (9765)—Silver by Locksley (4240) —by Stanhope (5315)—by Marquis (22292)—descended from Mr. Jobling's stock of Styford.

ROSE SOVEREIGN—Roan cow, calved Dec. 30, 1867, bred by T. Barnes, Westland, Moynalty, Ireland, got by Royal Sovereign (25044), out of Red Rose by The Druid (18981)—Rosy McHale by Dr. McHale (15887)—Moss Rose by Hopewell (10332)—Rosebud by Hamlet (8126)—Red Rose 3d by Rosebury (5011)—Red Rose 2d by Ury (5536)—Red Rose 1st by 2d Comet (5101)—by Prince George (2464)—by Frederick (7023)—by Kearney's Bull (4144).

Recorded under dam, E. H. B., vol. 18, p. 685.

HONEYSUCKLE—Red and white heifer, calved May 5, 1868, bred by T. Barnes, Westland, got by Royal Duke (25014), out of Sweetbrier by Nimrod (13388)—Charlotte by Selim (6454)—Rebecca by Rex (6385)—Fair Maid of Athens by Sir Thomas Fairfax (5196)—Medora by Ambo (1636)—Blossom by Memnon (2295)— Own Sister to Isabella by Pilot (496)—White Cow by Agamemnon (9)—by Mr. Burrell's Bull of Burdon (1768).

Recorded under dam, E. H. B., vol. 18, p. 743, and A. H. B., vol. 10, p. 567.

ROSEDALE 3D—Roan heifer, calved Feb. 24, 1869, bred by Rev. J. Storer, Hellidon, Daventry, got by Royal Buckingham (20718), out of Rosy by Master Belleville (11795)—Red Rose by Vanguard (10994)—Dinah by Diamond (5918)—Strawberry by Young Matchem (2282)—Rosamond by Jack Tar (1133)—by Pilot (496)— by Young Albion (15).

Recorded under dam, E. H. B., vol. 19, p. 715, and A. H. B., vol. 10, p. 827. Returned to England Aug. 4, 1877, and sold Sept. 4, 1877, at public sale to J. C. Toppin for 62 guineas.

COUNTESS OF YARBOROUGH—Roan heifer, calved Aug. 2, 1868, bred by Messrs. Dudding, Panton, Wragby, Lincolnshire, got by Baron Rosedale (21239), out of Countess of Wragby by Sir Roger (16991)—Columbine by Lambton (9273)—Cactus by Gen. Washington (6036)—Cranberry by Plenipo (4724)—Young Cowslip by Ratify (2481)—Cowslip by Wellington (680)—by Favorite (252)—by Punch (531).

Recorded under dam, E. H. B., vol. 19, p. 459, and A. H. B., vol. 10, p. 458.

MADRIGAL 18TH—Red roan cow, calved Nov. 16, 1867, bred by T. T. Drake, Shardeloes, Bucks., got by Wizard (25468), out of Madrigal 8th by Final Hope (17848)—Madrigal 4th by Yellow Jack (15526)—Madrigal by Vanguard (10994)—Mellow Tone by Exquisite (8048)—Music Note by Leonard (4210)—Martin Bell by Pilgrim (4701)—Minstrelsy by Prince Comet (1342)—by Count (170)—by Constellation (163)—by Young Favorite (255).

Recorded, E. H. B., vol. 18, p. 601, under dam, and A. H. B., vol. 10, p. 672.

LADY SOLWAY—Roan heifer, calved Jan. 2, 1869, bred by J. Beattie, Newby House, Annan, got by Baronet (27930), out of Julia by Duke (14419)—Juno by Mack Turk (14872)—Nelly by Pride (10631)—Bonny by The Peer (5455)—Nell by Henry (2113)—by son of George (2057)—by George (2057)—by Togston (5487)—bred by Mr. Laing, of Longhaughton.

Recorded, A. H. B., vol. 10, p. 629.

BOOTH'S LANCASTER—Roan heifer, calved April 4, 1868, bred by G. R. Barclay, Keavil, Dumfrieshire, Fife, got by Baron Booth (21212), out of Anne of Lancaster by Lord Raglan (13244)—Lancaster 25th by Matadore (11800)—Lancaster 16th by The Marquis (10938)—Lancaster 12th by Will Honeycomb (5660)—Lancaster 10th by George 3d (7038)—Lancaster 9th by Spectator (2688)—by Albion (1619)—by Lancaster (360)—by son of Windsor (698)—by Comet (155).

Recorded under dam, E. H. B., vol. 18, p. 375, and A. H. B., vol. 10, p. 418. Sold to W. S. King, Minneapolis, Minn.

PHILLIS 8TH—Roan cow, calved Dec. 25, 1867, bred by Hugh Aylmer, West Dereham Abbey, Norfolk, got by Prince Christian (22581), out of Phillis 4th by Hildebrand (18068)—Phillis by Homer (14714)—Young Polly by Cardigan (12556)—Polly by Young Rufus (13649)—by Constitution (12634)—by Young Comet (1853)—descended from Jolly's Bull (4115).

Recorded under dam, E. H. B., vol. 18, p. 656, and A. H. B., vol. 10, p. 768.

LADY DUDDING—Red and white, calved March 8, 1871, bred by Messrs. Dudding, Panton, got by Standard Bearer (30055), out of Countess of Yarborough by Baron Rosedale (21539)—Countess of Wragby by Sir Roger (16991)—Columbine by Lambton (9273) —Cactus by Gen. Washington (6036)—Cranberry by Plenipo (4724), &c., as in dam, above.

A. H. B., vol. 11, p. 765.

POTENTILLA—Red, calved June 18, 1867, bred by Mr. J. Christy, Boynton Hall, Chelmsford, got by Duke of Grafton (21594), out of Primula by Victor (15458)—Polyanthus by Orestes (15027) —Oxslip by Progress (11950)—Cowslip by Roan Robin (10721)— Countess of Liverpool by Mars (7212)—Lady Liverpool by Earl of Liverpool (3677)—Actress by Arthur (3039)—Ruby by Imperial (4068)—Red Lady by Juniper (1145)—Red Rose by Columella (904) —by Shakespeare (1429)—by Blyth Comet (85)—by Neswick (1266) —by Favorite (1033).

Recorded, A. H. B., vol. 10, p. 776.

WINSOME EYES 3D—Roan heifer, calved Feb. 11, 1869, bred by W. R. Bromet, Cocksford, Tadcaster, got by 5th Duke of Wharfdale (26033), out of Winsome Eyes by Rose Duke (22760)—Mild Eyes by Stout (17048)—Young Red Eyes by Oxford Duke (15036) —Red Eyes by Duke of Richmond (7996)—Wild Eyes 23d by 2d Cleveland Lad (3408)—Wild Eyes 9th by Duke of Northumberland (1940)—Wild Eyes 3d by Belvedere (1706)—Wild Eyes by Emperor (1975)—by Wonderful (700)—by Cleveland (145)—by Butterfly

(104)—by Hollon's Bull (313)—by Mowbray's Bull (2342)—by Masterman's Bull (422)—descended from Michael Dobison's stock.

Recorded under dam, E. H. B., vol. 19, p. 790, and A. H. B., vol. 10, p. 898. Returned to England July, 1873, by ship Sarmatian, for Earl Dunmore.

WILD FLOWER 6TH—Red heifer, calved Jan. 3, 1868, bred by T. Atherton, Chapel House, Speke, got by Imperial Oxford (18084), out of Wild Flower by Cherry Duke 2d (14265)—Wildair 2d by Marquis of Speke (13307)—Wildair by Santiago (12047)—Wild Eyes 8th by Duke of Northumberland (1940)—Wild Eyes 3d by Belvedere (1706)—Wild Eyes by Emperor (1975)—by Wonderful (700)—by Cleveland (145)—by Butterfly (104)—by Hollon's Bull (313)—by Mowbray's Bull (2342)—by Masterman's Bull (422)—descended from the stock of Michael Dobison.

Recorded, A. H. B., vol. 10, p. 897.

DUCHESS 2D—Red and white, calved July 24, 1867, bred by H. Aylmer, West Dereham Abbey, Norfolk, got by Norfolk Thorndale Duke (24666), out of Red Duchess by Red Knight (16809)—Roseleaf by Whittington (12299)—Lady Welbourn by Lord Lowther (7164)—Eliza by Panton Favorite (4646)—Miss Eliza by grandson of Grazier (1085)—Elizabeth by son of Grazier (1085)—Bessy by son of Vulcan (667)—Betty by Quaker (1349)—by 1st Stonehill Bull (3798).

E. H. B., vol. 18, p. 683, under dam, and A. H. B., vol. 10, p. 480, as Duchess 2d of Dereham Abbey. Sold to Geo. Brown, Bow Park.

DUCHESS 3D—Roan, calved Aug. 6, 1868, bred by H. Aylmer, West Dereham Abbey, Norfolk, got by Prince Christian (22581), out of Red Duchess by Red Knight (16809)—Roseleaf by Whittington (12299)—Lady Welbourn by Lord Lowther (7164)—Eliza by Panton Favorite (4646)—Miss Eliza by grandson of Grazier (1085)—Elizabeth by son of Grazier (1085)—Bessy by son of Vulcan (667)—Betty by Quaker (1349)—by 1st Stonehill Bull (3798).

E. H. B., vol. 18, p. 683, under dam, and A. H. B., vol. 10, p. 480, as Duchess 3d of Dereham Abbey. Sold to Geo. Brown, Bow Park.

FLORA—Red and white, calved Sept. 7, 1868, bred by H. Aylmer, West Dereham Abbey, Norfolk, got by Prince Christian (22581), out of Flounce by Alderman 2d (17292)—Filigree by First Fruits (16048)—Flirt by Kirklevington (11639)—Flounce by Broughton Hero (6811)—Frill by Rockingham (2550)—Fancy by Remus (2524)—White Rosette by Juniper (1144)—Rosette by White Comet (1582)—Young Rose by Grandson of Favorite (2073)—Old Rose by Cattley's Gray Bull (1798).

Recorded under dam, E. H. B., vol. 18, p. 500, and A. H. B., vol. 10, p. 528. Recorded, A. H. B., vol. 10, p. 533, as Flora of Dereham Abbey.

DUCHESS FOURTH OF DEREHAM ABBEY—White, calved Nov. 20, 1870, bred by H. Aylmer, got by Royal Broughton (27352). out of Duchess 3d of Dereham Abbey by Prince Christian (22581), &c., as above.

Recorded, A. H. B., vol. 10, p. 480.

DUCHESS 101st—Red and white, calved July 26, 1868, bred by Captain Gunter, Wetherby Grange, Yorkshire, got by 4th Duke of Thorndale (17750), out of Duchess 84th by Archduke (14099) —Duchess 72d by 4th Duke of Oxford (11387)—Duchess 67th by Usurer (9763)—Duchess 59th by 2d Duke of Oxford (9046)— Duchess 56th by 2d Duke of Northumberland (3646)—Duchess 51st by Cleveland Lad (3407)—Duchess 41st by Belvedere (1706) —Duchess 32d by 2d Hubback (1423)—Duchess 19th by 2d Hubback (1423)—Duchess 12th by The Earl (646)—Duchess 4th by Ketton 2d (710)—Duchess 1st by Comet (155)—by Favorite (252) —by Daisy Bull (186)—by Favorite (252)—by Hubback (319)—by James Brown's Red Bull (97).

Recorded under dam, E. H. B., vol. 18, p. 461. Returned to England by ship Sarmatian, July, 1873, for Lord Dunmore. A. H. B., vol. 10, p. 477.

DUCHESS 103d—Roan, calved Aug. 3, 1868, bred by Captain Gunter, Wetherby Grange, got by 4th Duke of Thorndale (17750), out of Duchess 92d by 4th Duke of Oxford (11387)—Duchess 84th

by Archduke (14099)—Duchess 72d by 4th Duke of Oxford (11387)
—Duchess 67th by Usurer (9763)—Duchess 59th by 2d Duke of
Oxford (9046)—Duchess 56th by 2d Duke of Northumberland
(3646)—Duchess 51st by Cleveland Lad (3407)—Duchess 41st
by Belvedere (1706)—Duchess 32d by 2d Hubback (1423)—
Duchess 19th by 2d Hubback (1423)—Duchess 12th by The Earl
(646)—Duchess 4th by Ketton 2d (710)—Duchess 1st by Comet
(155)—by Favorite (252)—by Daisy Bull (186)—by Favorite (252)
—by Hubback (319)—by J. Brown's Red Bull (97).

Recorded under dam, E. H. B., vol. 18, p. 461, and A. H. B.,
vol. 10, p. 477.

DUCHESS OF HILLHURST—White, calved Nov. 30, 1870,
bred by Captain Gunter, got by 8th Duke of York (28480), out of
Duchess 103d, above.

Sold to Lord Dunmore, and landed at Liverpool Nov. 14, 1871.

DUCHESS OF HILLHURST 2D—Roan, calved Dec. 16, 1870,
bred by Captain Gunter, got by 8th Duke of York (28480), out of
Duchess 101st, above.

Sold to Lord Dunmore. Landed at Liverpool Nov. 14, 1871.

ROSA LOUISA—Roan, calved May 3, 1867, bred by R. S.
Bruere, Braithwaite Hall, Middleham, Yorks., got by Royal Booth
(22772), out of Rosa Sybilla by Baron Booth (21212)—Rosewreath
by Windsor (14013)—Rose Garland by Baron Warlaby (7813)—
Garland by The Silkie Laddie (10947)—Damsel by Rouge (5012)—
Strawberry by Shipton (2620)—Damsel by Cleveland (3404)—Rose
by Danby (3550)—by son of Studley Grange (1483)—by son of
Duke (225)—by Midas (4470)—by Nailer (4528)—by Ambo (746).

Recorded under dam, E. H. B., vol. 18, p. 697, and A. H. B.,
vol. 10, p. 820.

JOAN OF ARC—Roan, calved Oct. 18, 1868, bred by G. Garne,
Churchill Heath, Chipping Norton, Oxon. got by Plymouth Candidate (22531), out of Jubilee by Cynric (19542)—Juna by Brigadier (14193)—Venus by Grand Duke (10284)—Manganese by
Barmpton (8900)—Metal by Lord Warden (7167)—Titania by

Orontes (4623)—Zinc by Guardian (3947)—Roguery by Mercury (2301)—Pageant by Monarch (2324)—No. 13 Chilton Sale by St. Albans (2584)—No. 4 Chilton Sale by Jupiter (342)—Sir Oliver Cow by Sir Oliver (605)—Raspberry by Trunnell (659)—Strawberry by Favorite (252)—Lily by Favorite (252)—Miss Lax by Dalton Duke (188)—Lady Maynard by R. Alcock's Bull (19)—by J. Smith's Bull (608)—by Jolly's Bull (337).

Recorded under dam, E. H. B., vol. 18, p. 544, and A. H. B., vol. 10, p. 586.

LADY HIGHTHORN—Roan, calved Aug. 15, 1868, bred by R. Plummer, Carlton, Husthwaite, Thirsk, got by Lord Abbot (20140), out of Dora 7th by The Duke (18982)—Dora by Gone Away (10279)—Dairymaid by Napoleon (10552)—by Hampden (3965)—by Plenipo (4724)—Miss Scarthe by Sir John (16985)—Omega by Rex (1375)—Primrose by Baron (58)—Lily by a bull of Mr. Mason's—by Falstaff (250)—by Irishman (329).

Recorded, A. H. B., vol. 10, p. 614.

PHILLIS 9th—White, calved Jan. 8, 1869, bred by H. Aylmer, West Dereham Abbey, Norfolk, got by Prince Christian (22581), out of Phillis 4th by Hildebrand (18068)—Phillis by Homer (14714)—Young Polly by Cardigan (12556)—Polly by Young Rufus (13649)—by Constitution (12634)—by Young Comet (1853)—descended from Jolly's Bull (4115).

Recorded, A. H. B., vol. 10, p. 768.

BRITANNIA 19th—Roan, calved March 26, 1868, bred by H. Aylmer, West Dereham Abbey, Norfolk, got by Prince Christian (22581), out of Britannia 16th by Hildebrand (18068)—Busy Bee by Brigand (12494)—Britannia 11th by Lord John (11731)—by Albion (7771)—by Eclipse (3685)—by Punch (7386)—Pink by Prince Paul (4827)—Britannia by Monarch (2324)—No. 8 Chilton Sale by Dr. Syntax (220)—Charles Cow by Charles (127)—St. John Cow by St. John (572)—Chilton Cow by Chilton (136)—Nymph by White Bull (421)—Lily by Favorite (252)—Miss Lax by Dalton

Duke (188)—Lady Maynard by R. Alcock's Bull (19)—by J. Smith's Bull (608)—by Jolly's Bull (337).

Under dam, E. H. B., vol. 18, p. 406, and A. H. B., vol. 10, p. 425.

JESSIE HOPEWELL—Roan cow, calved Oct. 12, 1866, bred by H. Aylmer, West Dereham Abbey, Norfolk, got by General Hopewell (17953), out of Jewess by Rifleman (18704)—Jewel by Meteor (18391)—Judith 3d by Rebellion (10684)—Judith by Orontes (4623) —Rosamond by Alabaster (1616)—No. 54 Chilton Sale by Monarch (2324)—No. 25 Chilton Sale by Satellite (1420)—No. 2 Chilton Sale by Cato (119)—by Jupiter (342)—by George (273)—by Chilton (136) —by Irishman (329)—by B (45).

Under dam, E. H. B., vol. 17, p. 542, and A. H. B., vol. 10, p. 585.

LORD WETHERBY 12337—Roan bull, calved Nov. 30, 1870, bred by T. Atherton, Chapel House, Speke, got by 2d Duke of Wetherby (21618), out of Wild Flower 6th by Imperial Oxford (18084), &c., as in dam, above.

IMPERIAL CÆSAR 17365 [1573]—Red and white, calved Jan. 16, 1871, bred by Hugh Aylmer, West Dereham Abbey, Norfolk, got by Royal Broughton (27352), out of Duchess 2d of Dereham Abbey by Norfolk Thorndale Duke (24666)—Red Duchess by Red Knight (16809)—Roseleaf by Whittington (12299)—Lady Welbourn by Lord Lowther (7164)—Eliza by Panton Favorite (4646)—Miss Eliza by grandson of Grazier (1085)—Elizabeth by a son of Grazier (1085)—Bessy by a son of Vulcan (667)—Bess by Vulcan (667)—Betty by Quaker (1349)—by 1st Stonehill Bull (3798).

WELCOME LADY—Roan heifer, calved Dec. 30, 1870, bred by J. B. Booth, Killerby Hall, Yorkshire, got by Banner Bearer (11326), out of Lady of the Lake by Knight Errant (18154)— Forest Queen by Royal Buck (10750)—Hecuba by Hopewell (10332) —Helen by Hamlet (8126)—by Leonard (4210).

A. H. B., vol. 11, p. 1143. Returned to England Aug. 4, and sold, Sept. 4, 1877, to J. B. Booth for 226 guineas. This cow was calved at Hillhurst after dam reached there.

CANDIDATE'S DUCHESS 3D—White, calved April 9, 1871, bred by D. R. Davies, got by Royal Chester 12847, out of Candidate's Duchess 2d by Grand Duke of Essex 4th (24068), &c., as in dam, above.

Vol. 11, p. 541, A. H. B.

PRINCESS CHRISTIAN—Roan cow, calved Oct. 25, 1867, bred by H. Aylmer, West Dereham Abbey, Norfolk, got by Prince Christian (22581), out of Queen of the Herd by Royal Counsellor (20725)—Queen of the May by Lord of the Harem (16430)—Sally by Gloster's Grand Duke (12949)—Silky by Bristol (11205)—Champion by Vagabond (9765)—Silver by Locksley (4210)—Silver by Stanhope (5315)—Silver by Marquis (22292)—bred from Mr. Jobling's stock of Styford.

Recorded, E. H. B., vol. 18, p. 678, under dam, and A. H. B., vol. 10, p. 786.

FAIRY GEM (twinned with Fairy Pearl)—Roan heifer, calved May, 1869, bred by J. B. Booth, Killerby Hall, got by K. C. B. (26492), out of Fairy Queen by Valasco (15443)—Forest Queen by Royal Buck (10750)—Hecuba by Hopewell (10332)—Helen by Hamlet (8126)—by Leonard (4210).

Recorded, E. H. B., vol. 19, p. 505, and A. H. B., vol. 10, p. 512.

REGA—Red and white heifer, calved Sept. 5, 1870, bred by R. S. Bruere, Braithwaite Hall, calved the property of M. H. Cochrane, Hillhurst, Can., got by Regal Booth (27262), out of Rosa Louisa by Royal Booth (22772)—Rosa Sybilla by Baron Booth (21212)—Rose Wreath by Windsor (14013)—Rose Garland by Baron Warlaby (7813)—Garland by The Silky Laddie (10947)—Damsel by Rogue (5012)—Strawberry by Shipton (2620)—Damsel by Cleveland (3404)—Rose by Danby (3550)—by a son of Studley Grange (1488)—by a son of Duke (225)—by Midas (4470)—by Nailer (4528)—by Ambo (746).

A. H. B., vol. 10, p. 809.

BRITISH LADY—Roan, calved September 6, 1870, bred by H. Aylmer, West Dereham Abbey, got in England by Royal Boughton

(27352), out of Britannia 19th by Prince Christian (22581)—Britannia 16th by Hildebrand (18068)—Busy Bee by Brigand (12494)—Britannia 11th by Lord John (11731)—by Albion (7771)—by Eclipse (3685)—by Punch (7386)—Pink by Prince Paul (4827)—Britannia by Monarch (2324)—No. 8 Chilton Sale by Dr. Syntax (220)—Charles Cow by Charles (127)—St. John Cow by St. John (572)—Chilton Cow by Chilton (136)—Nymph by White Bull (421)—Lily by Favorite (252)—Miss Lax by Dalton Duke (188)—Lady Maynard by R. Alcock's Bull (19)—by J. Smith's Bull (608)—by Jolly's Bull (337).

Vol. 10, p. 425, A. H. B.

Aug. 29, 1870. M. H. COCHRANE.

By Ship European, from Liverpool.

COWS.

BLUEBELL—Roan cow, calved Feb. 4, 1863, bred by A. & A. Mitchell, Alloa, got by Knight Errant (18154), out of Barbelle by Cardigan .(12556)—Barmaid by Lord Fanny (13187)—Beauty by Noble (4578)—Betsy by Newton (2367)—by Baronet (1686)—by Reformer (2502)—by Margrave (2263)—by Leopold (2199)—by Hector (2103)—by Surly (2715)—by Traveler (655)—by Colonel (152).

E. H. B., vol. 17, p. 386 (illustrated). and A. H. B., vol. 10, p. 417.

MISS BLITHE—Red and white, calved Feb. 28, 1869, bred by A. & A. Mitchell, Alloa, got by Lord Blithe (22126), out of Queen of the Isle by Sir Samuel (15302)—Lady of the Lake by Prince Arthur (13497)—Sonsie by Hudibras (10339)—1st Sonsie by Howard (6085)—3d Sonsie by Adam (2920)—Modesty by Belshazzar (1703)—Tulip by Camden (1776)—by Noble Henry (2374)—by Young Alexander (2977).

A. H. B., vol. 10, p. 721.

BADDOW ROSE—Roan heifer, calved Nov. 8, 1867, bred by R. H. Crabb, Baddow, Chelmsford, got by Manhattan (26802), out of Lady Ducie 3d by Guelder Rose (19910)—Lady Ducie 2d by Tally Ho (20927)—Lady Ducie by Grand Duke (12965)—Lady Jane by Red Roan Kirtling (10691)—Lady Ann by Pam (6272)—Countess by Vanguard (5545)—Dodona by Alabaster (1616)—No. 6 Chilton Sale by Dr. Syntax (220)—Charles Cow by Charles (127)—Henry Cow by Henry (301)—Lydia by Favorite (252)—Nell by Mason's White Bull (421)—Fortune by Bolingbroke (86)—by Foljambe (263)—by Hubback (319)—bred by Mr. Maynard.

A. H. B., vol. 10, p. 385, and E. H. B., vol. 18, p. 562, under dam.

FAREWELL.—Red and white heifer, calved April 25, 1868, bred by J. B. Booth, Killerby Hall, got by Brigade Major (21312), out of Virtue by Valasco (15443)—Lady Georgiana by Knight Errant (18154)—Georgie by Prince George (13510)—Hopeful by Hopewell (10332)—by Warrior (12287).

Recorded under dam, E. H. B., vol. 19, p. 771.

KILLERBY QUEEN—Roan heifer, calved July, 1867, bred by J. B. Booth, Killerby Hall, got by Brigade Major (21312), out of Clara by Fitz Clarence (14552)—Georgie by Prince George (13510)—Hopeful by Hopewell (10332)—by Warrior (12287).

Recorded under dam, E. H. B., vol. 19, p. 445, and A. H. B., vol. 10, p. 598. Returned to England Aug. 4, and sold Sept. 4, 1877, to J. Torr for 41 guineas.

MILLINER—Roan cow, calved Aug., 1867, bred by J. B. Booth, Killerby Hall, got by Brigade Major (21312), out of Lady Percy by Percival (20486)—Lady Day by Man Friday (13290)—Anna by The Irishman (5446)—by Sir Richard (5175)—by son of Booth's White Bull—by Seymour's Red Bull.

Recorded under dam, E. H. B., vol. 18, p. 576, and A. H. B., vol. 10, p. 712.

GOODY TWO-SHOES—Red heifer, calved May 29, 1869, bred by G. S. Foljambe, Osberton Hall, Worksop, Notts., got by Lord Lyons (26677), out of Ladyslipper by May Duke (16553)—Cenerentolia by Foig-a-Ballagh (8082)—Clementina by Clementi (3399)

—Farewell by Young Matchem (4422)—Flora by Isaac (1129)—by Young Pilot (4702)—by Pilot (496)—by Julius Cæsar (1143).

Recorded A. H. B., vol. 11, p. 682. Sold to W. T. Hughes, Lexington, Ky.

MAIDEN—Red, calved Nov. 6, 1870, bred by J. B. Booth, got by K. C. B. (26492), out of Milliner by Brigade Major (21312), &c., as above.

LADY GRATEFUL—Roan cow, calved Jan. 12, 1864, bred by T. C. Booth, Warlaby, Northallerton, got by Lord of the Valley (14837), out of Lady Blithe by Windsor (14013)—Blithe by Hopewell (10332)—Bliss by Leonard (4210)—Young Broughton by Young Matchem (2282)—Broughton by Jerry (4097)—by Young Pilot (4702)—by Pilot (496)—by son of Apollo (36).

She was twinned with Lady Gratitude. Recorded under dam, E. H. B., vol. 16, p. 519, and A. H. B., vol. 10, p. 613.

LADY BOOTH—Red and white heifer, calved April 25, 1869, bred by T. C. Booth, Warlaby, got by British Crown (21322), out of Lady Jane by Lord of the Valley (14837)—Lady Mirth by Sir Samuel (15302)—Lady Blithe by Windsor (14013)—Blithe by Hopewell (10332)—Bliss by Leonard (4210)—Young Broughton by Young Matchem (2282)—Broughton by Jerry (4097)—by Young Pilot (4702)—by Pilot (496)—by son of Apollo (36).

Recorded under dam, E. H. B., vol. 19, p. 582, and A. H. B., vol. 10, p. 606.

LADY OF THE LAKE—Red and white cow, calved Nov. 19, 1862, bred by John B. Booth, Killerby Hall, Catterick, Yorkshire, got by Knight Errant (18154), out of Forest Queen by Royal Buck (10750)—Hecuba by Hopewell (10332)—Helen by Hamlet (8126)—by Leonard (4210).

Recorded under dam, E. H. B., vol. 15, p. 497, and A. H. B., vol. 10, p. 622.

QUEEN OF BEAUTY—Red and white heifer, calved April 4, 1868, bred by J. B. Booth, Killerby Hall, got by Knight Errant (18154), out of Queen of the Glen by Valasco (15443)—Forest

Queen by Royal Buck (10750)—Hecuba by Hopewell (10332)—Helen by Hamlet (8126)—by Leonard (4210).

Recorded under dam, E. H. B., vol. 18, p. 677, and A. H. B., vol. 10, p. 795. Returned to England Aug. 4, 1877, and sold at public sale Sept. 4, 1877, to J. B. Booth for 120 guineas.

BULLS.

ROYAL COMMANDER 10914 (29857)—Roan, calved July 31, 1869, bred by T. C. Booth, Warlaby, Northallerton, got by Commander-in-Chief (21451), out of Prudence by Gen. Hopewell (17953)—Modesty 2d by Lord of the Valley (14837)—Modesty by Buckingham (3239)—Monica by Raspberry (4875)—White Strawberry by Rockingham (2551)—Strawberry 2d by Young Alexander (2977)—Strawberry by Pilot (496)—Halnaby by Booth's Lame Bull (359)—by Easby (232)—by Suwarrow (636).

Returned to England Aug. 14, 1875, by ship Polynesian, landing at Liverpool Aug. 30, for H. Aylmer.

BOOTH'S MARKSMAN (28059)—Roan, calved Sept. 2, 1869, bred by R. S. Bruere, Braithwaite Hall, Middleham, got by Booth's Kinsman (25658), out of Vernal Star by The Sutler (23061)—Venus Star by Prince George (13510)—Vesper by King Arthur (13110)—Vesper by Morning Star (6223)—Primrose by Roland (2556)—Strawberry by Priam (2452)—by Matchem (2281)—by son of Peter (487).

ROYAL RICHARD 15415—Red and white bull, calved Jan. 15, 1870, bred by Mr. Barnes, Westland, Ireland, got by King Richard 8469, out of Isabella Sovereign by Royal Sovereign (22802)—Isabella by British Prince (14197)—Sweetbrier by Nimrod (13388)—Charlotte by Selim (6454)—by Rex (6385)—by Sir Thomas Fairfax (5196)—by Ambo (1636)—by Memnon (2295)—by Pilot (496)—by Agamemnon (9)—by Burrell's Bull (1768).

Oct., 1870. M. H. COCHRANE,

Of Hillhurst, Compton, Can. By Ship European, from Liverpool.

WILD EYES DUCHESS—Red, calved Feb. 3, 1865, bred by C. R. Saunders, Nunwick Hall, Penrith, got by 9th Grand Duke (19879), out of Wild Eyes 19th by Lablache (16353)—Wild Eyes 18th by Solon (13766)—Wild Eyes 16th by 2d Duke of Oxford (9046)—Wild Eyes 15th by 4th Duke of Northumberland (3649)— by Duke of Northumberland (1940)—by Belvedere (1706)—by Emperor (1975)—by Wonderful (700)—by Cleveland (145)—by Butterfly (104)—by Hollon's Bull (313)—by Mowbray's Bull (2342)— by Masterman's Bull (422)—descended from the stock of M. Dobison.

E. H. B., vols. 21, p. 687, and 19, p. 786. and A. H. B., vol. 10, p. 896. Returned to England in July, 1873, to Earl of Dunmore.

LYNDALE WILD EYES—Roan, calved April 13, 1871, bred by Mr. Saunders, calved Mr. Cochrane's, got by Earl of Eglinton (23832), out of Wild Eyes Duchess, above.

E. H. B., vol. 21, p. 687, under dam, and A. H. B., vol. 11, p. 856.

WATERLOO 38TH—Red heifer, calved March 12, 1869, bred by C. R. Saunders, Nunwick Hall, Penrith, got by Earl of Eglinton (23832), out of Waterloo 32d by 9th Grand Duke (19879)— Waterloo 31st by 3d Grand Duke (16182)—Waterloo 18th by 2d Grand Duke (12961)—Waterloo 13th by 3d Duke of Oxford (9047) —Waterloo 9th by 2d Cleveland Lad (3408)—Waterloo 6th by Duke of Northumberland (1940)—Waterloo 3d by Norfolk (2377)—Waterloo Cow by Waterloo (2816)—by Waterloo (2816).

A. H. B., vol. 10, p. 890.

LADY WORCESTER—Roan, calved April 8, 1865, bred by J. Harward, Winterfold, Kidderminster, got by Charleston (21400), out of Clear Star by Marton Duke (22307)—Bright Star by Red Duke (18676)—Bright Eyes by 3d Duke of York (10166)—Wild Eyes 23d by 2d Cleveland Lad (3408)—by Duke of Northumberland (1940)—by Belvedere (1706)—by Emperor (1975)—by Wonderful (700)—by Cleveland (145)—by Butterfly (104)—by Hollon's Bull (313)—by Mowbray's Bull (2342)—by Masterman's Bull (422) —from the stock of M. Dobison.

A. H. B., vol. 10, p. 632, and E. H. B., vol. 17, p. 423, under dam. Returned to England in July, 1873, to Earl of Dunmore.

STAR QUEEN—Roan, calved June 23, 1866, bred by R. S. Bruere, Braithwaite Hall, Middleham, Yorks., got by The Sutler (23061), out of Star of Windsor by Windsor (14013)—Vesper by King Arthur (13110)—Vesper by Morning Star (6223)—by Roland (2556)—by Priam (2452)—by Matchem (2281)—by son of Peter (487).

A. H. B., vol. 10, p. 856, and E. H. B., vol. 17, p. 748, under dam.

GRAND DUKE OF GORDON 11216 (28757) [1496]—Roan, calved April 21, 1869, bred by W. Torr, Aylesby Manor, got by Blinkhoolie (23428), out of Geneva by Breastplate (19337)—Genuine Gold by British Prince (14197)—Genuine Gem by Vanguard (10994)—Glisten by Vanguard (10994)—by Baron Warlaby (7813) —by Londesbro' (6142)—by Ranunculus (2479)—by Remus (4932) —by Prince Comet (1342)—by Count (170)—by Constellation (163) —by Young Favorite (255).

LORD ABRAHAM 11223 (29056)—Roan, calved April 25, 1869, bred by W. Torr, Aylesby Manor, got by Breastplate (19337), out of Lady Zillah by Prince of Warlaby (15107)—Lady Eve by Dr. McHale (15887)—Lady Mary Bountiful by Baron Warlaby (7813) —Lady Bountiful by Usurer (9763)—Belinda by Ranunculus (2479) —Sylph by Sir Walter (2637)—by Hotspur (1117)—by Coxcomb (928)—by Midas (435)—by Comet (155)—by R. Colling's son of Favorite (252)—by same son of Favorite (252)—by Hubback (319).

PRINCE NICHOLAS 11235 (29647)—Roan, calved Feb. 5, 1869, bred by Hugh Aylmer, West Dereham Abbey, Norfolk, got by Prince Christian (22581), out of Queen of the Gems by Prince Imperial (22594)—Guava 2d by Red Knight (16809)—Guava by War Eagle (15483)—Gem by Sweet William (7571)—Juniper by

Mahomed (6170)—Red Lady by Juniper (1145)—by Columella (904)—by Shakespeare (1429)—by Blyth Comet (85)—by Neswick (1266)—by Favorite (1033).

VESPER STAR—Red and white, calved May 12, 1871, bred by R. S. Bruere, Braithwaite Hall, got by Sir Windsor Broughton (27507), out of Star Queen by The Sutler (23061)—Star of Windsor by Windsor (14013)—Vesper by King Arthur (13110)—Vesper by Morning Star (6223)—by Roland (2556)—by Priam (2452)—by Matchem (2281)—by son of Peter (487).

A. H. B., vol. 11, p. 1126. Calved after dam reached Hillhurst. Returned to England and sold Sept. 4, 1877, to W. T. Crosbie for 1,000 guineas.

July 20, 1871. M. H. COCHRANE.
By Ship European, from Liverpool.
COWS.

ACACIA—Red and white, calved May 10, 1864, bred by G. Bland, Coleby Hall, Lincoln, got by Knight Errant (18154), out of Amethyst by Magna Charta (16486)—Applin by Lord Raglan (14849)—Amaryllis by Burgomaster (12513)—by Baron of Ravensworth (7811)—by Lycurgus (7180)—by Zenith (5702)—by Guardian (3947)—by Firby (1040)—by Ivanhoe (1131)—by Regent (544) —by Blyth Comet (85).

A. H. B., vol. 11, p. 448.

LADY LUCY THORNDALE—Roan, calved June 20, 1864, bred by J. Clayden, Littlebury, Essex, got by 3d Duke of Thorndale (17749), out of Lady Bird by Count de Gourcy (17632)—Luxury by Cheltenham (12588)—Red Rose by Horatio (10335)—by 3d Duke of Northumberland (3647)—by Velocipede (5552)—by Sir Thomas (2636)—by Marske (418)—by Comet (155)—by Tom (652) —by Son of Favorite (1033)—by Hutton's Bull (323)—by Barningham (56).

E. H. B., vol. 17, p. 574, and A. H. B., vol. 11, p. 775.

ROSEDALE—Red, calved Sept. 9, 1869, bred by W. Derham, Palmer's Green, Middlesex, got by Bismarck (25637), out of Lady

Lucy Thorndale by 3d Duke of Thorndale (17749)—Lady Bird by Count de Gourcy (17632)—Luxury by Cheltenham (12588)—Red Rose by Horatio (10335)—by 3d Duke of Northumberland (3647) —by Velocipede (5552)—by Sir Thomas (2636)—by Marske (418) —by Comet (155)—by Tom (652)—by Son of Favorite (1033)—by Hutton's Bull (323)—by Barningham (56).

A. H. B., vol. 11, p. 1058.

WATERLOO ROSE—Roan, calved Dec. 19, 1866, bred by R. Chaloner, King's Fort, Ireland, got by Royal Sovereign (22802). out of Wellingtonia by British Prince (14197)—Waterloo Rose by Blood Royal (14169)—Water Bride by Brideman (12493)—Water Nymph by Vanguard (10994)—Water Witch by 4th Duke of Northumberland (3649)—by Norfolk (2377)—by Waterloo (2816)—by Waterloo (2816).

E. H. B., vol. 17, p. 786, under dam, and A. H. B., vol. 11, p. 1141.

BELINDA OXFORD—Roan, calved Nov. 10, 1869, bred by C. Barnett, Stratton Park, Bedfordshire, got by Brockley Oxford (25688), out of Pheasant by Provost (24878)—Water Wave 17th by Wandering Willie (19110)—Clove by Chieftain (14267)—Historia by Homer (13038)—Parisian by Prince John (9508)—Viola by Vampire (6632)—White Rose by Senator (2610)—Red Rose by Columella (904)—by Shakespeare (1429)—by Blyth Comet (85)—by Neswick (1266)—by Favorite (1033).

A. H. B., vol. 11, p. 492, and E. H. B., vol. 19, p. 667, under dam. Sold to C. E. Coffin, Muirkirk, Md.

FANNY 29TH—Red, calved Jan. 14, 1870, bred by J. Meadows. Thornville, Wexford, got by Prince of the Realm (22627), out of Fanny 14th by Fugleman (14580)—Fanny 3d by Chieftain (7898) —Fanny 1st by Mayo (16555)—Mary by Peel (4672)—by Duke (3633).

A. H. B., vol. 11, p. 642.

PRIMROSE 9TH—Roan, calved April 14, 1870, bred by J. Meadows, got by Prince of the Realm (22627), out of Primrose 6th

by First Fiddle (19749)—Primrose 4th by Fugleman (14580)—
Primrose 2d by Midsummer (14951)—Primrose by Abraham Parker
(9856)—Peony by Ernest (5987)—Passion Flower by Melbourne
(4448)—Passion by Newton (2367)—by Reformer (2502).

A. H. B., vol. 11, p. 999.

ROYAL DUCHESS 2D—Roan heifer, calved March 26, 1870,
bred by C. A. Barnes, Charleywood, Herts., got by Lord Wallace
(24473), out of Ladylove by Cock of the Walk (15782)—Looey
by Marmaduke (14897)—Lelia by Count of Gloucester (12650)—
Lizzy by 4th Duke of York (10167)—by Cramer (6907)—by Cato
(6836)—by Helicon (2107)—by Matchem (2281)—by Sir Alexander
(591)—by Stephen (1456)—by Western Comet (689)—by Charge's
Gray Bull (872)—by Favorite (252)—by Bartle (777)—descended
from Studley White Bull (627).

A. H. B., vol. 11, p. 1070, and E. H. B., vol. 20, p. 606, under
dam. Sold to Edward Iles, of Springfield, Ill.

WATERLOO ROSE 2D—Roan heifer, calved May 6, 1869, bred
by D. McIntosh, Havering Park, Essex, got by King Richard
(26523), out of Waterloo Rose by Royal Sovereign (22802)—Wellingtonia by British Prince (14197)—Waterloo Rose by Blood
Royal (14169)—Water Bride by Brideman (12493)—Water Nymph
by Vanguard (10994)—Water Witch by 4th Duke of Northumberland (3649)—by Norfolk (2377)—by Waterloo (2816)—by Waterloo
(2816).

E. H. B., vol. 19, p. 777, under dam, and A. H. B., vol. 11, p.
1141.

WEEPING WILLOW—Roan heifer, calved April 12, 1871,
bred by Wm. Torr, Aylesby Manor, Lincolnshire, got by Manfred
(26801), out of Water Cress by Dr. McHale (15887)—Water Maid
by Vanguard (10994)—Water Witch by 4th Duke of Northumberland (3649)—by Norfolk (2377)—by Waterloo (2816)—by Waterloo
(2816).

E. H. B., vol. 20, p. 812, under dam, and A. H. B., vol. 11, p.
1142.

PET GWYNNE—Roan heifer, calved Jan. 22, 1868, bred by P. Riall, Old Conna Hill, Wicklow, got by British Sailor (23471), out of Polly Gwynne by Killerby Lad (20052)—Pauline Gwynne by Paul Potter (16688)—White Moll Gwynne by Cadet (12521)—Modesty 5th by Young Usurer (10985)—Modesty by Sir Thomas Fairfax (5196)—by Wallace (5586)—by Wellington (2824)—by Marmion (406)—by Merlin (430)—by Layton (366)—by Phenomenon (491)—by Favorite (252)—by Favorite (252)—by Hubback (319)—by Snowdon's Bull (612)—by Waistell's Bull (669)—by Masterman's Bull (422)—by Studley Bull (626).

E. H. B., vol. 18, p. 659, under dam, and A. H. B., vol. 11, p. 980.

OXFORD DUCHESS—Roan, calved April 16, 1869, bred by Rev. R. B. Kennard, Marnhull Rectory, Dorsetshire, got by Oxford Duke (27019), out of Ada by Duke of Montrose (23771)—Juliet by Wonder (21126)—Ethelinda by Marmaduke (14897)—Electra by Lovemore (10476)—Dewberry by Hurricane (4061)—Daffodil by Eclipse (3684)—Bright Eyes by Milton (8315)—by Young Comet (3436)—by Macgregor (2235).

A. H. B., vol. 11, p. 972. Sold to E. Iles, of Springfield, Ill.

FRENCH ASTER—Red heifer, calved June 20, 1868, bred by J. Christy, Boynton Hall, Essex, got by Duke of Grafton (21594), out of German Aster by Royal Arch (18749)—China Aster by Archduke 2d (15588)—Cream by Young Weathercock (15495)—Roan Crocus by Gen. Elliott (10266)—Red Crocus by Young Locksley (22111)—by Sherborne (10805)—by Prince (4772)—by Stanhope (5315)—by son of Sir Kenneth (1450)—by Pilot (1319)—by Thorpe (1515)—by Waistell's Bull (1567).

E. H. B., vol. 18, p. 511, under dam, and A. H. B., vol. 11, p. 667. Sold to S. R. Steator, Cleveland, Ohio.

ANEMONE—Red and white heifer, calved June 27, 1869, bred by J. Christy, Boynton Hall, Essex, got by Duke of Grafton (21594), out of German Aster by Royal Arch (18749)—China Aster by Archduke 2d (15588)—Cream by Young Weathercock (15495)—Roan

Crocus by Gen. Elliott (10266)—Red Crocus by Sherborne (10805)—Crocus by Young Locksley (4240)—Crocus by Prince (4772)—Hawthorn by Stanhope (5315)—Beauty by son of Sir Kenneth (1450)—by Pilot (1319)—by Thorpe (1515)—by Waistell's Bull (1567).

A. H. B., vol. 11, p. 469.

PORTULACA—Red heifer, calved Aug. 27, 1869, bred by J. Christy, Boynton Hall, Essex, got by Duke of Grafton (21594), out of Primula by Victor (15458)—Polyanthus by Orestes (15027)—Oxlip by Progress (11950)—Cowslip by Roan Robin (10721)—Countess of Liverpool by Mars (7212)—Lady Liverpool by Earl of Liverpool (3677)—Actress by Arthur (3039)—Ruby by Imperial (4068)—Red Lady by Juniper (1145)—Red Rose by Columella (904)—by Shakespeare (1429)—by Blyth Comet (85)—by Neswick (1266)—by Son of Favorite (1033).

A. H. B., vol. 11, p. 990.   Sold to C. E. Coffin, Muirkirk, Md.

AMERICAN ASTER—Red, calved Dec. 17, 1871, bred by J. Christy, got by Rosolio (32346), out of French Aster by Duke of Grafton (21594), &c., as in French Aster, above.

A. H. B., vol. 12, p. 594.

PHYLLIS GWYNNE—Red heifer, calved Jan. 12, 1869, bred by P. Riall, Old Conna Hill, Wicklow, got by British Sailor (23471), out of Polly Gwynne by Killerby Lad (20052)—Pauline Gwynne by Paul Potter (16688)—White Moll Gwynne by Cadet (12521)—Modesty 5th by Young Usurer (10985)—Modesty by Sir Thomas Fairfax (5196)—by Wallace (5586)—by Wellington (2824)—by Marmion (406)—by Merlin (430)—by Layton (366)—by Phenomenon (491)—by Favorite (252)—by Favorite (252)—by Hubback (319)—by Snowdon's Bull (612)—by Waistell's Bull (669)—by Masterman's Bull (422)—by Studley Bull (626).

A. H. B., vol. 11, p. 981.

GLOSSY'S WOODBINE—Red heifer, calved March 17, 1869, bred by W. Bolton, The Island, County Wexford, got by Woodranger (27834), out of Ida Glossy by Gen. Clyde (19839)—Red Glossy by

Highland Laddie (13026)—Glossy 2d by Duke of Bedford (11378)
—Glossy by Eugene Aram (10210)—Amelia by 2d Comet (5101)—
by Corrector (1876)—by Tyro (2781)—by Grazier (1085)—by Jupiter (342)—by Sir Oliver (605)—by Trunnell (659)—by Favorite
(252)—by Favorite (252)—by Dalton Duke (188)—by R. Alcock's
Bull (19)—by J. Smith's Bull (608)—by Jolly's Bull (337).

E. H. B., vol. 19, p. 549, under dam, and A. H. B., vol. 11,
p. 679.

GLOSSY 5TH—Red heifer, calved February 12, 1870, bred by
W. Bolton, got by Gray Gauntlet (19908), out of Ida Glossy by Gen.
Clyde (19839)—Red Glossy by Highland Laddie (13026)—Glossy
2d by Duke of Bedford (11378)—Glossy by Eugene Aram (10210)—
Amelia by 2d Comet (5101)—by Corrector (1876)—by Tyro (2781)
—by Grazier (1085)—by Jupiter (342)—by Sir Oliver (605)—by
Trunnel (659)—by Favorite (252)—by Favorite (252)—by Dalton
Duke (188)—by R. Alcock's Bull (19)—by J. Smith's Bull (608)—
by Jolly's Bull (337).

E. H. B., vol. 19, p. 549, under dam, and A. H. B., vol. 11,
p. 679.

LADY AMELIA—Red heifer, calved Sept. 23, 1868, bred by
C. A. Barnes, got by Charles 2d (23539), out of Lady Anne by Lord
Chancellor (20160)—British Queen by Sir Charles (16948)—Miss
Amelia by Marquis of Bute (11788)—Miss Beauford by Red Roan
Kirtling (10691)—Celia by 3d Duke of Northumberland (3647)—
Cornflower by Bashaw (1692)—Columbine by Helmsman (2109)—
Columbine by Columella (904)—Charlottina by Regent (544)—
Charlotte Palatine by Palatine (478)—Charlotte by Palmflower
(480)—Crimson by Patriot (486)—Young Milbank by Driffield
(223)—Milbank by C. Holmes' Bull (314).

A. H. B., vol. 11, p. 755.

NEGUS—Roan heifer, calved Jan. 27, 1870, bred by Messrs.
Garne, Broadmoor, Gloucestershire, got by Royal Benedict (27348),
out of Neatness by Brigadier (23456)—Nemophila by Cynric (19542)
—Nectarine by Gen. Pelissier (14605)—Necklace by Uncle Tom

(13912)—Princess by Fitz Hardinge (8073)—by Raffler (7391)—by Consul (1868)—by son of Speculation (1472).

E. H. B., vol. 19, p. 649, under dam, and A. H. B., vol. 11, p. 955, both as Nellie Booth; vol. 13, p. 831.

VERNAL STAR—Red and white cow, calved April 22, 1866, bred by R. S. Bruere, got by The Sutler (23061), out of Venus Star by Prince George (13510)—Vesper by King Arthur (13110)—Vesper by Morning Star (6223)—Primrose by Roland (2556)—by Priam (2452)—by Matchem (2281)—by son of Peter (487).

A. H. B., vol. 11, p. 1125, and E. H. B., vol. 18, p. 763.

BULLS.

THE DOCTOR 13021—Red bull, calved March 9, 1870, bred by W. Derham, Palmer's Green, Middlesex, got by Bismarck (25637), out of Colleen Bawn by Lord Red Rose (22205)—Carlina by Valasco (15443)—Charlotte by Royal Buck (10750)—Dora by Leonard (4210)—Dorothy by Roland (2556)—by Priam (2452).

CHERUB 11505—Red bull, calved April 1, 1870, bred by Lord Sudeley, Toddington, Gloucestershire, got by Baron Booth (21212), out of Seraphina 13th by John o' Gaunt (16322)—Seraphina 7th by Duke of Sussex (12772)—Seraphina 2d by Sweet William (7571)—Seraphina by Earl of Essex (6955)—Sapphire by Stratton (5336)—Ruby by Fanatic (1996)—Rufe by Red Rover (4902)—by Rufus (2577)—by Emperor (1014).

Sold to Edward Iles, of Springfield, Ill.

WAR BANNER 13102—Roan bull, calved Feb. 5, 1871, bred by W. Torr, Aylesby Manor, Lincolnshire, got by Manfred (26801), out of Warlike by Breastplate (19337)—Warfare by The Druid (18981)—War Lady by British Prince (14197)—Water Lady by Baron Warlaby (7813)—Water Witch by 4th Duke of Northumberland (3649)—Waterloo 3d by Norfolk (2377)—by Waterloo (2816)—by Waterloo (2816).

BREADALBANE 11429—Roan bull, bred by William Torr, Aylesby Manor, calved March 24, 1871, got by Lord Napier (26688), out of Blink Bonny by Booth Royal (15673)—Bright Dawn by

Vanguard (10994)—Bright Phœbus by Crown Prince (10087)—Blanche 2d by Zadig (8796)—Blanche by Auld Robin Gray (6753)—White Rose by James Crisp's Bull—Rose by Burley (1766)—Young Anna by Isaac (1169)—Anna by Pilot (496)—Ariadne by Albion (14)—Bright Eyes by Booth's Lame Bull (359)—by Shipton (587)—by son of Suwarrow (636)—by Son of Twin Brother to Ben (88)—by Twin Brother to Ben (660).

CENTRAL PACIFIC 13219—Roan bull, calved Nov. 10, 1871, bred by G. Bland, Coleby Hall, Lincoln, Eng., got by Bismarck (25637), out of Acacia by Knight-Errant (18154)—Amethyst by Magna Charta (16486)—Applin by Lord Raglan (14849)—Amaryllis by Burgomaster (12513)—Acacia by Baron of Ravensworth (7811)—Penance by Lycurgus (7180)—Dimity by Zenith (5702)—by Guardian (3947)—by Firby (1040)—by Ivanhoe (1131)—by Regent (544)—by Blyth Comet (85).

1872. M. H. COCHRANE.
Per Steamship Vicksburg.

INNOCENCE—White heifer, calved Jan. 10, 1870, bred by R. Stratton, Burderop, Wilts., got by James 1st (24202), out of Minerva by 8th Duke of York (23808)—Europa by Windsor Castle (21118)—Lilla by Hermit (14697)—Euridice 2d by Lord of the Manor (14836)—Euridice by Red Duke (8694)—Euribia by Hero of the West (8150)—Modish by Kenilworth (7118)—Madame by Lottery (4280)—Premium by Phœnix (6290).

A. H. B., vol. 12, p. 852, under dam, and E. H. B., vol. 19, p. 629.

VILLAGE ROSE—Red heifer, calved Nov. 21, 1870, bred by R. Stratton, got by James 1st (24202), out of April Rose by Warwick (19120)—March Rose by Young Windsor (17241) —Christmas Rose by His Highness (14708)—Salthrop Rose 4th by Lord of the Manor (14836)—Salthrop Rose 1st by Waterloo (11025)—Young Moss Rose by Lottery (4280)—Moss Rose by Phœnix (6290).

A. H. B., vol. 12, p. 1277, and E. H. B., vol. 20, p. 395, under dam, as of October.

ADMIRAL JACK—Roan, calved April 1, 1873, bred by R. Stratton, got by Jack Frost (31425), out of Village Rose by James 1st (24202)—April Rose by Warwick (19120), &c., as above.

LOUIS LE GRAND 17610—Roan bull, calved Sept. 25, 1872, bred by Mr. S. Wiley, Brandsby, York, got by Warrior (32800), out of Louisa by Earl of Derby (21638)—Luna by Prince George (13510)—Aurora by King Arthur (13110)—Astra by Sylvan King (13819)—Stella by the Silky Laddie (10947)—Vesper by Morning Star (6223)—Primrose by Roland (2556)—Strawberry by Priam (2452)—by Matchem (2281)—by son of Peter (487).

LADY BIRD—Roan heifer, calved Nov. 4, 1870, bred by S. Wiley, Brandsby, York, got by Breastplate (19337), out of Luna by Prince George (13510)—Aurora by King Arthur (13110)—Astra by Sylvan King (13819)—Stella by The Silky Laddie (10947)—Vesper by Morning Star (6223)—Primrose by Roland (2556)—Strawberry by Priam (2452)—by Matchem (2281)—by son of Peter (487).

A. H. B., vol. 12, p. 896, and E. H. B., vol. 19, p. 609, under dam.

Spring of 1872. M. H. COCHRANE,

Of Hillhurst, Compton, Prov. Quebec, Can. Per Steamship Corinthian.

WAVE SWELL—Roan cow, calved June 26, 1867, bred by Wm. Torr, Aylesby Manor, Lincolnshire, got by Prince of Warlaby (15107), out of Wave Princess by British Prince (14197)—Wave Maid by Vanguard (10994)—Water Queen by Baron Warlaby (7813)—Water Witch by 4th Duke of Northumberland (3649)—by Norfolk (2377)—by Waterloo (2816)—by Waterloo (2816).

Under dam, E. H. B., vol. 18, p. 778.

FAIR NAPIER—Roan heifer, calved Sept. 11, 1870, bred by W. Torr, Aylesby Manor, got by Lord Napier (26688), out of Fair Jute by Breastplate (19337)—Fair Dane by Fitz Clarence (14552)—Flower Nymph by Vanguard (10994)—Flower Girl by Londesbro' (6142)—Flora of Farnesfield by Rinaldo (4949)—by Sir Thomas

(2636)—by Sir Alexander (591)—by Marske (418)—by North Star (459)—by Wellington (680)—by Favorite (252)—by Ben (70).

Under dam, E. H. B., vol. 19, p. 502, and A. H. B., vol. 12, p. 779.

VERNAL STAR—Red and white, calved April 22, 1866, bred by R. S. Bruere, Braithwaite Hall, Yorks., got by The Sutler (23061), out of Venus Star by Prince George (13510)—Vesper by King Arthur (13110)—Vesper by Morning Star (6223)—Primrose by Roland (2556)—Strawberry by Priam (2452)—by Matchem (2281)—by son of Peter (487).

A. H. B., vol. 12, p. 1270, and E. H. B., vol. 17, p. 771, under dam.

REGAL OR SILVER STAR—Roan, calved Jan. 5, 1872, bred by R. S. Bruere, got by Booth's Royal Signet (28061), out of Vernal Star, above.

A. H. B., vol. 14, p. 819. Returned to England, by Ship Sardinian, Aug. 4, 1877, and sold at public sale Sept. 4, 1877, to A. Darby for 460 guineas.

FORGET-ME-NOT—Roan, calved Jan. 25, 1873, bred by W. Torr, got by Lieutenant-General (31600), out of Fair Napier by Lord Napier (26688)—Fair Jute by Breastplate (19337), &c., as in Fair Napier, above.

Returned to England by Ship Sardinian from Quebec Oct. 16, 1875, for A. H. Browne, Doxford, Chathill, Northumberland.

LOUISA—Roan, calved Feb. 14, 1867, bred by S. Wiley, Brandsby, York, got by Earl of Derby (21638), out of Luna by Prince George (13510)—Aurora by King Arthur (13110)—Arthur by Sylvan King (13819)—Stella by Silky Laddie (10947)—Vesper by Morning Star (6223)—Primrose by Roland (2556)—Strawberry by Priam (2452)—by Matchem (2281)—by son of Peter (487).

A. H. B, vol. 12, p. 978, and E. H. B., vol. 18, p. 598, under dam.

Aug., 1873. M. H. COCHRANE.

By Ship Canadian, from Liverpool.

PARTRIDGE—Red heifer, calved Sept. 1, 1870, bred by Mr. G. Garne, Churchill Heath, Oxon., got by Royal Butterfly 20th (25007), out of Panacea by General Pelissier (14605)—Pane by Bashaw (12449)—Panic by Colchicum (8963)—Paleface by Harold (8131)—by Lord John (4259)—by Young Consul (6893)—by Newnham (2365)—by Satellite (1420)—by Jupiter (343)—by Sir Oliver (605)—by Trunnell (659)—by Favorite (252)—by Favorite (252)—by Dalton Duke (188)—by R. Alcock's Bull (19)—by J. Smith's Bull (608)—by Jolly's Bull (337).

E. H. B., vol. 19, p. 659, under dam, and A. H. B., vol. 14, p. 772.

JOHN PEEL 17436—Roan bull, calved Dec. 14, 1873, bred by G. Garne, got by 3d Lord of Warwickshire (28524), out of Partridge by Royal Butterfly 20th (25007), &c., as in Partridge, above.

Aug. 13, 1874. M. H. COCHRANE.

By Ship Phœnician, from Glasgow.

RED BESS—Red cow, calved Aug. 20, 1871, bred by Hugh Aylmer, West Dereham Abbey, Norfolk, got by Royal Broughton (27352), out of Roseleaf 4th by Prince Christian (22581)—Roseleaf 2d by Hildebrand (18068)—Roseleaf by Whittington (12299)—Lady Welbourn by Lord Lowther (7164)—Eliza by Panton Favorite (4646)—by grandson of Graziar (1085)—by son of Vulcan (667)—by Vulcan (667)—by Quaker (1349)—by 1st Stonehill Bull (3798).

E. H. B., vol. 20, p. 738, under dam, and A. H. B., vol. 15, p. 850.

GAIETY—Red cow, calved Feb. 21, 1870, bred by G. Garne, Churchill Heath, Oxon., got by Royal Benedict (27348), out of Garland by Duke of Towneley (21615)—Gazelle by Gondomar (17985)—Genteel by Bashaw (12449)—Gentle by Colchicum (8963)—Meg Merrilies by Harold (8131)—by son of Anthony (1640)—by a bull of Mr. Champion's, of Blyth.

E. H. B., vol. 19, p. 525, under dam, and A. H. B., vol. 14. p. 548.

RANGER PRINCE 21756—Red, calved Aug. 16, 1873, bred by W. G. Garne, Broadmoor, or J. Houlton, Aldsworth, got by Ranger (21755), out of Pippin by Cynric (19542)—Pica by Royal Oak (16870)—Picola by Gen. Pelissier (14605)—Picotee by Royal (13636) —Pink by Marchmont (9367)—Young Pye by Young Consul (6893) —Pye by Newnham (2365)—by Satellite (1420)—by Jupiter (342) —by Sir Oliver (605)—by Trunnell (659)—by Favorite (252)—by Favorite (252)—by Dalton Duke (188)—by R. Alcock's Bull (19) —by J. Smith's Bull (608)—by Jolly's Bull (337).

Sold to James Peterson, of Monmouth, Ill., June 16, 1875, for $950.

GUSTAVUS 21591—Roan, calved Sept. 5, 1874, bred by G. Garne, got by Buccaneer (25693), out of Gaity by Royal Benedict (27348)—Garland by Duke of Towneley (21615)—Gazelle by Gondomar (17985)—Genteel by Bashaw (12449)—Gentle by Colchicum (8963)—Meg Merrilies by Harold (8131)—Edgecott Rose by son of Anthony (1640)—Old Rose by a bull of Mr. Champion's, of Blyth.

Aug. 19, 1874. M. H. COCHRANE.
Hillhurst, Compton, Can. By Ship Vicksburg, from Liverpool.

PRINCESS CHRISTIAN—Red and white, calved Dec. 26, 1871, bred by Mr. R. Welsted, Ballywalter, County Cork, Ireland, got by Prince Christian (22581), out of Queen Victoria by Sir James (16980)—Elfin Queen by Elfin King (17796)—British Queen by British Prince (14197)—Primrose by Orson (13432)—Trinkett by Duke of Cornwall (5947)—Trifle by Mowbray (4516)—Folly by Fergus (3782)—by Woodford (2854)—by Sir Walter (2637)—by Hotspur (1117)—by Coxcomb (928)—by Midas (435)—by Comet (155)—by R. Colling's son of Favorite (252)—by same son of Favorite (252)—by Hubback (319).

Under dam, E. H. B., vol. 20, p. 716.

PRINCESS SALLY 2D—Red and white, calved Feb. 26, 1872, bred by Mr. R. Welsted, got by Prince Christian (22581), out of Aunt Sally by Sir James (16980)—Aunt Jane by Uncle Tom

(13913)—Lady Iola by Baron Norton (11155)—Japonica by Emperor (10198)—Lady Jane by Albert 3d (6725)—Adelaide by Haddock (3991)—by Tarquin (2734)—by Red Simon (2499)—by Governor (1077)—by son of Wellington (680).

E. H. B., vol. 20, p. 397, under dam, and A. H. B., vol. 14, p. 796.

PAT MOLLOY 24259—Red and white, calved March 5, 1875, bred by R. Welsted, got by England's Glory (23889), out of Princess Sally 2d by Prince Christian (22581), &c., as above.

Calved in Canada after dam's arrival.

PRINCESS ADELAIDE—Roan, calved March 3, 1872, bred by Mr. R. Welsted, got by Prince Christian (22581), out of Agatha by Sir James (16980)—Amulet by Uncle Tom (13913)—Australia by Master Charlie (13312)—Grace Darling by Rex (6385)—by Rex (6385)—by Solomon (6313)—by Skelbrook (5206)—by Frickley (3849).

E. H. B., vol. 20, p. 385, under dam, and A. H. B., vol. 15, p. 833.

RUTH—White, calved April 12, 1875, bred by R. Welsted, got by England's Glory (23889), out of Princess Adelaide by Prince Christian (22581), &c., as above.

A. H. B., vol. 15, p. 833, under dam.

June 10, 1875. M. H. COCHRANE.
Per Ship Dominion.

GUINEVERE—Red, calved Sept. 20, 1873, bred by J. W. Phillips, got by 3d Duke of Clarence (23727), out of Graceful by Duke of Albany (25931)—Lucinda by Priam (18567)—Coral by Cardinal (11246)—Charmer 3d by Earl of Dublin (10178)—by Little John (4232)—by Caliph (1774)—by Sir Walter (2637)—by Hotspur (1117)—by Coxcomb (928)—by Midas (435)—by Comet (155)—by R. Colling's son of Favorite (252)—by same son of Favorite (252) —by Hubback (319).

Recorded, A. H. B., vol. 17, p. 12915.

July, 1875. M. H. COCHRANE.

By Ship Nova Scotian, July 7, from Liverpool, landing at Quebec, July 19.

PRINCESS—Red, calved April 6, 1874, bred by Messrs. F. Leney & Sons, Orpines, Wateringbury, Kent, got by 6th Duke of Oneida (30997), out of Princess Louise by Grand Duke of Kent (26289)—Princess Alice by British Prince (14197)—Duchess by Duke of Cambridge (12742)—Cold Cream by Earl of Dublin (10178)—Pansy by Gray Friar (9172)—Freckle by Fawsley (6004)—Furbelow by Little John (4232)—Erato by Marcellus (2260)—Beatrice by Caliph (1774)—Quickly by Swing (2721)—Alamode by Argus (759)—Valuable by Defender (194)—Violet by Petrarch (488)—by own brother to R. Colling's White Heifer—by Butterfly (104)—by Globe (278).

A. H. B., vol. 16, p. 12290, and E. H. B., vol. 21, p. 811, under dam.

SIDDINGTON 5TH—Roan cow, calved June 12, 1867, bred by Mr. E. Bowly, Siddington House, Cirencester, got by 7th Duke of York (17754), out of Siddington by 4th Duke of Oxford (11387)—Kirklevington 7th by Earl of Derby (10177)—Kirklevington 4th by Earl of Liverpool (9061)—Kirklevington by Duke of Northumberland (1940)—Nell Gwynne by Belvedere (1706)—Northallerton by son of 2d Hubback (2683)—a cow of Mr. Bates', descended from the stock of Mr. Maynard, of Eryholme.

A. H. B., vol. 15, p. 904, and E. H. B., vol. 18, p. 728, under dam.

LORD HILLHURST 23771—Roan, calved Dec. 26, 1875, bred by Mr. E. Bowly, Siddington House, got by Grand Duke of Thorndale (31298), out of Siddington 5th by 7th Duke of Oxford (17754), &c., as above.

Calved after dam reached Hillhurst, Canada.

GRAND DUCHESS OF BARRINGTONIA—Red and white, calved May 5, 1872, bred by Mr. R. E. Oliver, Sholebroke Lodge, Towcester, Northamptonshire, got by 18th Duke of Oxford (25995), out of Grand Duchess of Barrington by 7th Grand Duke (19877)

—Countess of Barrington 2d by 9th Duke of Oxford (17738)—
Countess of Barrington by 3d Grand Duke (16182)—Laurel by
Grand Turk (12969)—Lally by Earl of Derby (10177)—Olive Leaf
3d by Earl of Liverpool (9061)—Olive Leaf 2d by 2d Duke of
Cambridge (3638)—Olive Leaf by Belvedere (1706)—Lady Barrington by son of Herdsman (304)—Young Alicia by Wonderful
(700)—Old Alicia by Alfred (23)—by Young Favorite (6994).

A. H. B., vol. 15, p. 586, and E. H. B., vol. 20, p. 547, under
dam.

MARCHIONESS BARRINGTON—Roan, calved Nov. 11, 1875,
bred by Mr. R. E. Oliver, got by Grand Duke 22d (24062), out of
Grand Duchess of Barringtonia by 18th Duke of Oxford (25995),
&c., as above.

A. H. B., vol. 15, p. 586, under dam.   Calved after dam reached
Hillhurst, Canada.   Returned to England Aug. 4, and sold Sept.
4, 1877, to Sir W. H. Salt, Bart., for 800 guineas.

Oct. 14, 1875. M. H. COCHRANE.
By Ship Polynesian, from Liverpool.

WILD EYES LASSIE—Roan, calved July 3, 1871, bred by
Mr. I. Downing, Turner's Hill, Dudley, got by 3d Duke of Claro
(23729), out of Wild Eyes 24th by 4th Duke of Oxford (11387)—
Wild Eyes 22d by Wild Duke (19148)—Wild Eyes 20th by Lord Barrington 1st (13170)—Wild Eyes 16th by 2d Duke of Oxford (9046)
—Wild Eyes 15th by 4th Duke of Northumberland (3649)—by
Duke of Northumberland (1940)—by Belvedere (1706)—by Emperor (1975)—by Wonderful (700)—by Cleveland (145)—by Butterfly (104)—by Hollon's Bull (313)—by Mowbray's Bull (2342)—
by Masterman's Bull (422)—descended from the stock of M.
Dobison.

Recorded under dam, E. H. B., vol. 20, p. 822.

KIRKLEVINGTON 26TH—Roan heifer, calved Feb. 16, 1875,
bred by W. Ashburner, Netherhouse, Ulverston, got by Grand
Duke of Kent 2d (28759), out of Kirklevington 24th by 5th Duke

of Wharfdale (26033)—Kirklevington 17th by Lord Lally (22161) —Kirklevington 10th by Delhi (15865)—Kirklevington 8th by Gen. Canrobert (12926)—Kirklevington 7th by Earl of Derby (10177)—Kirklevington 4th by Earl of Liverpool (9061)—Kirklevington 1st by Duke of Northumberland (1940)—by Belvedere (1706)—by son of 2d Hubback (2683)—a cow of Mr. Bates', descended from the stock of Mr. Maynard, of Eryholme.

Recorded under dam, E. H. B., vol. 22, p. 301, and A. H. B., vol. 17, p. 12951.

May, 1876. M. H. COCHRANE.
By Ship Circassian, from Liverpool, arriving at Quebec, May 15, 1876.

LADY BARRINGTON 11TH—Roan, calved Nov. 30, 1873, bred by W. Ashburner, Ulverston, got by Grand Duke of Oxford (28764), out of Lady Barrington 9th (vol. 23, p. 497, E.) by Wild Duke (27808) —Lady Barrington 7th by Baron Tarves (17387)—Lady Barrington 5th by Kirklevington 3d (13120)—Lady Barrington 3d by Weathercock (9815)—Lady Barrington 10th by 2d Duke of Oxford (9046) —Lady Barrington 2d by Belvedere (1706)—Lady Barrington by son of Herdsman (304)—Young Alicia by Wonderful (700)—Old Alicia by Alfred (23)—by Young Favorite (6994).

A. H. B., vol. 17, p. 12957.

LALLY DUCHESS 2D—Roan, calved Jan. 7, 1874, bred by C. W. Harvey, Walton-on-the-Hill, Liverpool, got by Grand Prince of Claro (28781), out of Lally 11th by 5th Lord Wild Eyes (26762) —Lally 5th by Duke of Wetherby (17754)—Lally 2d by Malachite (18313)—Lally by Earl of Derby (10177)—Olive Leaf 3d by Earl of Liverpool (9061)—Olive Leaf 2d by 2d Duke of Cambridge (3638)—by Belvedere (1706)—by son of Herdsman (304)—by Wonderful (700)—by Alfred (23)—by Young Favorite (6994).

E. H. B., vol. 21, p. 548.

DUKE OF OXFORD 35TH 26350 (36530)—Red, calved Aug. 6, 1875, bred by the Duke of Devonshire, Holker Hall, got by 5th Duke of Wetherby (31033), out of Grand Duchess of Oxford

19th by Grand Duke 10th (21848)—Grand Duchess of Oxford 6th by Imperial Oxford (18084)—Grand Duchess of Oxford 4th by Grand Duke of Wetherby (17997)—Oxford 15th by 4th Duke of York (10167)—by 2d Duke of Northumberland (3646)—by Short Tail (2621)—by Matchem (2281)—by Young Wynyard (2859).

March 31, 1881. M. H. COCHRANE.

By Ship Texas, from Liverpool.

BELLFLOWER—Red heifer, calved May 21, 1879, bred by S. Campbell, got by Luminary (34715), out of Bellflower by British Prince (33226)—Bellflower by Nobleman (26967)—Bellflower by Statesman (15342)—Wall Flower by Lord Raine (13248)—Windsor Flower by Earl of Scarborough (9064)—Magic Flower by Magician (7185)—by Sir Thomas (2636)—by Eryholme (1018)—by Eclipse (236)—by Charge's Gray Bull (872)—by Paddock Bull (477)—by Brown's Red Bull (97).

A. H. B., vol. 25, p. 627.

NONPAREIL—Roan heifer, calved in June, 1879, bred by S. Campbell, got by Luminary (34715), out of Nonpareil 31st by British Prince (33226)—Nonpareil 29th by Duke (28342)—Nonpareil 25th by Dipthong (17681)—Nonpareil 24th by Lord Sackville (13249)—Nonpareil 23d by The Baron (13833)—Nonpareil 17th by Matadore (11800)—Nonpareil 10th by Prince Edward Fairfax (9506)—Countess of Lincoln by Diamond (5918)—Nonpareil 3d by Young Frederick (3836)—Nonpareil 2d by Commodore (1858) —Nonpareil by Tathwell Studley (5401)—by Blyth Comet (85).

CECILIA—Roan, calved March 2, 1880, bred by S. Campbell, got by Luminary (34715), out of Cecilia by Cæsar Augustus (25714) —Columbine by Sir Walter Scott (23922)—Camelia by Lancaster Comet (11663)—Cactus by Lord Sackville (13249)—Rose of Sharon by Plantagenet (11906)—Fancy by Billy (3151)—Jessie by Sovereign (7539)—Rose by Satellite (1420)—by Baronet (60)—by Cleveland (143)—by Symmetry (641).

21

BUSHBURY COUNTESS KIRKLEVINGTON 1st—Red and white, calved March 11, 1880, bred by H. Lovatt, Low Hill, Wolverhampton, got by Duke of Hillhurst (28401), out of Siddington Grand Duchess by Grand Duke of Thorndale 2d (31298)—Siddington 7th by Grand Duke of Thorndale (31297)—Siddington 5th by 7th Duke of York (17754)—Siddington 3d by 7th Duke of York (17754)—Kirklevington 7th by Earl of Derby (10177)—Kirklevington 4th by Earl of Liverpool (9061)—Kirklevington by Duke of Northumberland (1940)—by Belvedere (1706)—by son of 2d Hubback (2683)—a cow of Mr. Bates', descended from the stock of Mr. Maynard, of Eryholme.

E. H. B., vol. 27, p. 494.

BUSHBURY COUNTESS BARRINGTON 1st—Red, calved Sept. 13, 1880, bred by H. Lovatt, got by Duke of Oxford 39th (38173), out of Countess of Barrington 7th by Baron Barrington 4th (36006)—Lally 15th by 8th Duke of Geneva (28390)—Lally 8th by 7th Duke of York (17754)—Lally 3d by 4th Duke of Oxford (11387)—Lally by Earl of Derby (10177)—Olive Leaf 3d by Earl of Liverpool (9061)—by 2d Duke of Cambridge (3638)—by Belvedere (1706)—by son of Herdsman (304)—by Wonderful (700)—by Alfred (23)—by Young Favorite (6994).

E. H. B., vol. 27, p. 493.

ROSEBUD—Red heifer, calved Jan. 24, 1879, bred by S. Campbell, got by Golden Prince (38363), out of Rosebud by Sir Christopher (22895)—Rosebud by Gladstone (26256), &c., as below.

ROSEBUD—Red heifer, calved March 2, 1879, bred by S. Campbell, got by Luminary (34715), out of Rosebud by British Prince (33228)—Rosebud by Duke (28342)—Rosebud by Gladstone (26256), &c., as above.

Vol. 25, p. 627, A. H. B.

BESSIE—Roan heifer, calved March 4, 1879, bred by S. Campbell, got by Luminary (34715), out of Bessie 6th by Foljambe (33950)—Bessie 3d by British Prince (23468).

MINA—Roan heifer, calved in March, 1879, bred by S. Campbell, got by Golden Prince (38363), out of Mina 1st by Dipthong 3d (21547)—Mina by Beeswing (12456)—Crocus by Sir Arthur (12072) —Bashful by Young Ury (10984)—Likely by The Pacha (7612)— Helen by 2d Duke of Northumberland (3646)—by Sillery (5131)— by Carleton (843)—by Diamond (205)—by Diamond (205).

MAID OF PROMISE—Red heifer, calved April 2, 1879, bred by S. Campbell, got by Luminary (34715), out of Maid of Promise by Scotland's Pride (25100)—Lady Kitty by Lord Lincoln (34583) —Rind Mix by Old England (24681)—Luna 2d by Earl of Rosse (17782)—Luna by California (12528)—Jessica by Bang (17346)— by Sir Thomas Fairfax (5196).

Vol. 25, p. 627, A. H. B.

AIRDRIE'S LADY JOCELYN—White heifer, calved Jan. 19, 1878, bred by Geo. Fox, Elmhurst Hall, Litchfield, got by 24th Duke of Airdrie (36460), out of Lady Jocelyn 4th by Janitor (24204) —Lady Jocelyn by 7th Duke of York (17754)—Lady Jane by Duke of Gloster (11382)—Jardine by Lord Warden (7167)—by Fawsley (6004)—by Warden (5595)—by Javelin (4093)—by Blyth (797)— by Wellington (684)—by Phenomenon (491)—by Favorite (252)— by Favorite (252)—by Favorite (252)—by Hubback (319)—by Snowdon's Bull (612)—by Waistell's Bull (669)—by Masterman's Bull (422)—by Studley Bull (626).

Vol. 25, p. 453, E. H. B., under dam.

BESSIE—Roan heifer, calved April 30, 1878, bred by R. H. Masfen, got by Duke of Milcote 4th (36523), out of Queen Bess by Broomstick (28099)—Queen Victoria by Cherry Duke (25752)— Queen Anne by Duke of Geneva 2d (21591)—Queen of Airdrie by 2d Duke of Airdrie (19600)—Queen of Hearts by May Duke (13320) —Queen Bess by 3d Duke of Oxford (9047)—Queen Anne by Freeman (10244)—by Hamlet (8128)—by Short Tail (2621)—by Emperor (1974)—by Young Lancaster (361)—by St. Albans (2584)— by Lawnsleeves (365)—bred by Messrs. James, of Stamford.

Vol. 25, p. 580, E. H. B., under dam.

ROSEBUD—Red heifer, calved June 20, 1878, bred by S. Campbell, Kinnellar, Aberdeenshire, Scotland, got by Novelist (52952), out of Rosebud by Dipthong 3d (21547)—Rosebud by British Prince (33226)—Rosebud by Duke (28342)—Rosebud by Gladstone (26256)—Thalia by Earl of Aberdeen (12800)—Myrtle by Balmoral (9920)—by Dannecker (7949)—by Ury (17157)—by Heriot (4017)—by Gray Diomed (2076)—by Juniper (1144).

Vol. 25, p. 628, A. H. B.

LORD ABERDEEN 52417—Red and white, calved May 12, 1881, bred by S. Campbell, got by Gladstone (43286), out of Rosebud by Novelist (52952), &c., as in dam, next above.

BEATRICE—Roan, calved Oct. 4, 1878, bred by Her Majesty the Queen, Prince Consort's Shaw Farm, Windsor, got by King Rufus (34351), out of Benedicta by Royal Benedict (27348)—Bluebell by Cynric (19542)—Royal Blossom by Royal Oak (16870)—Blooming by Bashaw (12449)—Bloom by Royal (13636)—Young Blossom by Magnet (10488)—by Fitz Hardinge (8073)—by Elevator (6969)—by Consul (1868)—by Gazer (7030).

Vol. 25, p. 305, E. H. B., under dam.

EMPRESS—Roan, calved April 17, 1876, bred by W. Bradburn, Wednesfield, Straffordshire, got by Phosphate (32064), out of Helen by Paris (29523)—Pheasant by Provost (24878)—Waterwave 17th by Wandering Willie (19110)—by Chieftain (14267)—by Homer (13038)—by Prince John (9508)—by Vampire (6632)—by Senator (2610)—by Columella (904)—by Shakespeare (1429)—by Blyth Comet (85)—by Neswick (1266)—by Favorite (1033).

Vol. 26, p. 567, E. H. B.

HOPEFUL—White cow, calved July 27, 1877, bred by T. Nash, Featherstone, Wolverhampton, got by Duke (38113), out of Helen by Paris (29523), &c., as in Empress, above.

FAIRY ROSE—Red cow, calved May 15, 1877, bred by R. H. Masfen, Prudeford, Wolverhampton, got by Duke of Milcote 4th (36523), out of Queen of the Roses by Broomstick (28099)—Telluria

9th by King of the Roses (22043)—Telluria 4th by Beau of Oxford (21254)—Telluria 3d by Romulus (15185)—Telluria 2d by Horatio (10335)—Taglioni by Lord George (10439)—by Orontes (4623)—by Guardian (3947)—by Mercury (2301)—by Monarch (2324)—by St. Albans (2584)—by Jupiter (342)—by Sir Oliver (605)—by Trunnell (659)—by Favorite (252)—by Favorite (252)—by Dalton Duke (188)—by R. Alcock's Bull (19)—by J. Smith's Bull (608)—by Jolly's Bull (337).

HELIOTROPE—Roan, calved April 15, 1881, bred by T. Nash, Featherstone, Wolverhampton, got by Lord Claro (41845), out of Hopeful by Duke (38113)—Helen by Paris (29523), &c., as in Empress.

QUEEN OF TRUMPS—White, calved March 20, 1881, bred by R. H. Masfen, got by Watchman 2d (47227), out of Bessie by Duke of Milcote 4th (36523)—Queen Bess by Broomstick (28099), &c., as in dam, above.

QUEEN OF CONNAUGHT—Red and white, calved Oct. 22, 1881, bred by Her Majesty the Queen, got by Duke of Connaught (44663), out of Beatrice by King Rufus (34351)—Benedicta by Roaly Benedict (27348), &c., as in dam, above.

EMPEROR OF THE FRENCH 48160—Red, calved Jan. 14, 1882, bred by W. Bradburn, Wednesfield, got by Lord France (45086), out of Empress by Phosphate (32064)—Helen by Paris (29523), &c., as in dam, above.

Aug. 13, 1881. M. H. COCHRANE.
By Ship Scandinavian, from Liverpool.

SIR LEWIS (45614) — Red, calved March 2, 1880, bred by H. Aylmer, West Dereham Abbey, Norfolk, got by Sir Wilfred (37484), out of Cassandra by Royal Monk (35392)—Celeste by Royal Broughton (27352)—Christine by Prince of the Realm (22627)—Charmian by Valasco (15443)—by British Boy (11206)—by Hamlet (8126)—by Leonard (4210)—by Buckingham (3239)—from the stock of Sir M. W. Ridley, Bart.

February 8, 1883. M. H. COCHRANE.
By Ship Montreal, from Liverpool to Halifax.

GRAND DUKE OF BARRINGTON 3D (46444)—Red, calved Oct. 30, 1881, bred by H. Lovatt, Low Hill, got by Grand Duke 37th (43307), out of Lady Ellen Barrington by Lord Stanley (24467)—Grand Duchess of Barrington by Grand Duke 7th (19877) —Countess of Barrington 2d by 9th Duke of Oxford (17738)— Countess of Barrington by Grand Duke 3d (16182)—by Grand Turk (12969)—by Earl of Derby (10177)—by Earl of Liverpool (9061)—by 2d Duke of Cambridge (3638)—by Belvedere (1706)— by son of Herdsman (304)—by Wonderful (700)—by Alfred (23)— by Young Favorite (6994).

Oct., 1875. M. H. COCHRANE and S. BEATTIE.
Shipped by Polynesian, Oct. 14, 1875, from Liverpool, for M. H. Cochrane and Simon Beattie.

WILD EYES LASSIE 2D—Roan, calved May, 1875, bred by J. P. Foster, Killhow, Carlisle, got by 22d Duke of Oxford (31000), out of Wild Eyes Lassie by 3d Duke of Claro (23729)—Wild Eyes 24th by 4th Duke of Oxford (11387)—Wild Eyes 22d by Wild Duke (19148)—Wild Eyes 20th by Lord Barrington 1st (13170)— Wild Eyes 16th by 2d Duke of Oxford (9046)—Wild Eyes 15th by 4th Duke of Northumberland (3649)—by Duke of Northumberland (1940)—by Belvedere (1706)—by Emperor (1975)—by Wonderful (700)—by Cleveland (145)—by Butterfly (104)—by Hollon's Bull (313)—by Mowbray's Bull (2342)—by Masterman's Bull (422) —descended from Mr. Dobison's stock.

Recorded under dam, E. H. B., vol. 22, p. 418; exported Oct., 1876; landed at Liverpool Nov. 8, 1876. Sold to J. P. Foster, her breeder, for $4,500, June 14, 1876.

SONATA—Red and white cow, calved Feb. 25, 1871, bred by Lord Skelmersdale, Latham House, Ormskirk, Lancashire, got by Cherry Grand Duke 2d (25758), out of Sonora by Romulus Butterfly (18741)—Sincerity by Mameluke (13289)—Sweetheart 4th by Cardinal (11246)—Sweetheart by Accordion (5708)—Charmer by

Little John (4232)—by Caliph (1774)—by Sir Walter (2637)—by Hotspur (1117)—by Coxcomb (928)—by Midas (435)—by Comet (155)—by R. Colling's son of Favorite (252)—by same son of Favorite (252)—by Hubback (319).

Recorded E. H. B., vol. 20, p. 769, under dam, and A. H. B., vol. 18, p. 13930. Sold June 14, 1876, to S. R. Streator, Cleveland, Ohio, for $900.

WATERLOO 30TH—Red and white cow, calved April 20, 1871, bred by R. B. Hetherington, Park Head, Silloth Cumberland, got by Grand Duke of Lightburne 2d (26291), out of Waterloo 22d by Speculator (13775)—Waterloo 18th by Bosquet (14183)—Waterloo 15th by The Hero (10934)—Waterloo 13th by 3d Duke of Oxford (9047)—Waterloo 9th by 2d Cleveland Lad (3408)—by Duke of Northumberland (1940)—by Norfolk (2377)—by Waterloo (2816)—by Waterloo (2816).

Recorded, E. H. B., vol. 21, p. 970. Sold June 14, 1876, to S. W. Jacobs, of West Liberty, Iowa, for $710. A. H. B., vol. 17, p. 13203.

JUNO—Roan, calved March 14, 1872, bred by W. Ashburner, got by Duke of Harlock (30962), out of Jantja 11th by Thorndale Rushden (25311)—Jantja 8th by Hayman (16245)—Jonquil by Aaron Smith (12331)—Jantja by Lycurgus (7180)—by Sweet William (5368)—by Javelin (4093)—by Blyth (797)—by Wellington (684)—by Phenomenon (491)—by Favorite (252)—by Favorite (252)—by Favorite (252)—by Hubback (319)—by Snowdon's Bull (612)—by Waistell's Bull (669)—by Masterman's Bull (422)—by Studley Bull (626).

Recorded, E. H. B., vol. 20, p. 565, under dam. Sold June 10, 1876, to A. L. Stebbin, Port Huron, Mich., for $670.

OXFORD J. 30517—Roan bull, calved Dec. 14, 1875, bred by W. Ashburner, got by Oxford Beau 4th (34964), out of Juno by Duke of Harlock (30962), &c., as in Juno, above.

AZALEA—Red and white heifer, calved March 1, 1873, bred by R. B. Hetherington, got by Grand Duke of Lightburne 2d (26291), out of Alice by Lord of the Valley (29184)—Athena by Duke of Darlington (21586)—Acacia by Count de Gourcy (17632)—Asia by 2d Grand Duke (12961)—Apricot by Fusileer (11499)—by 3d Duke of York (10166)—by 2d Cleveland Lad (3408)—by Duke of Cleveland (1937)—by Belvedere (1706)—cow of Mr. Bates', Kirklevington.

Recorded, A. H. B., vol. 17, p. 12786, as bred by Mr. A. Brodgen. Sold to Col. LeG. B. Cannon, Shelburn, Vt.

GRAND DUCHESS OF BARRINGTON 3D—Red and white heifer, calved Jan. 29, 1874, bred by H. J. Sheldon, Brailes House, Warwickshire, got by 2d Duke of Collingham (23730), out of Grand Duchess of Barrington 2d by Duke of Brailes (23724)—Grand Duchess of Barrington by 7th Grand Duke (19877)—Countess of Barrington 2d by 9th Duke of Oxford (17738)—Countess of Barrington by Grand Duke 3d (16182)—Laurel by Grand Turk (12969) —Lally by Earl of Derby (10177)—by Earl of Liverpool (9061)—by 2d Duke of Cambridge (3638)—by Belvedere (1706)—by son of Herdsman (304)—by Wonderful (700)—by Alfred (23)—by Young Favorite (6994).

Recorded, E. H. B., vol. 21, p. 924, under her dam. Sold June 14, 1876, to Lord Feversham, England, for $1,700. A. H. B., vol. 16, p. 12095.

PRINCESS VICTORIA 11TH—Roan heifer, calved March 27, 1874, bred by Lord Skelmersdale, got by 1st Duke of Oneida (30996), out of Princess Victoria 5th by Lord Oxford 2d (20215)— Princess Victoria 2d by Lord Oxford 2d (20215)—Princess Alice by Gen. Canrobert (12927)—Princess by Earl of Derby (10177)—by 2d Cleveland Lad (3408)—by 2d Earl of Darlington (1945)—by a son of 2d Hubback (2683)—a cow bought of Mr. Bates.

Recorded, E. H. B., vol. 21, p. 931, under dam, and A. H. B., vol. 17, p. 13112. Sold to Col. LeG. B. Cannon, of Shelburn, Vt., for $1,375.

OXFORD DUCHESS—Roan, calved April 21, 1870, bred by Messrs. W. Horswell & Sons, Burns Hall, Lewdown, Devonshire, got by Baron Oxford 2d (23376), out of Duchess of Northumberland by Duke (15908)—Lady Barrington 2d by Weathercock (9815)—Lady Barrington 10th by 2d Duke of Oxford (9046)—by Belvedere (1706)—by son of Herdsman (304)—by Wonderful (700)—by Alfred (23)—by Young Favorite (6994).

Recorded, E. H. B., vol. 21, p. 774.

LORD BRIGHT EYES 23742—Red, calved March 12, 1875, bred by Earl Dunmore, Dunmore, Sterling, got by 3d Duke of Hillhurst (30975), out of Lady Worcester 2d by Charleston (21400)—Clear Star by Marton Duke (22307)—Bright Star by Red Duke (18676)—Bright Eyes by 3d Duke of York (10166)—by 2d Cleveland Lad (3408)—by Duke of Northumberland (1940)—by Belvedere (1706)—by Emperor (1975)—by Wonderful (700)—by Cleveland (145)—by Butterfly (104)—by Hollon's Bull (313)—by Mowbray's Bull (2342)—by Masterman's Bull (422)—descended from M. Dobison's stock.

Sold to Wm. Miller, Brougham, Ont., Can., for $1,075.

SONATA 2d—Roan heifer, calved Nov. 11, 1875, bred by Lord Skelmersdale, Latham House, got by Baron Oxford 4th (25580), out of Sonata by Cherry Grand Duke 2d (25758)—Sonora by Romulus Butterfly (18741)—Sincerity by Mameluke (13289)—Sweetheart 4th by Cardinal (11246)—Sweetheart by Accordion (5708)—Charmer by Little John (4232)—Graceful by Caliph (1774)—Sylph by Sir Walter (2637)—by Hotspur (1117)—by Coxcomb (928)—by Midas (435)—Rachel by Comet (155)—Russell by R. Colling's son of Favorite (252)—by same son of Favorite (252)—by Hubback (319).

Recorded, A. H. B., vol. 18, p. 13920, by the name of Royal Charmer 5th. Sold at sale of Messrs. Cochrane, Hope & Beattie, June 14, 1876, to S. R. Streator, Cleveland, Ohio, for $500. She was calved after dam reached M. H. Cochrane, Compton, Can. Owned by Bow Park Association.

WILD EYES 33d—Roan heifer, calved July 18, 1874, bred by W. Ashburner, got by Grand Duke of Kent 2d (28759), out of Wild Eyes 31st by Grand Duke of Cambridge 3d (31285)—Wild

Eyes 29th by Knight of the Harem (24278)—Wild Eyes 27th by Gainford 5th (12913)—Wild Eyes 26th by 2d Cleveland Lad (3408) —Wild Eyes 5th by Short Tail (2621)—Wild Eyes by Emperor (1975)—by Wonderful (700)—by Cleveland (145)—by Butterfly (104)—by Hollon's Bull (313)—by Mowbray's Bull (2342)—by Masterman's Bull (422)—descended from the stock of M. Dobison.

Recorded, A. H. B., vol. 16, p. 12389. Sold to Major Greig, Toronto, Canada, June 14, 1876, for $2,100. E. H. B., vol. 23, p. 316, under dam.

ACOMB BELLE—Roan heifer, calved Oct. 1, 1874, bred by Sir W. C. Trevelyan, Bart., Wallington, Northumberland, got by Oxford Beau 3d (32013), out of Anemone by Grand Duke of Lightburne 2d (26291)—Athena by Duke of Darlington (21586)—Acacia by Count de Gourcy (17632)—Asia by 2d Grand Duke (12961)— Apricot by Fusileer (11499)—by 3d Duke of York (10166)—by 2d Cleveland Lad (3408)—by Duke of Cleveland (1937)—by Belvedere (1706)—a cow of Mr. Bates'.

Recorded under dam, E. H. B., vol. 21, p. 968. Bought by S. W. Jacobs, West Liberty, Iowa, June 14, 1876, for $575. Recorded, A. H. B., vol. 19, p. 14353.

LADY ACOMB 4TH—Red heifer, calved March 3, 1875, bred by Mr. T. Gow, Camba, Newcastle-on-Tyne, got by Oxford Beau 4th (34964), out of Lady Acomb by 2d Earl of Walton (19672)— Amy by Duke of Darlington (21586)—Anemone by Duke of Kent (19619)—Acacia by Count de Gourcy (17632)—Asia by 2d Grand Duke (12961)—Apricot by Fusileer (11499)—by 3d Duke of York (10166)—by 2d Cleveland Lad (3408)—by Duke of Clevelend (1937) —by Belvedere (1706)—a cow of Mr. Bates'.

Recorded under dam, E. H. B., vol. 22, p. 436, and A. H. B., vol. 16, p. 12131. Bought by T. L. McKeen, of Easton, Pa., for $600.

LORD LIGHTBURNE 27087—Roan, calved Feb. 7, 1876, bred by R. B. Hetherington, Park Head, Silloth, calved M. H. Cochrane's, Compton, Can., got by Lord Radstock (34656), out of

Azalea by Grand Duke of Lightburne 2d (26291)—Alice by Lord of the Valley (29184)—Athena by Duke of Darlington (21586)—Acacia by Count de Gourcy (17632)—Asia by 2d Grand Duke (12961) —Apricot by Fusileer (11499)—by 3d Duke of York (10166)—by 2d Cleveland Lad (3408)—by Duke of Cleveland (1937)—by Belvedere (1706)—a cow of Mr. Bates'.

Sold with his dam to Col. LeG. B. Cannon, Shelburn, Vt.

WATERLOO'S OXFORD—Red, calved Jan. 28, 1876, bred by R. B. Hetherington, got by Oxford Beau 4th (34964), out of Waterloo 30th by Grand Duke of Lightburne 2d (26291), &c., as in dam, above.

A. H. B., vol. 20, p. 16256.

ADMIRAL SIR ISAAC COFFIN.

For Massachusetts Agricultural Society in 1823 and 1824.

ADMIRAL (1608)—Roan, calved 1821, bred by Mr. Wetherell, got by North Star (460), dam by Comet (155)—by Wellington (678)—by Danby (190).

ANNABELLA (vol. 5, p. 39, E.)—Red and white, bred by Mr. Wetherell, calved in 1820, got by Major (398), out of Ada by Denton (198)—Aurora by Comet (155)—by Henry (301)—by Danby (190).

EMMA (vol. 5, p. 319, E.)—Red and white, calved in 1824, bred by Mr. Derby, got in England by Wellington (683), out of Annabella.

BLANCHE—White, imported in 1823 by Sir Isaac Coffin, got by a son of Comet (155). See Harlem Comet 71.

SNOWDROP—Got by Fitz Favorite (1042), out of Blanche by a son of Comet (155). See 2483.

1875. WM. COLLUM.

Haysville, Waterloo Co., Ont.

LIBERATOR [5573]—Roan, calved Feb. 20, 1874, bred by Robt. Bruce, Newton of Struthers, Tarves, Scotland, got by Baron Killerby (27949), out of Lucy by Hydra (19995)—Jamestina by Picotee (15063)—Bellona by Robert (10704)—Amazon by Duke 3d (17697)—Averne by Bucephalus (6784)—by Crusader (934)—by Sultan (1485)—by Mars (411)—by North Star (458).

AGGIE BUCKINGHAM—Red and white, calved Feb. 11, 1876, bred by A. Cruickshank, got by Lord Irwin (29123), out of Airy Buckingham by Master of Arts (26867), &c., as below.

C. H. B., vol. 4, p. 3.

AIRY BUCKINGHAM—Roan, calved March 6, 1871, bred by A. Cruickshank, got by Master of Arts (26867), out of Ada Buckingham by Lord Raglan (13244)—Alice Buckingham by The Baron (13833)—Miss Buckingham by Dr. Buckingham (14405)—Arabella by Robin o' Day (4973)—Picotee by Premier (6308)—by Unicorn (8725)—by Young Satellite (8538)—by Valentine (661)—bred by Mr. Rennie.

C. H. B., vol. 4, p. 5.

DOROTHY—Roan, calved May 29, 1873, bred by John Law, New Keig, White House, Scotland, got by Shuttlecock (27942), out of Viscountess by Prince Louis (20560)—Countess by Report (10704)—Bet by The Pacha (7612)—Strawberry by 2d Duke of Northumberland (3646)—Margaret by Mahomed (6170)—Mary Anne by Sillery (5131)—Miss Gibson by Carleton (843)—Dora by Diamond (205)—Kitty by Diamond (205).

C. H. B., vol. 4, p. 98, and A. H. B., vol. 23, p. 17815.

VISCOUNTESS 2D—Red and white, calved April 2, 1876, bred by John Law, New Keig, White House, Aberdeen, Scotland, got by Lord Irwin (29123), out of Dorothy by Shuttlecock (27942), &c., as in Dorothy, above.

C. H. B., vol. 4, p. 587.

1874. LESLIE COMBS, JR.,

Of Lexington, Ky., October 28, 1874, from Liverpool, by Steamship Egypt, to New York, N. Y.

ROSARY MONK (35316)—Red and white, calved Aug. 15, 1872, bred by Wm. Torr, Aylesby Manor, got by Royal Prince (27384), out of Rosary Link by Breasptlate (19337)—Rosary Bead by Dr. McHale (15887)—Rose Butterfly by Master Butterfly 2d (14918)—Rosabel by Fanatic (8054)—by Auld Robin Gray (6753)—by Scrip (2604)—by Burley (1766)—by Isaac (1129)—by Pilot (496)—by Albion (14)—by Lame Bull (359)—by Shipton (587)—by son of Suwarrow (636)—by Son of Twin Brother to Ben (88)—by Twin Brother to Ben (660).

ANNA 5TH—Roan, calved December 26, 1869, bred by Rev. J. Storer, Hellidon, Northamptonshire, got by Earl of Rosedale (26072), out of Anna 3d by Mantalini Prince (22276)—Anna 2d by Brilliant Star (17450)—Princess Julia by Lord of Rainsber (13211)—Dowager Queen by Sir Frederick Fairfax (5152)—Adelaide by Albert (727)—Anna by Pilot (496)—by Albion (14)—by Lame Bull (359)—by Shipton (587)—by son of Suwarrow (636)—by Son of Twin Brother to Ben (88)—by Twin Brother to Ben (660).

Under dam, E. H. B., vol. 19, p. 395.

ROSY MORN—Red cow, calved in June, 1871, bred by Donald Fisher, Pitlockrie, Perthshire, got by Brother Windsor (25690), out of Early Dawn by Fashion (21724)—New-Year's Morn by Baltic (12431)—Lady Marguerite by Master Charlie (13312)—Florence by Rex (6385)—Fair Maid of Athens by Sir Thomas Fairfax (5196)—Medora by Ambo (1636)—Blossom by Memnon (2295)—Sister to Isabella by Pilot (496)—White Cow by Agamemnon (9)—by Burrell's Bull of Burdon (1768).

E. H. B., vol. 21, p. 706.

DAIRY MAID—Roan, calved in May, 1868, bred by D. Fisher, got by Scottish Chief (22850), out of Diamond by The Chieftain (20942)—Damsel by Lord Hopewell (18239)—Datura by Fitz Clarence (14552)—Duchess by Duke of Cambridge (12742)—Cold

Cream by Earl of Dublin (10178)—Pansy by Gray Friar (9172)—Freckle by Fawsley (6004)—Furbelow by Little John (4232)—Erato by Marcellus (2260)—Beatrice by Caliph (1774)—Quickly by Swing (2721)—Alamode by Argus (759)—Valuable by Defender (194)—Violet by Petrarch (488)—by own brother to R. Colling's White Heifer—by Butterfly (104)—by Globe (278).

E. H. B., vol. 20, p. 470. E. H. B. says calved 1869—1868 is correct.

DUCHESS OF KNIGHTLY—Roan, calved in May, 1871, bred by D. Fisher, got by Great Hope (24082), out of Damsel by Lord Hopewell (18239)—Datura by Fitz Clarence (14552)—Duchess by Duke of Cambridge (12742)—Cold Cream by Earl of Dublin (10178)—Pansy by Gray Friar (9172)—Freckle by Fawsley (6004)—Furbelow by Little John (4232)—Erato by Marcellus (2260)—Beatrice by Caliph (1774)—Quickly by Swing (2721)—Alamode by Argus (759)—Valuable by Defender (194)—Violet by Petrarch (488)—by own brother to R. Colling's White Heifer—by Butterfly (104)—by Globe (278).

Under dam, E. H. B., vol. 20, p. 473.

COLD CREAM 6TH—Roan heifer, calved in May, 1873, bred by D. Fisher, got by Fawsley Prince (31150), out of Cold Cream 2d (vol. 20, p. 452, E.) by Prince of Saxe Cobourg (20576)—Comely by Buckingham (17471)—Cold Cream by Earl of Dublin (10178)—Pansy by Gray Friar (9172)—Freckle by Fawsley (6004)—Furbelow by Little John (4232)—Erato by Marcellus (2260)—Beatrice by Caliph (1774)—Quickly by Swing (2721)—Alamode by Argus (759)—Valuable by Defender (194)—Violet by Petrarch (488)—by own brother to R. Colling's White Heifer—by Butterfly (104)—by Globe (278).

PRIME MINISTER—Roan, calved in Feb., 1875, bred by D. Fisher, got by Valentine Vox (32752), out of Dairymaid by Scottish Chief (22850)—Diamond by The Chieftain (20942)—Damsel by Lord Hopewell (18239)—Datura by Fitz Clarence (14552)—Duchess by Duke of Cambridge (12742)—Cold Cream by Earl of Dublin (10178)—Pansy by Gray Friar (9172)—Freckle by Fawsley (6004)—

Furbelow by Little John (4232)—Erato by Marcellus (2260)—
Beatrice by Caliph (1774)—Quickly by Swing (2721)—Alamode
by Argus (759)—Valuable by Defender (194)—Violet by Petrarch
(488)—by own brother to R. Colling's White Heifer—by Butterfly
(104)—by Globe (278).

Calved after dam reached Kentucky.

DAME KNIGHTLEY—Roan heifer, calved Jan. 12, 1875, bred
by D. Fisher, got by Valentine Vox (32752), out of Duchess of
Knightley by Great Hope (24082)—Damsel by Lord Hopewell
(18239)—Datura by Fitz Clarence (14552)—Duchess by Duke of
Cambridge (12742)—Cold Cream by Earl of Dublin (10178)—
Pansy by Gray Friar (9172)—Freckle by Fawsley (6004)—Furbelow by Little John (4232)—Erato by Marcellus (2260)—Beatrice by
Caliph (1774)—Quickly by Swing (2721)—Alamode by Argus
(759)—Valuable by Defender (194)—Violet by Petrarch (488)—by
own brother to R. Colling's White Heifer—by Butterfly (104)—by
Globe (278).

COLD CREAM 8TH—Roan heifer, calved in March, 1874, bred
by Donald Fisher, got by Valentine Vox (32752), out of Cold Cream
2d (vol. 20, p. 452, E.) by Prince of Saxe Cobourg (20576)—Comely
by Buckingham (17471)—Cold Cream by Earl of Dublin (10178)—
Pansy by Gray Friar (9172)—Freckle by Fawsley (6004)—Furbelow
by Little John (4232)—Erato by Marcellus (2260)—Beatrice by
Caliph (1774)—Quickly by Swing (2721)—Alamode by Argus
(759)—Valuable by Defender (194)—Violet by Petrarch (488)—by
own brother to R. Colling's White Heifer—by Butterfly (104)—by
Globe (278).

LIONESS—Red, calved Jan. 4, 1871, bred by Mr. Blackwell,
Tausly, Derbyshire, got by Jupiter (24228), out of Lady Leoine by
Lord Cobham (20164)—Lioness by Sir Edmund Lyons (15284)—
Loyal by Triumph (8717)—Lydia by Matchless (4428)—Laura by
Broughton (2868)—Lily by Roman (2559)—by Columella (904)—
by Albion (14)—by Cinnamon (139)—by Neswick (1266).

S. H. R., vol. 5, p. 523.

LADY FARNLEY—Roan heifer, calved Sept. 21, 1874, bred by Mr. Thorn, got by Flag of Ireland (28613), out of Lioness by Jupiter (24228), &c., as in dam Lioness, above.

1818 or 1819. CORNELIUS COOLIDGE.
Boston, Mass.

CŒLEBS 349—Roan, bred by Mr. Mason, Chilton, England, got by Hercules, a son of Comet (155).

FLORA—Bred by Mr. Mason, of Chilton, England, imported in 1818, at about six months old, got by Lafon's son of Comet (155).

JOSEPH COPE.
Pennsylvania.

YORKSHIREMAN 189 (5700)—Roan, calved Aug. 28, 1838, bred by Thomas Bates, Kirklevington, got by Short Tail (2621), out of Blanche by Belvedere (1706)—Lupin by Belvedere (1706)—Tulip by Lancaster (360)—Ruby by Petrarch (488)—by Major (497) —Stranger by Chapman's Son of Punch (122)—Old Roaney by Dickson's Grandson of Punch (213)—by Checks (132)—Sockburn Sal by J. Coates' Bull (148).

1838 or 1839. MR. COPES or MR. J. BURTON.
Marshalltown, Pa.

TRUELOVE—Red and white, calved Nov. 3, 1835, bred by W. S. Gill, Grimston, near Tadcaster, got by Sir John (2628), out of Sweetheart by Forester (1055)—2d Strawberry by Newton (1271) —Strawberry by a son of Comet (155)—by a son of Major (397)— by Charge's Gray Bull (872)—by Favorite (252)—by Bartle (777). E. H. B., vol. 5, p. 1020.

DONNA MARIA—Roan, calved May 5, 1833, bred by Mr. Paley, got by Buckingham (1755), out of Lucky by Corinthian

Tom (921)—Lady by Young Dimple (971)—Lady by Young Comet (905)—Cherry by Favorite—Old Cherry by Goldfinder (1075).

E. H. B., vol. 5, p. 277.

ROBIN GREY (4968)—Roan, calved June 9, 1836, bred by Mr. Allison, got by Imperial (4068), out of Bessie by Helmsman (2109) —Betty by Columella (904)—Lady Betty by Shakespeare (1479)— Roan Lady by Blyth Comet (85)—by Neswick (1266)—by Favorite (1033).

ROBIN GREY Jr. (4969)—Roan, got in England, calved Jan. 1, 1840, bred by Mr. Gill, got by Robin Grey (4968), out of Donna Maria by Buckingham (1755), &c., as in dam, above.

1863. EZRA CORNELL,
Of Ithaca, N. Y. To New York, N. Y.

FIDGET 5TH—Roan, calved April 8, 1860, bred by Thos. Atherton, Chapel House, got by Delhi (15865), out of Fidget 4th by 4th Duke of York (10167)—Fidget 2d by Duke of Northumberland (1940)—Fidget by 2d Earl of Darlington (1945)—Fletcher by a son of Young Wynyard (2859)—descended from J. Brown's Red Bull (97).

A. H. B., vol. 6, p. 266.

KIRKLEVINGTON 11TH—Roan, calved Jan. 18, 1861, bred by C. W. Harvey, Walton-on-the-Hill, got by Delhi (15865), out of Kirklevington 7th by Earl of Derby (10177)—Kirklevington 4th by Earl of Liverpool (9061)—Kirklevington 1st by Duke of. Northumberland (1940)—Nell Gwynne by Belvedere (1706)—Northallerton by Son of 2d Hubback (2683).

A. H. B., vol. 6, p. 303.

KIRKLEVINGTON 5860—Red, calved Feb. 26, 1864, bred by C. W. Harvey, got by 4th Duke of Oxford (11387), out of Kirklevington 11th by Delhi (15865)—Kirklevington 7th by Earl of Derby (10177)—Kirklevington 4th by Earl of Liverpool (9061)— Kirklevington by Duke of Northumberland (1940)—Nell Gwynne by Belvedere (1706)—Northallerton by Son of 2d Hubback (2683).

23

DUKE OF LIVERPOOL (19627)—White, calved Nov. 15, 1862, bred by E. Cornell, at Liverpool, got by Duke of Wetherby (17753), out of Fidget 5th by Delhi (15865), &c., as in dam Fidget 5th, above.

FIDGET 6TH—Red, calved April 20, 1864, bred by Thos. Atherton, got by 4th Duke of Oxford (11387), out of Fidget 5th by Delhi (15865), &c., as in dam Fidget 5th, above.

A. H. B., vol. 7, p. 338.

ERASTUS CORNING and W. H. SOTHAM.
New York.

COLUMBUS (5869)—Calved Dec. 13, 1841, bred by Mr. Lovell, Edgecott Lodge, near Banbury, got by Baronet (6762), out of Cassandra by Anthony (1640)—by Edgcott (1953)—by Audley (3055)—by Son of Merlin (6522).

PRINCE 841—Red and white, begotten in England by Baronet (6762) (owned by Sir Charles Knightley), out of Wilddame by Anthony (1640)—Witch by Magnet Jr. (2242)—by a grandson of Merlin (2302)—by Son of Merlin (6522).

WILDDAME—Roan, calved March 11, 1837, bred by W. Lovell, got by Anthony (1640), out of Witch by Magnet Jr. (2242)—by a grandson of Merlin (2302)—by Son of Merlin (6522).

E. H. B., vol. 5, p. 1113.

MARY—Red and white, calved in 1838, bred by Mr. W. Lovell, got by Mortimer, out of Mabel by Anthony (1640)—by Edgcott (1953)—by Son of Merlin (6522).

E. H. B., vol. 5, p. 592, under dam.

MABEL—Red and white, calved in 1835, bred by Mr. W. Lovell, got by Anthony (1640), dam by Edgcott (1953)—by Son of Merlin (6522).

E. H. B., vol. 5, p. 592.

CHERRY—Calved in 1834, bred by Mr. Lovell, got by Anthony (1640), dam by Edgcott (1953)—by Audley (3055)—by Son of Merlin (6522).

E. H. B., vol. 5, p. 170.

ASHLY (3045)—Red and white, calved in 1840, bred by Mr. Lownds, got in England by Young Rubens (5026), out of Princess by Henry (4008)—Beauty by Fitz Form (2024)—White Princess by Cupid (938)—Young Princess by Lionel (1171)—Princess by Favorite (256)—Elvira by Phenomenon (491)—Princess by Favorite (252)—by Favorite (252), &c.

CLEOPATRA—Red and white, calved 1836, bred by Mr. Lovell, got by Anthony (1640), dam by Edgcott (1953)—by Audley (3055)—by Son of Merlin (6522).

E. H. B., vol. 5, p. 188.

PET—Roan, calved 1832, bred by Mr. Lovell (see 5th vol., E. H. B., p. 769), got by Anthony (1640), dam by Edgcott (1953) —by Audley (3055)—by Son of Merlin (6522).

E. H. B., vol. 5, p. 769.

VENUS—White, bred by Mr. Lovell, Edgecott Lodge, calved 1834, got by Anthony (1640), out of roan cow Vanity.

A. H. B., vol. 1, p. 233.

1854. R. G. CORWINE,
Of Lebanon, Ohio. Summer of 1854.

SCOTTISH BLUE BELLE—Roan, calved April 5, 1852, bred by James Douglas, got by Molecatcher (10537), out of Blue Belle by Capt. Shaftoe (6833)—Daisy by Tomboy (5494)—by Priam (4758)—Cora, sold at sale of Mr. Denton to Mr. Wetherell, of Yorkshire.

A. H. B., vol. 2, p. 555.

EDITH 2D—Red and white, calved March 7, 1850, bred by James Douglas, got by Fitz Adolphus Fairfax (9124), out of Edith Fairfax by Sir Thomas Fairfax (5196)—Kirton by Billy

(3151)—Jessie by Sovereign (5285)—Rose by Satellite (1420)—by Baronet (60)—by Cleveland (144)—by Symmetry (641).

A. H. B., vol. 2, p. 360.

CRUSADER (12667)—White, calved July 5, 1852, bred by James Douglas, got by Crusade (7938), out of Crocus by Daniel (5907)—Woodlass by Homer (2134)—by Stamford (8629).

This bull was the joint property of Mr. Corwine and the Shakers of Lebanon, Ohio, who jointly imported him.

H. J. COWDEN.
Ripley, N. Y.

HAROLD 2D 1638—White, calved Jan., 1850, bred by Robert Golding, Hunton, Kent, got by Harold (10300), out of Elfrida by Hengist (10315)—Modish by Little John (4232)—Urania by Marcellus (2260)—Modish by Caliph (1774)—Brenda by Norman (2379) —Eleanor by Mentor (426)—Eliza by Badsworth (47)—bred by Sir John Ramsden.

1874. J. R. CRAIG,
Of Edmonton, Can., by Ship Texas, from Liverpool, Aug. 5, 1874.

LADY LE MOOR—Roan, calved Feb. 15, 1868, bred by Mr. T. Maynard, Marton le Moor, Yorkshire, got by Young Lord Abbot (31609), out of Fawn by Grand Duke of Wetherby (17997)—Fawn by Cornborough (14327)—Fly by Prince of Wales (6348)—by Vauxhall (5550)—by Brilliant (1741)—by Red Rover (2495).

E. H. B., vol. 20, p. 603; C. H. B., vol. 3, p. 556, and A. H. B., vol. 17, p. 12972.

WATERLOO J.—Roan heifer, calved Feb. 1, 1872, bred by Sir W. C. Trevelyan, Bart., Wallington, Northumberland, got by Lord Waterloo (24475), out of Jessy by Duke of Waterloo (21616)—Julia by Ranter (18666)—Lady Jersey by 7th Duke of York (17754) —Lady Jane by Duke of Gloster (11382)—Jardine by Lord Warden (7167)—by Fawsley (6004)—by Warden (5595)—by Javelin (4093) —by Blyth (797)—by Wellington (684)—by Phenomenon (491)—

by Favorite (252)—by Favorite (252)—by Favorite (252)—by Hubback (319)—by Snowden's Bull (612)—by Waistell's Bull (669)—by Masterman's Bull (422)—by Studley Bull (626).

Under dam, E. H. B., vol. 20, p. 571, and C. H. B., vol. 3, p. 805.

EUPHEMIA—Red, calved March, 1871, bred by R. Stratton, Burderop, Swindon, Wilts, got by James 1st (24202), out of Minerva by 8th Duke of York (23808)—Europa by Windsor Castle (21118)—Lilla by Hermit (14697)—Eurydice by Lord of the Manor (14836)—Eurydice by The Red Duke (8694)—Euribia by Hero of the West (8150)—Modish by Kenilworth (7118)—Madam by Lottery (4280)—by Phœnix (6290).

C. H. B., vol. 3, p. 457.

EUPHEMIA 2D—Roan heifer, calved Nov. 9, 1874, bred by R. Stratton, Burderop, Swindon, Wilts, got by Protector (32221), out of Euphemia by James 1st (24202)—Minerva by 8th Duke of York (23808), &c., as above.

C. H. B., vol. 3, p. 457.

- - - - - - -

1875. ALBERT CRANE.

Shipped by Polynesian from Liverpool, Oct. 14, 1875, for Albert Crane, Durham Park. Kansas.

STATIRA 13TH—Red and white, calved March 5, 1870, bred by Mr. T. Comber, Myddleton Hall, Warrington, Lancashire, got by 12th Duke of Oxford (19633), out of Statira 7th by 4th Duke of Thorndale (17750)—Statira 5th by May Duke (13320)—Statira 1st by Duke of Gloster (11382)—Stately by Balco (9918)—Statice by Sir Launcelot (5166)—by Major (4345)—by Ganthorp (2049)—by Don Juan (1923)—by Shylock (2622).

Recorded, A. H. B., vol. 16, p. 12358, and E. H. B., vol. 19, p. 738, under dam.

DELIGHT—Roan heifer, calved Feb. 13, 1873, bred by J. P. Foster, Killhow, Carlisle, got by 17th Duke of Oxford (25994), out of Dorothy by 3d Duke of Wharfdale (21619)—Dora by Duke of

Geneva (19614)—Duchess 1st by Master Rembrandt (16545)—Duchess Nanny by Jasper (11609)—Duchess Nancy by 2d Duke of Oxford (9046)—Nettle by 2d Duke of Northumberland (3646)—Nell Gwynne by Belvedere (1706)—Northallerton by Son of 2d Hubback (2683)—a cow of Mr. Bates', descended from the stock of Mr. Maynard, of Eryholme.

Recorded, A. H. B., vol. 16, p. 12030, and E. H. B., vol. 21, p. 715, under dam.

STATIRA'S OXFORD 27964—Red bull, calved March 16, 1876, bred by Sir W. C. Trevelyan, Bart., Wallington, got by Oxford Beau 4th (34964), out of Statira 13th by 12th Duke of Oxford (19633)—Statira 7th by 4th Duke of Thorndale (17750)—Statira 5th by May Duke (13320)—Statira 1st by Duke of Gloster (11382) —Stately by Balco (9918)—Statice by Sir Launcelot (5166)—by Major (4345)—by Ganthorp (2049)—by Don Juan (1923)—by Shylock (2622).

This bull was calved after dam reached Durham Park, Kans.

TELLURIA WASSAIL—Red cow, calved Oct. 17, 1872, bred by W. Torr, Aylesby Manor, Grimsby, Lincolnshire, got by Duke of York (23804), out of Telluria Cup by British Crown (21322)—Telluria Royal by Booth Royal (15673)—Telluria 2d by Horatio (10335)—Taglioni by Lord George (10439)—Telluria by Orontes (4623)—by Guardian (3947)—by Mercury (2301)—by Monarch (2324)—by St. Albans (2584)—by Jupiter (342)—by Sir Oliver (605)—by Trunnell (659)—by Favorite (252)—by Favorite (252)—by Dalton Duke (188)—by R. Alcock's Bull (19)—by J. Smith's Bull (608)—by Jolly's Bull (337).

Recorded, E. H. B., vol. 20, p. 788, under dam, and A. H. B., vol. 16, p. 12366.

KIRKLEVINGTON DUCHESS 15th—Roan heifer, calved Oct. 25, 1873, bred by J. Fawcett, Scaleby Castle, Carlisle, got by 2d Duke of Gloster (28392), out of Kirklevington Duchess 5th by 2d Duke of Claro (21576)—Duchess of Kent by Lord Liverpool (22168)—Kirlevington Duchess 14th by 4th Duke of Oxford (11387)

—Kirklevington 7th by Earl of Derby (10177)—Kirklevington 4th by Earl of Liverpool (9061)—Kirklevington 1st by Duke of Northumberland (1940)—Nell Gwynne by Belvedere (1706)—Northallerton by Son of 2d Hubback (2683)—a cow of Mr. Bates', from the stock of Mr. Maynard, Eryholme.

Recorded, A. H. B., vol. 16, p. 12128, and E. H. B., vol. 21, p. 700, under dam. Exported, by ship Polynesian, from Montreal, July, 1878. Landed in Liverpool, Aug. 5, 1878.

SERAPHINA 26TH—Roan heifer, calved Feb. 16, 1874, bred by J. Fawcett, Scaleby Castle, Carlisle, got by 8th Duke of York (28480), out of Seraphina 20th by 6th Grand Duke (19876)—Seraphina 19th by Imperial Oxford (18084)—Seraphina 11th by May Duke (13320)—Seraphina 7th by Duke of Sussex (12772)—Seraphina 2d by Sweet William (7571)—Seraphina by Earl of Essex (6955)—Sapphire by Stratton (5336)—Ruby by Fanatic (1996).

E. H. B., vol. 21, p. 702, under dam, and A. H. B., vol. 16, p. 12349.

LADY MARY 6TH—White heifer, calved Feb. 28, 1874, bred by J. Fawcett, Scalesby Castle, Carlisle, got by 8th Duke of York (28480), out of Lady Mary 2d by Earl of Gloster (21644)—Mary Jane 2d by Lord Ravensworth (20222)—Mary Jane by Gen. Canrobert (12927)—Mary by Earl of Derby (10177)—a cow of Mr. Thompson's, Kirklevington, bred from Mr. Bates' bulls.

Recorded, A. H. B., vol. 16, p. 12157, and E. H. B., vol. 21, p. 701, under dam.

PANIC 27414—White (with roan ears) bull, calved May 21, 1876, bred by J. Fawcett, Scalesby Castle, Carlisle, Cumberland, got by 8th Duke of York (28480), out of Lady Mary 6th by 8th Duke of York (28480)—Lady Mary 2d by Earl of Gloster (21644)—Mary Jane 2d by Lord Ravensworth (20222)—Mary Jane by Gen. Canrobert (12927)—Mary by Earl of Derby (10177)—a cow of Mr. Thompson's, bred from Mr. Bates' bulls.

This bull was begotten in England and calved at Mr. Crane's place, Durham Park, Kans.

KNIGHT OF THE CRESCENT 26996—Roan, calved Nov. 10, 1875, bred by William Torr, Aylesby Manor, Great Grimsby, Lincolnshire, got by Knight of the Shire (26552), out of Telluria Wassail by Duke of York (23804)—Telluria Cup by British Crown (21322)—Telluria Royal by Booth Royal (15673)—Telluria 2d by Horatio (10335)—Taglioni by Lord George (10439)—Telluria by Orontes (4623)—Zinc by Guardian (3947)—Roguery by Mercury (2301)—Pageant by Monarch (2324)—No. 13 Chilton Sale by St. Albans (2584)—No. 4 Chilton Sale by Jupiter (342)—Sir Oliver Cow by Sir Oliver (605)—Raspberry by Trunnell (659)—Strawberry by Favorite (252)—Lily by Favorite (252)—Miss Lax by Dalton Duke (188)—Lady Maynard by R. Alcock's Bull (19)—by Jacob Smith's Bull (608)—by Jolly's Bull (337).

This bull was begotten in England and calved after dam's reaching America.

1853. W. B. CREW.
Toronto, Can.

EMILY 2D—Red roan, calved Dec. 16, 1850, got by Roan Duke (8486), out of Emily by The Chieftain (2893)—Enemone by Kiveton Reformer (4164)—Elegance by Topper (2768)—Ellen by Ackland (713)—by Blaize (75)—by Alfred (23)—by Butterfly (104)—by Suwarrow (636).

C. H. B., vol. 1, p. 284.

Oct., 1872. C. S. C. CRISPIGNY & SEVIER.
To Missouri or Kansas.

LISTLESS—Roan, calved Oct. 24, 1870, bred by Thos. Morris, Mausamore Court, got by Charleston (21400), out of Listless 4th by Kent Oxford (20047)—Listless by Samson (12045)—Linette by Belus (8879)—Lily by Vincent (5567)—Laura by White's Bull—bred by Mr. Strickland.

A. H. B., vol. 14, p. 662.

LORD LYTTON 17575—Red, calved April 27, 1873, bred by Thos. Morris, Mausamore Court, got by Earl Lally (28492), out of Listless by Charleston (21400), &c., see Listless, above.

LIKELY—Red, calved Sept. 20, 1870, bred by Thos. Morris, Mausamore Court, got by Charleston (21400), out of Likely 3d by Seaweed (22856)—Likely by MacDonald (13268)—Little Paulina by Shamrock (10803)—Locket by Stewart (17042)—Lady by Vincent (5567)—Laura by White's Bull—bred by Mr. Strickland.

A. H. B., vol. 14, p. 654.

LOVELY—Roan, calved March 1, 1873, got by Earl Lally (28492)—out of Likely by Charleston (21400), &c., as above.

A. H. B., vol. 13, p. 743.

DUKE OF HAZELCOTE 19TH 26253 (30967)—Roan, calved June 19, 1871, bred by Col. Kingscote, Wotton-under-Edge, got by 3d Duke of Clarence (23727), out of Honeymoon by Caleb (15718)—Honeydew by Viceroy (13945)—Helen by Oregon (8371)—Honeysuckle by Premier (7344)—by Bellerophon (3119)—by Alderman (2976)—by Waterloo (2816)—by Young Wynyard (2859)—by Son of Simon (590)—by Styford (103).

June, 1871. J. I. DAVIDSON,

Of Balsam, Ontario, Can. From Glasgow, by Ship Glenoffer, to Montreal, Canada.

OAK WREATH—Red heifer, calved Feb. 20, 1869, bred by A. Cruickshank, Sittyton, Aberdeen, Scotland, got by Allan (21172), out of Oak Apple by Lancaster Comet (11663)—Oak Leaf by The Baron (13833)—Aroma by Matadore (11800)—Admah by Fitz Adolphus Fairfax (9124)—Brokenhorn by Fitz Leonard (7010)—Princess by Sovereign (7539)—Queen, bred by Mr. Robertson, of Ladykirk.

C. H. B., vol. 2, p. 692; E. H. B., vol. 19, p. 652, under dam, and A. H. B., vol. 14, p. 762.

24

ROSE BLOSSOM—Red heifer, calved April 20, 1870, bred by A. Cruickshank, Sittyton, Aberdeen, Scotland, got by Senator (27441), out of Ring Dove by Lord Raglan (13244)—Roseberry by Somerset (10858)—Ruby by Guy Fawkes (7062)—Little Red Rose by Pedestrian (7321)—by Lucien (2228)—by Raby (2473).

C. H. B., vol. 2, p. 760, and A. H. B., vol. 12, p. 1203.

MATCHLESS 15th—Red, calved Feb. 20, 1870, bred by A. Cruickshank, got by Champion of England (17526), out of Matchless 12th by Lord Raglan (13244)—Matchless 10th by Lord Stanley (16454)—May Day by Master Butterfly 2d (14918)—Mayflower by The Baron (13833)—May Rose by Van Dunck (10992)—Matchless 2d by Fairfax Royal (6987)—Matchless by Holkar (4041)—Premium by George (2057)—by Togston (5487)—bred by Mr. Laing.

C. H. B., vol. 2, p. 642, and A. H. B., vol. 15, p. 746.

MATCHLESS 16th—Roan, calved April 20, 1870, bred by A. Cruickshank, got by Senator (27441), out of Matchless 10th by Lord Stanley (16454), &c., as in Matchless 15th, above.

C. H. B., vol. 2, p. 643, and A. H. B., vol. 12, p. 1019.

WATER WITCH—Red, calved Feb. 16, 1870, bred by A. Cruickshank, Sittyton, got by Scotland's Pride (25100), out of Water Nymph by Viceroy (19054)—Mermaid by Benedict Balco (14159)—Virago by Inheritor (13065)—Bloomer by Seafield (9616)—Rosamond by Sultan (5349)—Rose by Plenipo (4725)—Thorn by Abbott (2899).

C. H. B., vol. 2, p. 822, and A. H. B., vol. 14, p. 903, and vol. 12, p. 1286.

WATER MAID—Roan, calved Feb. 20, 1872, bred by A. Cruickshank, got by Dazzler (30862), out of Water Witch by Scotland's Pride (25100), &c., as in Water Witch, above.

C. H. B., vol. 3, p. 806, and vol. 14, p. 903.

Aug., 1873. J. I. DAVIDSON,

Of Balsam, Canada. August, 1873, from Glasgow to Montreal.

MYSIE 37TH—Roan, calved March 27, 1872, bred by A. Cruickshank, got by Senator (27441), out of Mysie 29th by Lord Raglan (13244)—Mysie 3d by Grand Duke (10284), &c., as in Mysie 35th.

C. H. B., vol. 3, p. 663.

OMELETTE ALIAS ORANGE BLOSSOM 18TH—Red and white, calved April 30, 1872, bred by A. Cruickshank, got by Breadalbane (28073), out of Orange Blossom 8th by Sir Walter Scott (22923)—Orange Blossom by Dr. Buckingham (14405)—Queen of Scotland by Matadore (11800)—Edith Fairfax by Sir Thomas Fairfax (5196)—Fancy by Billy (3151)—Jessie by Sovereign (7539)—Rose by Satellite (1420)—by Baronet (60)—by Cleveland (144)—by Symmetry.

C. H. B., vol. 3, p. 679, and A. H. B., vol. 21, p. 16631.

OMELETTE 2D ALIAS ORANGE BLOSSOM 20TH—Red and white, calved Jan. 14, 1874, bred by A. Cruickshank, got by Lord Lansdowne (29128), out of Omelette alias Orange Blossom 18th by Breadalbane (28073), &c., as in Orange Blossom 18th.

C. H. B., vol. 3, p. 680, and A. H. B., vol. 21, p. 16631.

CROWN PRINCE 22502 [2929]—Roan, calved Jan. 27, 1874, bred by A. Cruickshank, Sittyton, got by Lord Lansdowne (29128), out of Mysie 37th by Senator (27441)—Mysie 29th by Lord Raglan (13244)—Mysie 3d by Grand Duke (10284)—Mysie by Kelly 2d (9265)—Molly by the Pacha (7612)—Margaret by Mahomed (6170)—Mary Anne by Sillery (5131)—Miss Gibson by Carleton (843)—Dora by Diamond (205)—Kitty by Diamond (205).

1874. JAMES I. DAVIDSON.

BUTTERFLY 36TH—Roan, calved April 2, 1871, bred by A. Cruickshank, got by Cæsar Augustus (25704), out of Butterfly 10th by Grand Monarque (21867)—Butterfly 3d by The Baron (13833)—Butterfly by Matadore (11800)—Buttercup by Report (10704)—Bounty by The Pacha (7612)—Strawberry by 2d Duke of Northumberland (3646)—Margaret by Mahomed (6170)—Mary Anne by

Sillery (5131)—Miss Gibson by Carleton (843)—Dora by Diamond (205)—Kitty by Diamond (205).

A. H. B., vol. 15, p. 466, and C. H. B., vol. 3, p. 385.

BUTTERFLY 43D—Red, calved March 2, 1873, bred by A. Cruickshank, got by Royal Duke of Gloster (29864), out of Butterfly 36th by Cæsar Augustus (25704), &c., as above.

A. H. B., vol. 14, p. 452, and C. H. B., vol. 3, p. 386.

BUTTERFLY 44TH—White, calved March 14, 1875, bred by A. Cruickshank, got by Viceroy (32764), out of Butterfly 36th by Cæsar Augustus (25704), &c., as above.

A. H. B., vol. 15, p. 466, and C. H. B., vol. 3, p. 386.

CHARMING—Roan, calved March 3, 1872, bred by A. Cruickshank, got by Breadalbane (28073), out of Charmer by Champion of England (17526)—Ceremony by The Baron (13833)—Clipper by Billy (3151)—Favorite by Danby (6918)—Keepsake by Tip-Top (7633)—Old Lady, bred by Mr. Mason, of Chilton.

A. H. B., vol. 14, p. 462, and C. H. B., vol. 3, p. 392.

CORAL—Roan, calved April 1, 1873, bred by A. Cruickshank, got by Lord Lancaster (26666), out of Cornelia by Cæsar Augustus (25704)—Cicely by Lancaster Royal (18167)—Crocus by Jenny (11611)—Kitty by Somerset (10858)—Kate by Hawthorn (7071)—Kilmeny by The Peer (5455)—by George (2057)—by Togston (5487)—bred by Mr. Laing, of Longhoughton.

A. H. B., vol. 15, p. 489, and C. H. B., vol. 2, p. 401.

GOLDEN CROWN [4343]—Red, calved March 15, 1875, bred by A. Cruickshank, Sittyton, got by Red Gauntlet (32256), out of Golden Galaxy by Senator (27441)—Gold of Sheba by Caractacus (19397)—Golden Pippin by Champion of England (17526)—Gold Mint by The Baron (13833)—Pure Gold by 4th Duke (9037)—Star Pagoda by Duplicate Duke (6952)—The Mint by Robin O'Day (4973)—Brawith Bud by Sir Walter (2639)—by Young Jerry (8177)—by Roseberry (567)—by Roseberry (567)—by Constellation (163)—by Hastings (293)—by Leopold (372).

ACORN 2D—Red, calved April 10, 1873, bred by A. Cruickshank, got by Scotland's Pride (25100), out of Acorn by The Czar (20947)—Oak Leaf by The Baron (13833)—Aroma by Matadore (11800)—Admah by Fitz Adolphus Fairfax (9124)—Brokenhorn by Fitz Leonard (7010)—Princess by Sovereign (7539)—Queen, bred by Mr. Robertson, of Ladykirk.

A. H. B., vol. 15. p. 401, and C. H. B., vol. 3, p. 341.

AUTUMN LADY—Roan, calved Nov. 2, 1872, bred by A. Cruickshank, got by Senator (27441), out of Autumn Rose by Lord Raglan (13244)—Rose Leaf by Matadore (11800)—China Rose by Hudson (9228)—Carmine Rose by Fairfax Royal (6987)—Red Rose by Inkhorn (6091)—Moss Rose by Grazier (1085)—Cicely by Sampson—Marion by Wallace (1560).

A. H. B., vol. 19, p. 14379, and C. H. B., vol. 3, p. 355.

AUTUMN QUEEN—Red, calved Feb. 17, 1875, bred by A. Cruickshank, got by Ben Wyvis (30528), out of Autumn Lady, above.

A. H. B., vol. 25, p. 761, and C. H. B., vol. 3, p. 356.

FRAMEWORK (33964) [5254]—Roan, calved March 7, 1873, bred by A. Cruickshank, got by Lord Landsdowne (29128), out of Vellum by Grand Monarque (21867)—Violette by Lorenzo (20235) —by Dannecker (7949)—by The Chief (5425)—Eliza by Billy (3151) —Princess by Sovereign (7539)—Queen, bred by Mr. Rennie, of Phantassie.

CROWN PRINCE 2D [4301]—Roan, calved Feb. 13, 1875, bred by A. Cruickshank, got by Viceroy (32764), out of Mysie 36th by Scotland's Pride (25100)—Mysie 33d by Champion of England (17526)—Mysie 26th by Speculator (13775)—Mysie 6th by Red Knight (11976)—Mysie 2d by The Hero (10934)—Mysie by Kelly 2d (9265)—Molly by The Pacha (7612)—Margaret by Mahomed (6170)—Mary Anne by Sillery (5131)—Miss Gibson by Carleton (843)—Dora by Diamond (205)—Kitty by Diamond (205).

ACORN OF LINWOOD—Red, calved April 13, 1875, bred by A. Cruickshank, got by Framework (33964), out of Acorn 2d by Scotland's Pride (25100), &c., as in Acorn 2d.

A. H. B., vol. 16, p. 11945.

ORANGE BLOSSOM 19TH—Red, calved March 30, 1874, bred by A. Cruickshank, got by Royal Duke of Gloster (29864), out of Orange Blossom 8th by Sir Walter Scott (22922), &c., as in Orange Blossom 18th.

C. H. B., vol. 3, p. 679, and A. H. B., vol. 15, p. 806.

MISSIE 26TH—Red and white, calved April 16, 1868, bred by W. S. Marr, Aberdeen, Scotland, got by Prince Louis (27158), out of Missie 12th by Master Gunner (22316)—Missie 4th by Clarendon (14280)—Missie by Son of Duke 3d (17697)—Countess by The Pacha (7612)—Jessamine by Mahomed (6170)—Rose by Plenipo (4725)—Thorn by Abbot (2899).

Recorded, C. H. B., vol. 3, p. 648.

VILLAGE GIRL—Red, calved Jan. 31, 1872, bred by A. Cruickshank, got by Scotland's Pride (25100), out of Village Bride by Allan (21172)—Village Maid by Baronet (15614)—Village Belle by Champion of England (17526)—Vintage by Lord Bathurst (13173) —Lady Franklin by Matadore (11800)—The Vine by Fairfax Royal (6987)—Picotee by Premier (6308)—Sunflower by Unicorn (8725) —by Young Satellite (8538)—by Valentine (661)—bred by Mr. Rennie, of Phantassie—bred by Mr. Robertson, of Ladykirk.

A. H. B., vol. 14, p. 898, and C. H. B., vol. 3, p. 799.

VILLAGE LASS—Red, calved Sept. 17, 1874, bred by A. Cruickshank, got by Ben Wyvis (30258), out of Village Girl by Scotland's Pride (25100), &c., as in Village Girl, above.

A. H. B., vol. 14, p. 898, and C. H. B., vol. 3, p. 799.

VILLAGE VINE—Roan, calved March 6, 1873, bred by A. Cruickshank, got by Laudable (31587), out of Village Bride by Allan (21172), &c., as in Village Girl, above.

A. H. B., vol. 14, p. 898, and C. H. B., vol. 3, p. 800.

BARONET 54629 [4613]—Roan, calved March 15, 1875, bred by A. Cruickshank, got by Roan Gauntlet (35284), out of Butterfly 43d by Royal Duke of Gloster (29864), &c., as in Butterfly 43d, above.

RED LADY OF LINWOOD—Red, calved March 20, 1875, bred by A. Cruickshank, got by Framework (33964), out of Red Lady by Scotland's Pride (25100), &c., as in Red Lady, below.

A. H. B., vol. 16, p. 12306.

ORANGE BOY 30468—Red and white, calved May 1, 1875, bred by A. Cruickshank, got by Royal Duke of Gloster (29864), out of Rosemary by Breadalbane (28073), &c., as in dam.

MYSIE 35TH—Roan, calved April 16, 1871, bred by A. Cruickshank, got by Senator (27441), out of Mysie 27th by Harlequin (19922)—Mysie 3d by Grand Duke (10284)—Mysie by Kelly 2d (9265)—Molly by The Pacha (7612)—Margaret by Mahomed (6170)—Mary Anne by Sillery (5131)—Miss Gibson by Carleton (843)—Dora by Diamond (205)—Kitty by Diamond (205).

A. H. B., vol. 14, p. 747, and C. H. B., vol. 3, p. 663.

MYSIE 36TH—Red, calved Oct. 12, 1871, bred by A. Cruickshank, got by Scotland's Pride (25100), out of Mysie 33d by Champion of England (17526)—Mysie 26th by Speculator (13775)—Mysie 6th by Red Knight (11976)—Mysie 2d by The Hero (10934)—Mysie by Kelly 2d (9265)—Molly by The Pacha (7612), &c., as above.

A. H. B., vol. 15, p. 790, and C. H. B., vol. 3, p. 663.

RED LADY—Red, calved Feb. 27, 1873, bred by A. Cruickshank, got by Scotland's Pride (25100), out of Red Rose by Champion of England (17526)—Blush Rose by Lord Raglan (13244)—Province Rose by The Baron (13833)—Sharon's Rose by Plantagenet (11906)—Fancy by Billy (3151)—Jessie by Sovereign (7539)—Rose by Satellite (1420)—by Baronet (60)—by Cleveland (144)—by Symmetry (641).

A. H. B., vol. 15, p. 855, and C. H. B., vol. 3, p. 718.

ROSEMARY—Red and white. calved Feb. 17, 1872, bred by A. Cruickshank. got by Breadalbane (28073), out of Rose of May by Lord Raglan (13244)—Sharon's Rose by Plantagenet (11906). &c.. as in Red Lady. above.

C. H. B.. vol. 3, p. 750.

SOLEMNITY—Red, calved April 28. 1873, bred by A. Cruickshank, got by Lord Warden (31766), out of Sweet Briar by Ivanhoe (14735)—Splendor by Lord Sackville (13249)—Sympathy by Duke of Athol (10150)—Silence by Earl of Derby (10177)—Secret 3d by Duke of Sutherland (6945)—Secret 2d by Locomotive (4242)—Secret by Short Tail (2621)—White Rose by Gambier (2046)—White Rose by Young Wynyard (2859)—by bulls of Messrs. C. & R. Colling.

A. H. B., vol. 15, p. 907, and C. H. B., vol. 3, p. 778.

FLORA 6TH—Red, calved March 28, 1870, bred by A. Cruickshank, got by Golden Eagle (26267). out of Flora 2d by Baronet 2d (17363)—Flora by Marquis of Bute (18336)—by Count Fairfax (8991)—by South Durham (9676)—by Uptaker (5534)—by Miracle (2320)—by Matchem (2281)—by Fitz Remus (2025)—by Cato (119) —by Whitworth (695)—bred by Mr. Mason, of Chilton.

A. H. B., vol. 14, p. 540, and C. H. B., vol. 3, p. 479.

GOLDEN GALAXY—Roan, calved July, 1871, bred by A. Cruickshank, got by Senator (27441), out of Gold of Sheba by Caractacus (19397)—Golden Pippin by Champion of England (17526)—Gold Mint by The Baron (13833)—Pure Gold by Young 4th Duke (9037), &c., as in Golden Crown [4343], above.

A. H. B., vol. 14, p. 555, and C. H. B., vol. 3, p. 496.

GOVERNESS—Roan, calved March 21, 1872, bred by A. Cruickshank, got by Master of Arts (26867), out of Graceful by Baronet (15614)—Grandiflora by Lord Sackville (13249)—Flora by Fairfax Royal (6987)—Jessica by Premier (6308)—Venus by Saturn (5089) —Dairymaid by Favorite (6997)—by Grindon (3942)—bred by Mr. Rennie, of Phantassie.

A. H. B., vol. 15, p. 583, and C. H. B., vol. 3. p. 497.

GAIETY—Red, calved March 20, 1875, bred by A. Cruickshank, got by Viceroy (32764), out of Governess by Master of Arts (26867), &c., as in Governess, above.

C. H. B., vol. 3, p. 489.

ORANGE BLOSSOM OF LINWOOD—Red, calved April 25, 1875, bred by A. Cruickshank, got by Barmpton Royal (32996), out of Orange Blossom 19th by Royal Duke of Gloster (29864), &c., as in dam, above.

A. H. B., vol. 17, p. 13091.

COUNTESS—Roan, calved March 5, 1875, bred by A. Cruickshank, got by Ben Wyvis (30528), out of Coral by Lord Lancaster (26666), &c., as in Coral, above.

C. H. B., vol. 3, p. 403.

ETHEL BUCKINGHAM—Red and white, calved March 13, 1873, bred by A. Cruickshank, got by Lord Warden (31766), out of Edith Buckingham by Vice-President (23126)—Julia Buckingham by Lord Stanley (16454)—Lady Buckingham by Dr. Buckingham (14405)—Verdant by Exchequer (9721)—Prigg by Young Holkar (7090)—Tranquil by Billy (3151)—Prudence by Son of Northern Light (4586)—by a bull of Mr. Mason's—by Blucher (3175)—by Lawnsleeves (365).

A. H. B., vol. 15, p. 538, and C. H. B., vol. 3, p. 456.

EVA BUCKINGHAM—Red, calved March 14, 1875, bred by A. Cruickshank, got by Ben Wyvis (30528), out of Ethel Buckingham by Lord Warden (31766), &c., as above.

A. H. B., vol. 15, p. 540, and C. H. B., vol. 3, p. 458.

EVENING STAR—Red, calved March 1, 1873, bred by A. Cruickshank, got by Octavius (31997), out of Day Star by Scotland's Pride (25100)—Morning Star by Champion of England (17526)—Grandiflora by Lord Sackville (13249)—Flora by Fairfax Royal (6987)—Jessica by Premier (6308)—Venus by Saturn (5089)—Dairymaid by Favorite (6997)—by Grindon (3942)—bred by Mr. Rennie, of Phantassie—bred by Mr. Robertson, of Ladykirk.

Vol. 14, p. 523, A. H. B.

25

May 28, 1881. J. I. DAVIDSON.
By Ship Quebec, to Quebec.

ARTLESS—Red heifer, calved Sept, 13, 1879, bred by A. Cruickshank, Sittyton, Aberdeen, Scotland, got by Lord of the Isles (40218), out of Abarilla by Barmpton Prince (32995)—Arabella by Earl of Windsor (15968)—Lady by Bushranger (12516)—Mary by Mosstrooper (11827)—Geraldine by Robin O'Day (4973)—Angelica by Holkar (4041)—by Jopp's Bull (9256)—Kate of Darlington.

Vol. 27, p. 361, E. H. B., under dam, and vol. 24, p. 18422, and vol. 25, p. 674, A. H. B.

CISTUS—Red heifer, calved Sept. 26, 1879, bred by A. Cruickshank, got by Lord of the Isles (40218), out of Castus by Champion of England (17526)—Cicely by Lancaster Royal (18167)—Crocus by Jemmy (11611)—Kitty by Somerset (10858)—Kate by Hawthorn (7071)—Kilmeny by The Peer (5455)—by George (2057)—by Togston (5487)—bred by Mr. Laing, of Longhoughton.

JESSAMINE—Red and white heifer, calved Oct. 16, 1879, bred by A. Cruickshank, got by Lord of the Isles (40218), out of Juliet by Barmpton Prince (32995)—Joyful by Master of Arts (26867)—Jealousy by Champion of England (17526)—Josephine by The Baron (13833)—Flora by Fairfax Royal (6987)—Jessica by Premier (6308)—Venus by Saturn (5089)—Dairymaid by Favorite (6997)—by Grindon (3942)—bred by Mr. Rennie, of Phantassie.

Vol. 26, p. 392, E. H. B., under dam, and vol. 22, p. 17197, and vol. 24, p. 18570, A. H. B.

GLOXINIA—Red and white heifer, calved Nov. 13, 1879, bred by A. Cruickshank, got by Lord of the Isles (40218), out of Geranium by Pride of the Isles (35072)—Garland by Scotland's Pride (25100)—Graceful by Baronet (15614)—Grandiflora by Lord Sackville (13249)—Flora by Fairfax Royal (6987), &c., as in Jessamine, above.

Vol. 26, p. 392, E. H. B., under dam, and vol. 25, p. 1130, A. H. B.

GARDENIA—Red and white, calved Feb. 1, 1882, bred by A. Cruickshank, got by Antiquary (44314), out of Gloxinia by Lord of the Isles (40218), &c., as in dam, above.

Vol. 25, p. 836, A. H. B.

CINERARIA—Red heifer, calved Nov. 25, 1879, bred by A. Cruickshank, got by Viceroy (32764), out of Circassia by Champion of England (17526)—Cicely by Lancaster Royal (18167)—Crocus by Jemmy (11611)—Kitty by Somerset (10858)—by Hawthorn (7071)—by The Peer (5455)—by George (2057)—by Togston (5487)—bred by Mr. Laing.

Vol. 26, p. 391, E. H. B. Twinned with Convolvulus.

VICTORIA 66TH—Roan, calved Dec. 28, 1879, bred by A. Cruickshank, got by Lord of the Isles (40218), out of Victoria 55th by Lord Lancaster (26666)—Victoria 48th by Lord Lancaster (26666)—Victoria 39th by Champion of England (17526)—Victoria 29th by Red Knight (11976)—Victoria 19th by Lord John (11731)—Victoria 4th by Prince Albert (11933)—by Belzoni (783)—by Satellite (1420)—by Cato (119)—by Pope (514)—by Favorite (252)—by White Bull (421)—by Favorite (252)—by Dalton Duke (188)—by R. Alcock's Bull (19)—by J. Smith's Bull (608)—by Jolly's Bull (337).

Vol. 22, p. 17475.

GIRDLE—Red heifer, calved Feb. 22, 1880, bred by A. Cruickshank, got by Pride of the Isles (35072), out of Garnet by Prince Alfred (27107)—Guineas by Prince Imperial (22595)—Golden Chain by Lord Raglan (13244)—Gold Mint by The Baron (13833)—Pure Gold by Young 4th Duke (9037)—Star Pagoda by Duplicate Duke (6952)—The Mint by Robin O'Day (4973)—Brawith Bud (vol. 6, p. 276, E.) by Sir Walter (2639)—by Young Jerry (8177)—by Roseberry (567)—by Roseberry (567)—by Constellation (163)—by Hastings (293)—by Leopold (372).

Vol. 22, p. 17169.

SAXIFRAGE—Red heifer, calved March 6, 1880, bred by A. Cruickshank, got by Roan Gauntlet (35284), out of Selina by Grand

Duke of Gloster (26288)—Science by Baronet (15614)—Splendid by Lord Sackville (13249)—Sympathy by Duke of Athol (10150)—Silence by Earl of Derby (10177)—Secret 3d by Duke of Sutherland (6945)—by Locomotive (4242)—by Short Tail (2621)—by Gambier (2046)—by Young Wynyard (2859)—by bulls bred by Messrs. C. & R. Colling.

Vol. 22, p. 17445.

SEMPTRESS—Roan heifer, calved April 3, 1879, bred by A. Cruickshank, got by Barmpton (37763), out of Sycamore by Count Robert (30812)—Surmise by Champion of England (17526)—Superb by The Czar (20947)—Splendor by Lord Sackville (13249)—Sympathy by Duke of Athol (10150)—Silence by Earl of Derby (10177)—Secret 3d by Duke of Sutherland (6945)—by Locomotive (4242)—by Short Tail (2621)—by Gambier (2046)—by Young Wynyard (2859)—by bulls bred by C. & R. Colling.

Vol. 22, p. 17446, and vol. 25, p. 1131, A. H. B.

VICTORIA 63d—Roan, calved May 8, 1878, bred by A. Cruickshank, got by Pride of the Isles (35072), out of Victoria 45th by Cæsar Augustus (25704)—Victoria 42d by Forth (17866)—Victoria 32d by Prince Regent (16762)—Victoria 29th by Red Knight (11976)—Victoria 19th by Lord John (11731)—Victoria 4th by Prince Albert (11933)—Victoria 2d by Belzoni (783)—by Satellite (1420)—by Cato (119)—by Pope (514)—by Favorite (252)—by The White Bull (421)—by Favorite (252)—by Dalton Duke (188)—by R. Alcock's Bull (19)—by J. Smith's Bull (608)—by Jolly's Bull (337).

Vol. 26, p. 392, E. H. B., under dam, and vol. 25, p. 838, A. H. B.

EARL OF ABERDEEN 45992—Red, calved Feb. 25, 1881, bred by A. Cruickshank, got by Barmpton (37763), out of Silvery by Champion of England (17526)—Spicey 4th by Prince Louis (20560)—Spicey by Marmaduke (14897)—Saucebox by The Beau (12182)—by Lottery (10472)—by Fanatic (8054)—by Splendid

(5298)—by Belshazzar (1703)—by Belshazzar (1704)—by Snowball (2648)—by Prince (522)—by Favorite (256)—by a Bull of Mr. Colling.

DUKE OF LAVENDER (2006 B. A. H. B.)—Red, calved Sept. 9, 1880, bred by A. Cruickshank, got by Perfection (37185), out of Lavender 12th by Count Bickerstaffe 2d (25838)—Lavender 8th by Brian Boru (17440)—Lavender 7th by Friar Tuck (14578) —Lavender 3d by Eclipse (10186)—Lavender 2d by The Queen's Roan (7389)—by Will Honeycomb (5660)—by Spectator (2688)— by Albion (1619)—by Lancaster (360)—by Son of Windsor (698) —by Comet (155).

BARON VICTOR 42824—Red, calved Nov. 9, 1880, bred by A. Cruickshank, got by Barmpton (37763), out of Victoria 58th by Pride of the Isles (35072)—Victoria 43d by Champion of England (17526)—Victoria 36th by Baronet (15614)—Victoria 31st by Master Butterfly 2d (14918)—Victoria 29th by Red Knight (11976)— Victoria 19th by Lord John (11731)—Victoria 4th by Prince Albert (11933)—Victoria 2d by Belzoni (783)—by Satellite (1420)—by Cato (119)—by Pope (514)—by Favorite (252)—by White Bull (421)—by Favorite (252)—by Dalton Duke (188)—by R. Alcock's Bull (19)—by J. Smith's Bull (608)—by Jolly's Bull (337).

August I, 1882. JAMES I. DAVIDSON,

Of Balsam, Ont., Canada. By Steamship Buenos Ayrean to Quebec.

ORANGE BLOSSOM 31st—Red and white, calved Sept. 18, 1880, bred by A. Cruickshank, Sittyton, Aberdeenshire, got by Roan Gauntlet (35284), out of Orange Blossom 8th by Sir Walter Scott (22922)—Orange Blossom by Dr. Buckingham (14405)—Queen of Scotland by Matadore (11800)—Edith Fairfax by Sir Thomas Fairfax (5196)—Fancy by Billy (3151)—Jessie by Sovereign (7539) —Rose by Satellite (1420)—by Baronet (60)—by Cleveland (144)— by Symmetry (641).

E. H. B., vol. 28, p. 353, under dam, and A. H. B., vol. 25, p. 1095.

AURICULAR—Roan heifer, calved Sept. 24, 1880, bred by
A. Cruickshank, got by Roan Gauntlet (35284), out of Amaryllis
by Lord Lancaster (26666)—Azalea by Cæsar Augustus (25704)—
Anemone by Forth (17866)—Avalanche by Sir Samuel (15302)—
Angerona by Lemnos (13146)—Amy by Earl Stanhope (5966)—
Augusta by Trueblue (5522)—Albinia by Miracle (2321)—Alice by
Sir Henry (1446)—Young Madam by Count (170)—Young Venus
by Bracken (91)—Venus by Badsworth (47)—by Driffield (223)—
bred by Sir Geo. Strickland.

E. H. B., vol. 28, p. 353, under dam.

ACONITE—Red heifer, calved Nov. 9, 1880, bred by A. Cruick-
shank, got by Pride of the Isles (35072), out of Abarilla by
Barmpton Prince (32995)—Arabella by Earl of Windsor (15968)—
Lady by Bushranger (12516)—Mary by Mosstrooper (11827)—Ger-
aldine by Robin O'Day (4973)—Angelica by Holkar (4041)—
Glendronach by Jopp's Bull (9256)—Kate of Darlington.

E. H. B., vol. 27, p. 361, under dam, and A. H. B., vol. 25,
p. 1094.

MARSH VIOLET—Roan heifer, calved Nov. 19, 1880, bred by
A. Cruickshank, got by Pride of the Isles (35072), out of Rose of
Knowlmore by Knight of Knowlmore (22055)—Red Violet by
Allan (21172)—Violet by Lord Bathurst (13173)—Roseate by Mat-
adore (11800)—China Rose by Hudson (9228)—Carmine Rose by
Fairfax Royal (6987)—Red Rose by Inkhorn (6091)—Moss Rose
by Grazier (1085)—Cicely by Sampson—Marion by Wallace (1560).

E. H. B., vol. 28, p. 354, under dam, and A. H. B., vol. 25,
p. 838.

LAVENDER 31st—Roan heifer, calved January 7, 1881, bred
by A. Cruickshank, got by Barmpton (37763), out of Lavender
20th by Pride of the Isles (35072)—Lavender 12th by Count Bickerstaffe 2d (25838)—Lavender 8th by Brian Boru (17440)—Lavender
7th by Friar Tuck (14578)—Lavender 3d by Eclipse (10186)—
Lavender 2d by Queen's Roan (7389)—Lavender by Will Honey-

comb (5660)—by Spectator (2688)—by Albion (1619)—by Lancaster (360)—by Son of Windsor (698)—by Comet (155).

A. H. B., vol. 25, p. 1131.

LAVENDER 32D—Red heifer, calved February 9, 1881, bred by A. Cruickshank, got by Roan Gauntlet (35284), out of Lavender 15th by Lord Warden (31766)—Lavender 12th by Count Bickerstaffe 2d (25838)—Lavender 8th by Brian Boru (17440)—Lavender 7th by Friar Tuck (14578)—Lavender 3d by Eclipse (10186)—Lavender 2d by Queen's Roan (7389)—Lavender by Will Honeycomb (5660)—by Spectator (2688)—by Albion (1619)—by Lancaster (360)—by Son of Windsor (698)—by Comet (155).

A. H. B., vol. 25, p. 837.

LAVENDER 34TH—Red roan heifer, calved March 3, 1882, bred by A. Cruickshank, got by Roan Gauntlet (35284), out of Lavender 15th by Lord Warden (31766)—Lavender 12th by Count Bickerstaffe 2d (25838), &c., as in full sister, Lavender 32d.

GLADIOLUS—Red, calved Jan. 10, 1881, bred by A. Cruickshank, got by Pride of the Isles (35072), out of Golden Year by Viceroy (32764)—Golden Days by Lord Raglan (13244)—Pure Gold by Young 4th Duke (9037)—The Star Pagoda by Duplicate Duke (6952)—The Mint by Robin O'Day (4973)—Brawith Bud (vol. 6, p. 276, E.) by Sir Walter (2639)—by Young Jerry (8177)—by Roseberry (567)—by Roseberry (567)—by Constellation (163)—by Hastings (293)—by Leopold (327).

Vol. 25, p. 836, A. H. B.

LOVELY 36TH—Red and white, calved March 7, 1881, bred by A. Cruickshank, got by Perfection (37185), out of Lovely 12th by Scotch Rose (25099)—Lovely 9th by Windsor Augustus (19157)—Lovely 8th by Bosquet (14183)—Lovely by Kelly 2d (9265)—Lady Ythan by Robin O'Day (4973)—Lady by Favorite (9116)—Marion by Anthony (1640)—Miranda by Anthony (1640)—Merino by Edgcott (1953)—Matilda by Son of Merlin (6522)—White Cow by Acton (1607).

British-American H. B., vol. 2, p. 431.

VICTORIA 71st—Roan, calved March 12, 1881, bred by A. Cruickshank, got by Roan Gauntlet (35284), out of Victoria 63d by Pride of the Isles (35072)—Victoria 45th by Cæsar Augustus (25704)—Victoria 42d by Forth (17866)—Victoria 32d by Prince Regent (16762)—Victoria 29th by Red Knight (11976)—Victoria 19th by Lord John (11731)—Victoria 4th by Prince Albert (11933)—Victoria 2d by Belzoni (783)—Victoria by Satellite (1420)—No. 1 Mason's Sale by Cato (119)—Pope Cow by Pope (514)—Flora by Favorite (252)—Nymph by White Bull (421)—Lily by Favorite (252)—Miss Lax by Dalton Duke (188)—Lady Maynard by Alcock's Bull (19)—by Jacob Smith's Bull (608)—by Jolly's Bull (337).

Vol. 25, p. 839, under dam, and vol. 26, p. 507, both A. H. B.

GEAN BLOSSOM—Red, calved April 3, 1881, bred by A. Cruickshank, got by Perfection (37185), out of Garland by Scotland's Pride (25100)—Graceful by Baronet (15614)—Grandiflora by Lord Sackville (13249)—Flora by Fairfax Royal (6987)—Jessica by Premier (6308)—Venus by Saturn (5089)—Dairymaid by Favorite (6997)—by Grindon (3942)—bred by Mr. Rennie, of Phantassie—bred by Mr. Robertson, of Ladykirk.

British-American H. B., vol. 2, p. 431.

SORREL—Red, calved March 31, 1881, bred by A. Cruickshank, got by Roan Gauntlet (35284), out of Surname by Pride of the Isles (35072)—Surmise by Champion of England (17526)—Superb by The Czar (20947)—Splendor by Lord Sackville (13249)—Sympathy by Duke of Athol (10150)—Silence by Earl of Derby (10177)—Secret 3d by Duke of Sutherland (6945)—Secret 2d by Locomotive (4242)—Secret by Short Tail (2621)—White Rose by Gambier (2046)—White Rose by Young Wynyard (2859)—by bulls bred by R. & C. Colling.

Vol. 25, p. 838, A. H. B.

SPLEENWORT—Red and white, calved April 9, 1881, bred by A. Cruickshank, got by Royal Victor (43972), out of Sensation by Roan Gauntlet (35284)—Sybilla by Pride of the Isles (35072)—Sentiment by Prince Alfred (27107)—Science by Baronet (15614)

—Splendid by Lord Sackville (13249)—Sympathy by Duke of Athol (10150)—Silence by Earl of Derby (10177)—Secret 3d by Duke of Sutherland (6945)—Secret 2d by Locomotive (4242)—Secret by Short Tail (2621)—White Rose by Gambier (2046)—White Rose by Young Wynyard (2859)—by bulls bred by R. & C. Colling.

Vol. 28, p. 354, E. H. B., under dam. and vol. 26, p. 507, A. H. B.

BARON CAMPERDOWN (47389)—Red and white, calved Sept. 23, 1881, bred by A. Cruickshank, got by Roan Gauntlet (35284), out of Crescent by Gen. Windsor (28701)—Circassia by Champion of England (17526)—Cicely by Lancaster Royal (18167)—Crocus by Jemmy (11611)—Kitty by Somerset (10858)—Kate by Hawthorn (7071)—Kilmeny by The Peer (5455)—Premium by George (2057)—by Togston (5487).

LORD GLAMIS (48192)—Red, calved Sept. 20, 1881, bred by A. Cruickshank, got by Barmpton (37763), out of Garnish by Royal Duke of Gloster (29864)—Garnet by Prince Alfred (27107)—Guineas by Prince Imperial (22595)—Golden Chain by Lord Raglan (13244)—Gold Mint by The Baron (13833)—Pure Gold by Young 4th Duke (9037)—The Star Pagoda by Duplicate Duke (6952)—The Mint by Robin O'Day (4973)—Brawith Bud by Sir Walter (2639)—by Young Jerry (8177)—by Roseberry (567)—by Roseberry (567)—by Constellation (163)—by Hastings (293)—by Leopold (372).

LORD LOCHABER (1205 B. A. H. B.)—Red, calved Nov. 11, 1881, bred by A. Cruickshank, got by Barmpton (37763), out of Lovely 28th by Pride of the Isles (35072)—Lovely 12th by Scotch Rose (25099)—Lovely 9th by Windsor Augustus (19157)—Lovely 8th by Bosquet (14183)—Lovely by Kelly 2d (9265)—Lady Ythan by Robin O'Day (4973)—Lady by Favorite (9116)—Marion by Anthony (1640)—Mirandi by Anthony (1640)—Merino by Edgcott (1953)—Matilda by Son of Merlin (6522)—White Cow by Acton (1607).

VAN TROMP 54160 (48847)—Red, calved Dec. 11, 1881, bred by A. Cruickshank, got by Barmpton (37763), out of Victoria 45th by Cæsar Augustus (25704)—Victoria 42d by Forth (17866)—Victoria 32d by Prince Regent (16762)—Victoria 29th by Red Knight (11976)—Victoria 19th by Lord John (11731)—Victoria 4th by Prince Albert (11933)—Victoria 2d by Belzoni (783)—Victoria by Satellite (1420)—No. 1 Mason's Sale by Cato (119)—Pope Cow by Pope (514)—Flora by Favorite (252)—Nymyh by White Bull (421) —Lily by Favorite (252)—Miss Lax by Dalton Duke (188)—Lady Maynard by Alcock's Bull (19)—by Jacob Smith's Bull (608)—by Jolly's Bull (337).

BARON GLENCOE (47404)—Red, calved Dec. 14, 1881, bred by A. Cruickshank, got by Barmpton (37763), out of Graceful's Forth by Forth (17866)—Graceful by Baronet (15614)—Grandiflora by Lord Sackville (13249)—Flora by Fairfax Royal (6987)—Jessica by Premier (6308)—Venus by Saturn (5089)—Dairymaid by Favorite (6997)—by Grindon (3942)—bred by Mr. Rennie, of Phantassie —bred by Mr. Robertson, of Ladykirk.

JULIUS 56643 (48073)—Red, calved Dec. 19, 1881, bred by A. Cruickshank, got by Barmpton (37763), out of Juliet by Barmpton Prince (32995)—Joyful by Master of Arts (26867)—Jealousy by Champion of England (17526)—Josephine by The Baron (13833) —Flora by Fairfax Royal (6987)—Jessica by Premier (6308)—Venus by Saturn (5098)—Dairymaid by Favorite (6997)—by Grindon (3942)—bred by Mr. Rennie, of Phantassie—bred by Mr. Robertson, of Ladykirk.

VARNA 58341 (48848)—Red, calved April 13, 1881, bred by A. Cruickshank, Sittyton, got by Royal Victor (43972), out of Violet Bloom by Royal Violet (40649)—Village Pride 2d by Senator (27441)—Village Rose by Champion of England (17526)—Violet by Lord Bathurst (13173)—Roseate by Matadore (11800)—China Rose by Hudson (9228)—Carmine Rose by Fairfax Royal (6987)— Red Rose by Inkhorn (6091)—Moss Rose by Grazier (1085)—Cicely by Sampson—Marion by Wallace (1560).

AMHURST 49768 (47341)—Red, calved April 16, 1881, bred by A. Cruickshank, got by Royal Victor (43792), out of Amaranth by Barmpton (37763)—Amaryllis by Lord Lancaster (26666)—Azalea by Cæsar Augustus (25704)—Anemone by Forth (17866)—Avalanche by Sir Samuel (15302)—Angerona by Lemnos (13146)—Amy by Earl Stanhope (5966)—Augusta by True Blue (5522)—Albinia by Miracle (2321)—Alice by Sir Henry (1446)—Young Madam by Count (170)—Young Venus by Bracken (91)—Venus by Badsworth (47)—by Driffield (223)—bred by Sir George Strickland.

DUKE OF ALBANY (47709)—Roan, calved Aug. 19, 1881, bred by A. Cruickshank, got by Roan Gauntlet (35284), out of Acrimony by Barmpton (37763)—Abarilla by Barmpton Prince (32995)—Arabella by Earl of Windsor (15968)—Lady by Bushranger (12516)—Mary by Mosstrooper (11827)—Geraldine by Robin O'Day (4973)—Angelica by Holkar (4041)—Glendronach by Jopp's Bull (9256)—Kate of Darlington.

Sept., 1882. J. I. DAVIDSON,

Of Balsam, Canada. From Liverpool to Quebec, by Steamer Texas.

VIOLET EMPEROR 58381 (48876)—Red, calved March 20, 1882, bred by A. Cruickshank, got by Roan Gauntlet (35284), out of Russian Violet by Scotland's Pride (25100)—Red Violet by Allan (21772)—Violet by Lord Bathurst (13173)—Roseate by Matadore (11800)—China Rose by Hudson (9228)—Carmine Rose by Fairfax Royal (6987)—Red Rose by Inkhorn (6091)—Moss Rose by Grazier (1085)—Cicely by Sampson—Marion by Wallace (1560).

PREMIER EARL (48454)—Roan, calved March 10, 1882, bred by Amos Cruickshank, got by Barmpton (37763), out of Village Pride 2d by Senator (27441)—Village Rose by Champion of England (17526)—Violet by Lord Bathurst (13173)—Roseate by Matadore (11800), &c., as in Violet Emperor, above.

DUKE OF GUELDERS (47740)—Roan, calved March 26, 1882, bred by A. Cruickshank, got by Lamlash (45025), out of

Guelder Rose by Pride of the Isles (35072)—Evening Star by Royal Duke of Gloster (29864)—Morning Star by Champion of England (17526)—Grandiflora by Lord Sackville (13249)—Flora by Fairfax Royal (6987)—Jessica by Premier (6308)—Venus by Saturn (5089) Dairymaid by Favorite (6997)—by Grindon (3942)—bred by Mr. Rennie, of Phantassie—bred by Mr. Robertson, of Ladykirk.

BARON BRAWITH (47385)—Roan, calved Jan. 23, 1882, bred by A. Cruickshank, got by Roan Gauntlet (35284), out of Grizzle by Grand Vizier (34086)—Garnet by Prince Albert (27107)—Guineas by Prince Imperial (22595)—Golden Chain by Lord Raglan (13244)—Gold Mint by The Baron (13833)—Pure Gold by Young 4th Duke (9037)—The Star Pagoda by Duplicate Duke (6952)—The Mint by Robin O'Day (4973)—Brawith Bud by Sir Walter (2639)—by Young Jerry (8177)—by Roseberry (567)—by Roseberry (567)—by Constellation (163)—by Hastings (293)—by Leopold (372).

LORD NONPAREIL (48225)—Roan, calved March 3, 1882, bred by A. Cruickshank, got by Roan Gauntlet (35284), out of Nonpareil 15th by Royal Duke of Gloster (29864)—Nonpareil 13th by Gloster's Satellite (31258)—Nonpareil 12th by Lord Wharfdale (22231)—Nonpareil 3d by Laurel Hope (20105)—Nonpareil 9th by Cotherstone (12645)—Nonpareil 8th by Diamond (5918)—Nonpareil 3d by Adonis (2936)—by Commodore (1858)—Old Nonpareil by Tathwell Studley (5401)—Twine Tail by Blyth Comet (85).

BARON SECRET (47434)—Roan, calved March 10, 1882, bred by A. Cruickshank, got by Roan Gauntlet (35284), out of Sycamore by Count Robert (30812)—Surmise by Champion of England (17526)—Superb by The Czar (20947)—Splendor by Lord Sackville (13249)—Sympathy by Duke of Athol (10150)—Silence by Earl of Derby (10177)—Secret 3d by Duke of Sutherland (6945)—Secret 2d by Locomotive (4242)—Secret by Short Tail (2621)—White Rose by Gambier (2046)—White Rose by Young Wynyard (2859)—by bulls of R. & C. Colling.

June, 1883. JAMES I. DAVIDSON.

ORANGE BLOSSOM 33D—Roan, calved Dec. 17, 1881, bred by A. Cruickshank, got by Roan Gauntlet (35284), out of Orange Blossom 8th by Sir Walter Scott (22922)—Orange Blossom by Dr. Buckingham (14405)—by Matadore (11800)—by Sir Thomas Fairfax (5196)—by Billy (3151)—by Sovereign (7539)—by Satellite (1420)—by Baronet (60)—by Cleveland (144)—by Symmetry (641).

Under dam, E. H. B., vol. 28, p. 353.

LAVENDER 33D—Red, calved Dec. 31, 1881, bred by A. Cruickshank, got by Barmpton (37763), out of Lavender 16th by Lord Lansdowne (29128)—Lavender 11th by Count Bickerstaffe 2d (25838)—Lavender 9th by Brian Boru (17440)—Lavender 7th by Friar Tuck (14578)—Lavender 3d by Eclipse (10186)—Lavender 2d by Queen's Roan (7389)—by Will Honeycomb (5660)—by Spectator (2688)—by Albion (1619)—by Lancaster (360)—by son of Windsor (698)—by Comet (155).

A. H. B., vol. 26, p. 792.

LAVENDER 36TH—Red roan, calved Nov. 10, 1882, bred by A. Cruickshank, got by Roan Gauntlet (35284), out of Lavender 12th by Count Bickerstaffe 2d (25838)—Lavender 8th by Brian Boru (17440)—Lavender 7th by Friar Tuck (14578)—Lavender 3d by Eclipse (10186)—Lavender 2d by Queen's Roan (7389)—by Will Honeycomb (5660)—by Spectator (2688)—by Albion (1619)—by Lancaster (360)—by son of Windsor (698)—by Comet (155).

Vol. 26, p. 792, A. H. B.

NARCISSUS—Red, calved March 12, 1882, bred by A. Cruickshank, got by Barmpton (37763), out of Flora 7th by Royal Forth (25022)—Flora 6th by Golden Eagle (26267)—Flora 2d by Baronet 2d (17363)—Flora by Marquis of Bute (18336)—by Count Fairfax (8991)—by South Durham (9676)—by Uptaker (5534)—by Miracle (2320)—by Matchem (2280)—by Fitz Remus (2025)—by Cato (119).

LIDDESDALE (48146)—Red, calved May 17, 1882, bred by A. Cruickshank, got by Barmpton (37763), out of Lavender 20th by Pride of the Isles (35072)—Lavender 12th by Count Bickerstaffe 2d (25838)—Lavender 8th by Brian Boru (17440)—Lavender 7th by Friar Tuck (14578)—Lavender 3d by Eclipse (10186)—Lavender 2d by Queen's Roan (7389)—by Will Honeycomb (5660)—by Spectator (2688)—by Albion (1619)—by Lancaster (360)—by son of Windsor (698)—by Comet (155).

ORLANDO (48379)—Red, calved May 25, 1882, bred by A. Cruickshank, got by Perfection (37185), out of Orange Blossom 27th by Barmpton Prince (32995)—Orange Blossom 21st by Cæsar Augustus (25704)—Orange Blossom 14th by Knight of the Whistle (26558)—Orange Blossom 12th by Prince Imperial (22595)—Orange Blossom 2d by The Baron (13833)—by Dr. Buckingham (14405)—by Matadore (11800)—by Sir Thomas Fairfax (5196)—by Billy (3151)—by Sovereign (7539)—by Satellite (1420)—by Baronet (60)—by Cleveland (144)—by Symmetry (641).

SAPPHIRE—Red, calved Oct. 9, 1881, bred by A. Cruickshank, got by Barmpton (37763), out of Souchong by General Windsor (28701)—Souvenir by Royal Duke of Gloster (29864)—Superb by The Czar (20947)—Splendor by Lord Sackville (13249)—by Duke of Athol (10150)—by Earl of Derby (10177)—by Duke of Sutherland (6945)—by Locomotive (4242)—by Short Tail (2621)—by Gambier (2046)—by Young Wynyard (2859)—by bulls of Messrs. Colling.

Vol. 29, p. 402, E. H. B., under dam, and vol. 26, p. 793, A. H. B.

GOLDEN THISTLE—Roan, calved Nov. 4, 1881, bred by A. Cruickshank, got by Roan Gauntlet (35284), out of Golden Lady by Champion of England (17526)—Golden Princess by Lord Raglan (13244)—Gold Leaf by Lord Cardigan (13177)—Pure Gold by Young 4th Duke (9037)—by Duplicate Duke (6952)—by Robin O'Day (4973)—by Sir Walter (2639)—by Young Jerry (8177)—by Roseberry (567)—by Roseberry (567)—by Constellation (163)—by Hastings (293)—by Leopold (372).

Vol. 26, p. 792, A. H. B.

VICAR GENERAL (48855)—Roan, calved May 1, 1882, bred by A. Cruickshank, Sittyton, Aberdeen, Scotland, got by Barmpton (37763), out of Violet's Rose by Pride of the Isles (35072)—Violet's Pride by Scotland's Pride (25100)—Violet's Forth by Forth (17866)—Sweet Violet by Lord Stanley (16454)—Violet by Lord Bathurst (13173)—Roseate by Matadore (11800)—China Rose by Hudson (9228)—by Fairfax Royal (6987)—by Inkhorn (6091)—by Grazier (1085)—by Sampson—by Wallace (1560).

ABBOTSBURN (47312)—Roan, calved May 7, 1882, bred by A. Cruickshank, got by Roan Gauntlet (35284), out of Amaranth by Barmpton (37763)—Amaryllis by Lord Lancaster (26666)—Azalea by Cæsar Augustus (25704)—Anemone by Forth (17866)—Avalanche by Sir Samuel (15302)—Angerona by Lemnos (13146)—by Earl Stanhope (5966)—by True Blue (5522)—by Miracle (2321)—by Sir Henry (1446)—by Count (170)—by Bracken (91)—by Budsworth (47)—by Driffield (223)—bred by Sir G. Strickland.

VANCOUVER (48846)—Roan, calved May 11, 1882, bred by A. Cruickshank, got by Roan Gauntlet (35284), out of Victoria 48th by Lord Lancaster (26666)—Victoria 39th by Champion of England (17526)—Victoria 29th by Red Knight (11976)—Victoria 19th by Lord John (11731)—Victoria 4th by Prince Albert (11933)—Victoria 2d by Belzoni (783)—Victoria by Satellite (1420)—No. 1 Mason's Sale by Cato (119)—by Pope (514)—Flora by Favorite (252)—Nymph by White Bull (421)—Lily by Favorite (252)—Miss Lax by Dalton Duke (188)—Lady Maynard by Alcock's Bull (19)—by J. Smith's Bull (608)—by Jolly's Bull (337).

BARON LENTON—Roan, calved Dec. 12, 1882, bred by A. Cruickshank, got by Cumberland (46144), out of Lavender 16th by Lord Lansdowne (29128)—Lavender 11th by Count Bickerstaffe 2d (25838)—Lavender 9th by Brian Boru (17440)—Lavender 7th by Friar Tuck (14578)—Lavender 3d by Eclipse (10186)—Lavender 2d by Queen's Roan (7389)—Lavender by Will Honeycomb (5660)—by Spectator (2688)—by Albion (1619)—by Lancaster (360)—by son of Windsor (698)—by Comet (155).

Oct. 6, 1883. J. I. DAVIDSON,

Of Balsam, Ontario, Can. By Steamer Buenos Ayrean, Oct. 6, 1883.

MARY ANN OF LANCASTER 7TH—Red, calved Jan. 22, 1882, bred by N. Reid, Danestown, Bridge of Don, Scotland, got by Royal Lancaster (45535), out of Mary Ann of Lancaster 3d by The Hope of Britain (34179)—Mary Ann of Lancaster by Baronet (30448)—Anne of Lancaster by Lord Raglan (13244)—Lancaster 25th by Matadore (11800)—by The Marquis (10938)—by Will Honeycomb (5660)—by George 3d (7038)—by Spectator (2688)—by Albion (1619)—by Lancaster (360)—by son of Windsor (698)—by Comet (155).

COUNTESS—Roan, calved March 4, 1877, bred by Mr. Rennie, Conlie, Aberdeenshire, Scotland, got by Roan Gauntlet (35284), out of Duchess by Lord Privy Seal (16444)—Roseblush by Lord Sackville (13249)—Rosa by Procurator (10657)—Rose by 4th Duke of Northumberland (3649)—by Red Highflyer (2488)—by Pyramid (4812)—by Harry Lorrequer (3985)—by Blucher (84)—by Magnum Bonum (4322)—by Buston's son of Styford (103)—by son of Wetherell's Bull (690).

SCOTCH ROSE—Red, calved July 1, 1883, bred by A. Cruickshank, Sittyton, Aberdeenshire, got by Balfour, out of Countess by Roan Gauntlet (35284), &c., as in Countess, next above.

ALICE—Roan, calved March 13, 1879, bred by Mr. Rennie, Conlie, got by Roan Gauntlet (35284), out of Gipsey by Bolivar—Duchess by Lord Privy Seal (16444)—Roseblush by Lord Sackville (13249)—Rosa by Procurator (10657), &c., as in Countess, above.

SITTYTON ROSE—Roan, calved May 29, 1883, bred by A. Cruickshank, got by Balfour, out of Alice by Roan Gauntlet (35284), &c., as in Alice, next above.

DUCHESS OF GLOSTER 22D—Red, calved Jan. 9, 1879, bred by A. Cruickshank, got by Barmpton Prince (32995), out of 15th Duchess of Gloster by Master of Arts (26867)—Duchess of Gloster 10th by Royal Oak (22792)—Duchess of Gloster 9th by Lord Raglan (13244)—Duchess of Gloster 6th by Lord Carlies (14819)—

Duchess of Gloster by Duke of Oxford (11386)—Chance by Duke of Gloster (11382)—by Usurer (9763)—by Duke of Cornwall (5947) —by Morpeth (7254)—by Helicon (2107)—by Henwood (2114)— by Nestor (452)—by Harold (291)—by Meteor (432)—by Comet (155)—by Cupid (177).

WHIN BLOSSOM—Roan, calved March 25, 1880, bred by A. Cruickshank, got by Lord of the Isles (40218), out of Gratitude by Breadalbane (28073)—Golden Princess by Lord Raglan (13244)— Gold Leaf by Lord Cardigan (13177)—Pure Gold by Young 4th Duke (9037)—by Duplicate Duke (6952)—by Robin O'Day (4973) —by Sir Walter (2639)—by Young Jerry (8177)—by Roseberry (567)—by Roseberry (567)—by Constellation (163)—by Hastings (293)—by Hastings (293)—by Leopold (372).

ELIZA—Red, calved Jan. 15, 1881, bred by James Murray, Malberry, Aberdeenshire, got by Sherwood (48685), out of Jenny Lind by Golden Prince (31269)—Lady May by Lord Lieutenant (31688)—Elizabeth by Master Stanley (31885)—Lovely by Sir William (22924)—by Count Fairfax (8991)—by South Durham (9676) —by Uptaker (5534)—by Miracle (2320)—by Matchem (2280)—by Fitz Remus (2025)—by Cato (119)—by Whitworth (695)—bred by Mr. Mason, of Chilton.

ROYAL NONSUCH—Roan, calved Feb. 25, 1883, bred by A. Cruickshank, Sittyton, got by Roan Gauntlet (35284), out of Nonpareil 15th by Royal Duke of Gloster (29864)—Nonpareil 13th by Gloster's Satellite (31258)—Nonpareil 12th by Lord Wharfdale (22231)—Nonpareil 3d by Laurel Hope (20105)—Nonpareil 9th by Cotherstone (12645)—Nonpareil 8th by Diamond (5918)—Nonpareil 3d by Adonis (2936)—by Commodore (1858)—by Tathwell Studley (5401)—by Blyth Comet (85).

JULIUS CÆSAR—Red, calved Feb. 1, 1883, bred by A. Cruickshank, got by Cumberland (46144), out of Jonquil by Pride of the Isles (35072)—Juliet by Barmpton Prince (32995)—Joyful by Master of Arts (26867)—Jealousy by Champion of England (17526)— Josephine by The Baron (13833)—Flora by Fairfax Royal (6987)—

27

Jessica by Premier (6308)—Venus by Saturn (5089)—Dairymaid by Favorite (6997)—by Grindon (3942)—bred by Mr. Rennie, of Phantassie.

LOYAL SUBJECT—Roan, calved Feb. 8, 1883, bred by A. Cruickshank, got by Roan Gauntlet (35284), out of Lavender 15th by Lord Warden (31766)—Lavender 12th by Count Bickerstaffe 2d (25838)—Lavender 8th by Brian Boru (17440)—Lavender 7th by Friar Tuck (14578)—Lavender 3d by Eclipse (10186)—Lavender 2d by Queen's Roan (7389)—by Will Honeycomb (5660)—by Spectator (2688)—by Albion (1619)—by Lancaster (360)—by son of Windsor (698)—by Comet (155).

CONFESSOR—Red, calved March 24, 1883, bred by A. Cruickshank, got by Barmpton (37763), out of Costume by Bridesman (30586)—Cactus by Champion of England (17526)—Cicely by Lancaster Royal (18167)—Crocus by Jemmy (11611)—Kitty by Somerset (10858)—by Hawthorn (7071)—by The Peer (5455)—by George (2057)—by Togston (5487)—bred by Mr. Laing, of Longhoughton.

SPARTAN HERO—Red, calved March 23, 1883, bred by A. Cruickshank, got by Barmpton (37763), out of Souvenir by The Royal Duke of Gloster (29864)—Superb by The Czar (20947)—Splendor by Lord Sackville (13249)—Sympathy by Duke of Athol (10150)—by Earl of Derby (10177)—by Duke of Sutherland (6945) —by Locomotive (4242)—by Short Tail (2621)—by Gambier (2046) —by Young Wynyard (2859)—by bulls of Messrs. C. & R. Colling.

VIOLET KNIGHT—Red, calved March 26, 1883, bred by A. Cruickshank, got by Barmpton (37763), out of Village Pride 2d by Senator (27441)—Village Rose by Champion of England (17526)— Violet by Lord Bathurst (13173)—Roseate by Matadore (11800)— China Rose by Hudson (9228)—by Fairfax Royal (6987)—by Inkhorn (6091)—by Grazier (1085)—by Sampson—by Wallace (1560).

DOUBLE GLOSTER 55406—Red, calved Feb. 11, 1883, bred by A. Cruickshank, got by Barmpton (37763), out of Duchess of Gloster 24th by Lord of the Isles (40218)—Duchess of Gloster 21st by Barmpton Prince (32995)—Duchess of Gloster 13th by Grand

Duke of Gloster (26288)—Duchess of Gloster 12th by Champion of England (17526)—Duchess of Gloster 7th by Lord Raglan (13244)—Duchess of Gloster 2d by The Baron (13833)—Chance by The Duke of Gloster (11382), &c., as in Duchess of Gloster 22d, above.

BARON SENATOR—Roan, calved Dec. 12, 1882, bred by A. Cruickshank, got by Cumberland (46144), out of Lavender 16th by Lord Lansdowne (29128)—Lavender 11th by Count Bickerstaffe 2d (25838)—Lavender 9th by Brian Boru (17440)—Lavender 7th by Friar Tuck (14578), &c., as in Loyal Subject, above.

CIRCASSIAN CHIEFTAIN—Roan, calved Jan. 22, 1883, bred by A. Cruickshank, got by Perfection (37185), out of Circassia by Champion of England (17526)—Cicely by Lancaster Royal (18167), &c., as in Confessor, above.

VIOLET PRINCE—Red, calved March 26, 1883, bred by A. Cruickshank, got by Barmpton (37763), out of Village Pride by Senator (27441)—Village Rose by Champion of England (17526)— Violet by Lord Bathurst (13173), &c., as in Violet Knight, above.

May 26, 1884. J. I. DAVIDSON,
Of Balsam, Ontario. By Steamer Crecian, from Glasgow to Quebec.

BARON BARMPTON—Red, calved Feb. 22, 1883, bred by A. Cruickshank, Sittyton, Aberdeenshire, got by Roan Gauntlet (35284), out of Barmpton Spray (vol. 29, p. 400, E. H. B.) by Cæsar Augustus (25704)—Barmpton Flower by Allan (21172)— Butterfly's Joy by 2d Duke of Wharfdale (19649)—Butterfly's Pride by Royal Butterfly (16862)—Frederick's Pride by Frederick (11489) —Pride by 4th Duke of York (10167)—Princess Fairfax by Lord Adolphus Fairfax (4249)—Princess Royal by Thick Hock (6601)— Barmpton Rose by Expectation (1988)—by Belzoni (1709)—by Comus (1861)—by Denton (198).

AMETHYST—Roan, calved June 11, 1883, bred by A. Cruickshank, got by Perfection (37185), out of Abarilla (vol. 27, p. 361, E. H. B.) by Barmpton Prince (32995)—Arabella by Earl of Windsor (15968)—Lady by Bushranger (12516)—Mary by Mosstrooper

(11827)—Geraldine by Robin O'Day (4973)—Angelica by Holkar (4041)—Glendronach by Jopp's Bull (9256)—Kate of Darlington.

CLOUDBERRY—Roan, calved Jan. 14, 1883, bred by A. Cruickshank, got by Cumberland (46144), out of Cochineal by Bridesman (30586)—Carmine Rose (vol. 25, p. 409, E. H. B.) by Champion of England (17526)—Carmine by The Czar (20947)—Cressida by John Bull (11618)—Clipper by Billy (3151)—Favorite by Dandy (6918)—Keepsake by Tip Top (7633)—Old Lady, bred by Mr. Mason, of Chilton.

GENTIANELLA (Twinned with Gladys)—Red, caived March 12, 1883, bred by A. Cruickshank, got by Barmpton (37763), out of Grizzle (vol. 27, p. 361, E. H. B.) by Grand Vizier (34086) —Garnet by Prince Alfred (27107)—Guineas by Prince Imperial (22595)—Golden Chain by Lord Raglan (13244)—Gold Mint by The Baron (13833)—Pure Gold by Young 4th Duke (9037)—The Star Pagoda by Duplicate Duke (6952)—The Mint by Robin O'Day (4973)—Brawith Bud (vol. 6, p. 276, E.) by Sir Walter (2639)—by Young Jerry (8177)—by Roseberry (567)—by Roseberry (567)—by Constellation (163)—by Hastings (293)—by Leopold (372).

GLADYS—Red, calved March 12, 1883. Twinned with Gentianella, above.

GOLDEN SOCKS—Roan, calved Jan. 18, 1883, bred by A. Cruickshank, got by Roan Gauntlet (35284), out of Golden Lady (vol. 27, p. 361, E. H. B.) by Champion of England (17526)—Golden Princess by Lord Raglan (13244)—Gold Leaf by Lord Cardigan (13177)—Pure Gold by Young 4th Duke (9037)—The Star Pagoda by Duplicate Duke (6952)—The Mint by Robin O'Day (4973)—Brawith Bud by Sir Walter (2639)—by Young Jerry (8177)—by Roseberry (567)—by Roseberry (567)—by Constellation (163)—by Hastings (293)—by Leopold (372).

LOVELY 40TH—Red, calved April 24, 1883, bred by A. Cruickshank, got by Barmpton (37763), out of Lovely 12th (vol. 27, p. 362, E. H. B.) by Scotch Rose (25099)—Lovely 9th by Windsor

Augustus (19157)—Lovely 8th by Bosquet (14183)—Lovely by Kelly 2d (9265)—Lady Ythan by Robin O'Day (4973)—Lady by Favorite (9116)—Marion by Anthony (1640)—Miranda by Anthony (1640)—Merino by Edgcott (1953)—Matilda by Son of Merlin (6522)—White Cow by Acton (1607).

SONGSTRESS—Roan, calved Sept. 29, 1882, bred by A. Cruickshank, got by Staplehurst (47148), out of Spinster by Roan Gauntlet (35284)—Surname by Pride of the Isles (35072)—Surmise by Champion of England (17526)—Superb by The Czar (20947)—Splendor by Lord Sackville (13249)—Sympathy by Duke of Athol (10150)—Silence by Earl of Derby (10177)—Secret 3d by Duke of Sutherland (6945)—Secret 2d by Locomotive (4242)—Secret by Short Tail (2621)—White Rose by Gambier (2046)—White Rose by Young Wynyard (2859)—bulls bred by R. & C. Colling.

Vol. 29, p. 402, E. H. B., under dam.

SUNFLOWER 2D—Roan, calved April 12, 1882, bred by James Murray, Macterry, got by Marquis,* out of Sunflower by Golden Prince (31269)—Elizabeth by Master Stanley (31885)—Lovely by Sir William (22924)—by Count Fairfax (8991)—by South Durham (9676)—by Uptaker (5534)—by Miracle (3320)—by Matchem (2280) —by Fitz Remus (2025)—by Cato (119).

NOTE.—*Marquis, red, calved April 18, 1880, bred by James Murray, Macterry, got by Sherwood (48685), out of Helen by King Alfred (31471)— Elizabeth by Master Stanley (31885), as in pedigree of Sunflower 2d, above.

VICTORIA 75TH—Roan, calved Nov. 21, 1882, bred by A. Cruickshank, got by Barmpton (37763), out of Victoria 56th (vol. 27, p. 362, E. H. B.) by Bridesman (30586)—Victoria 47th by Lord Lansdowne (29128)—Victoria 39th by Champion of England (17526)—Victoria 29th by Red Knight (11976)—Victoria 19th by Lord John (11731)—Victoria 4th by Prince Albert (11933)—Victoria 2d by Belzoni (783)—Victoria by Satellite (1420)—No. 1 Mason's Sale by Cato (119)—Pope Cow by Pope (514)—Flora by Favorite (252)—Nymph by White Bull (421)—by Favorite (252) —Miss Lax by Dalton Duke (188)—Lady Maynard by R. Alcock's Bull (19)—by Jacob Smith's Bull (608)—by Jolly's Bull (337).

1831. CAPT. N. DE COST.

BRUTUS—Bred by Michael Ashcroft, near Liverpool. Pedigree lost.

CHARMER—Got by Don Juan (1923), out of Cherry by Wonderful (700)—by Alfred (23)—by Chilton's Red Bull.
See Lady Gay Spanker, vol. 2, p. 426, A. H. B.

HENRY DE GROOT.
New York.

JUDY ALIAS BETTY—See No. 2338.

W. DELANCY.
Westchester County, New York.

MOLLY—See 3586½. .

HARMER DENNY,
Of Pittsburg, Pa. Selected in England, by Rev. John A. Robertson. Imported in Ship Milo, of Portsmouth, N. H., Thompson, Master, landed at Philadelphia, Pa., Nov. 15, 1835.

YOUNG BUCKINGHAM (1758)—Red and white, calved March 24, 1834, bred by Mr. Rowlandson, of Newton Morrell, near Darlington, got by Buckingham (1756), out of Primrose by Scipio (1421)—Countess by Old Stephen (Mr. Charge's)—Lady Brough by Albion (14).

COWSLIP—Light roan, calved in Dec., 1832, got by Architect, out of Lady by Rob Roy—White Rose by George (Mr. Thornton's) —Burdon.

ARCHITECT—Got by Highflyer, out of Lady Brough by Duke.

Sold by Mr. Colling for 110 guineas, to Mr. Robinson, of Acklam.

1851. JOHN DOW.
Whitby.

YOUNG MARNOCK [436] ALIAS MARNOCK 1844—Roan, calved in May, 1850, bred by Wm. Stronach, Ardmeallie, Scotland, got by Young Pasha (11883), out of Flora by Commander (8976) —Catherine by Mahomed (6170)—by Jerry—bred by the late Mr. Hay, of Shethin, Scotland.

1853. LENIAT DOWLEY.
Brattleboro, Vt.

ALICE—Red, calved June 20, 1851, bred by J. S. Tanqueray, Hendon, got by Magician 2d (10486), out of Sally by Sheldon (8557)—Victoria by Mehemet Ali (7227)—Adelaide by Mahomet (4332)—White Flank 2d by William 4th (5663)—White Flank 1st by Forester (1055)—by Charge's Gray Bull (872)—by a son of Houghton (318)—by a bull of Col. Trotter's.

E. H. B., vol. 10, p. 572, under dam,.and A. H. B., vol. 2, p. 277.

SMILE—Roan, calved Jan. 29, 1849, bred by D. Pelham, Isle of Wight, got by Humber (7102), out of Laughter by Sweet William (5638)—Onion by Roman (2561)—Sage by William (2840) —Goose by Firby (1040)—by brother to Kalmico.

A. H. B., vol. 2, p. 556, and E. H. B., vol. 11, p. 698.

BELLE—Red, calved May 10, 1851, bred by W. D. Manning, Rothersthorpe, got by Monarch (7249), out of Barmaid by Hurricane (4061)—Spotted Boughton by Crusader (7939)—Bombazine by Regent (544)—bred by Mr. Edmonds, Boughton, Northamptonshire.

Vol. 2, p. 299, A. H. B.

MYRTLE—Red, calved May 9, 1848, bred by Robert Bell, Mosbro Hall, got by 4th Duke of York (10167), out of Mrs. Charlton by Bellman (7825)—by 2d Cleveland Lad (3408)—a cow from the herd of T. Bates.

Vol. 2, p. 489, A. H. B.

1873. JOHN DRYDEN,
Of Brooklin, Can. To Quebec.

BARMPTON ROYAL 31461 (32996)—Red roan, calved May 15, 1873, bred by A. Cruickshank, got by Scotland's Pride (25100), out of Butterfly's Delight by Allan (21172)—Butterfly's Joy by 2d Duke of Wharfdale (19649)—Butterfly's Pride by Royal Butterfly (16862)—Frederick's Pride by Frederick (11489)—Pride by 4th Duke of York (10167)—Princess Fairfax by Lord Adolphus Fairfax (4249)—Princess Royal by Thick Hock (6601)—Barmpton Rose by Expectation (1988)—by Belzoni (1709)—by Comus (1861)—by Denton (198).

Recorded as Royal Barmpton [3969] and as sired by Lord Lansdowne (29128).

COLUMBIA—Roan, calved May 21, 1873, bred by A. Cruickshank, got by Lord Lancaster (26666), out of Columbine by Sir Walter Scott (22922)—Camelia by Lancaster Comet (11663)—Cactus by Lord Sackville (13249)—Sharon's Rose by Plantagenet (11906)—Fancy by Billy (3151)—Jessy by Sovereign (7539)—Rose by Satellite (1420)—by Baronet (60)—by Cleveland (144)—by Symmetry (641).

C. H. B., vol. 3, p. 401.

Aug. 1880. JOHN DRYDEN,
Of Brooklin, Ont., Can. From Liverpool.

SUNBEAM—Roan heifer, calved Nov. 22, 1878, bred by A. Cruickshank, Sittyton, got by Royal Violet (40649), out of Songstress by Lord Lancaster (26666)—Superb by The Czar (20947)—Splendor by Lord Sackville (13249)—Sympathy by Duke of Athol (10150)—by Earl of Derby (10177)—by Duke of Sutherland (6945)—by Locomotive (4242)—by Short Tail (2621)—by Gambier (2046)—by Young Wynyard (2859)—by bulls of C. & R. Colling.

Brit.-Am. H. B., vol. 1, p. 434.

VIOLET BUD—Roan heifer, calved Feb. 24, 1879, bred by A. Cruickshank, got by Barmpton (37763), out of Violet's Rose by Pride of the Isles (35072)—Violet's Pride by Scotland's Pride

(25100)—Violet's Forth by Forth (17866)—Sweet Violet by Lord Stanley (16454)—Violet by Lord Bathurst (13173)—Roseate by Matadore (11800)—China Rose by Hudson (9228)—Carmine Rose by Fairfax Royal (6987)—by Inkhorn (6091)—by Grazier (1085)—by Sampson—by Wallace (1560).

E. H. B., vol. 26, p. 393, under dam.

VIOLET GEM—White, calved March 7, 1881, bred by A. Cruickshank, got by Royal Victor (43792), out of Violet Bud, above.

ORANGE BLOSSOM 30th—Red and white heifer, calved March 29, 1879, bred by A. Cruickshank, got by Pride of the Isles (35072), out of Orange Blossom 8th by Sir Walter Scott (22922)—Orange Blossom by Dr. Buckingham (14405)—Queen of Scotland by Matadore (11800)—by Sir Thomas Fairfax (5196)—by Billy (3151)—by Sovereign (5285)—by Satellite (1420)—by Baronet (60) —by Cleveland (144)—by Symmetry (641).

E. H. B., vol. 26, p. 392, under dam.

LANCASTER ROYAL 301 (Brit.-Am. H. B.)—Roan bull, calved Sept. 10, 1879, bred by A. Cruickshank, Sittyton, got by Lord of the Isles (40218), out of Lavender 18th by Lord Lancaster (26666)—Lavender 14th by Senator (27441)—Lavender 10th by Count Bickerstaffe 2d (25838)—Lavender 7th by Friar Tuck (14578)—Lavender 3d by Eclipse (10186)—Lavender 2d by The Queen's Roan (7389)—by Will Honeycomb (5660)—by Spectator (2688)—by Albion (1619)—by Lancaster (360)—by son of Windsor (698)—by Comet (155).

BARON SURMISE (45933)—Red and white bull, calved Oct. 25, 1879, bred by A. Cruickshank, got by Pride of the Isles (35072), out of Souvenir by Royal Duke of Gloster (29864)—Superb by The Czar (20947)—Splendor by Lord Sackville (13249)—Sympathy by Duke of Athol (10150)—by Earl of Derby (10177)—by Duke of Sutherland (6945)—by Locomotive (4242)—by Short Tail (2621)—by Gambier (2046)—by Young Wynyard (2859)—by bulls of C. & R. Colling.

28

BARMPTON VIOLET—Red heifer, calved March 20, 1879, bred by A. Cruickshank, got by Royal Violet 40649, out of Barmpton's Flower by Allan (21172)—Butterfly's Joy by 2d Duke of Wharfdale (19649)—Butterfly's Pride by Royal Butterfly (16862)—Frederick's Pride by Frederick (11489)—Pride by 4th Duke of York (10167)—Princess Fairfax by Lord Adolphus Fairfax (4249)—by Thick Hock (6601)—by Expectation (1988)—by Belzoni (1709)—by Comus (1861)—by Denton (198).

Brit.-Am. H. B., vol. 1, p. 430.

May 28, 1881. JOHN DRYDEN,

Of Brooklin, Ont., Can.   Per Steamer Texas, from Liverpool to Quebec.

CORN FLOWER—Roan heifer, calved August 26, 1880, bred by A. Cruickshank, Sittyton, Aberdeen, Scotland, got by Perfection (37185), out of Carmine by Meridian (38748)—Carmine Rose by Champion of England (17526)—Carmine by The Czar (20947)—Cressida by John Bull (11618)—Clipper by Billy (3151)—Favorite by Dandy (6918)—Keepsake by Tip Top (7633)—bred by Mr. Mason, Chilton.

E. H. B., vol. 27, p. 361, under dam, and Brit.-Am. H. B., vol. 1, p. 433.

VICTORIA 69TH—Roan heifer, calved Oct. 23, 1880, bred by A. Cruickshank, got by Barmpton (37763), out of Victoria 56th by Bridesman (30586)—Victoria 47th by Lord Lansdowne (29128)—Victoria 39th by Champion of England (17526)—Victoria 29th by Red Knight (11976)—Victoria 19th by Lord John (11731)—Victoria 4th by Prince Albert (11933)—by Belzoni (783)—by Satellite (1420)—by Cato (119)—by Pope (514)—by Favorite (252)—by White Bull (421)—by Favorite (252)—by Dalton Duke (188)—by R. Alcock's Bull (19)—by J. Smith's Bull (608)—by Jolly's Bull (337).

E. H. B., vol. 27, p. 362, under dam, and Brit.-Am. H. B., vol. 1, p. 434.

SULTANA—Roan heifer, calved Nov. 18, 1880, bred by A. Cruickshank, got by Pride of the Isles (35072), out of Souvenir by

Royal Duke of Gloster (29864)—Superb by The Czar (20947)—
Splendor by Lord Sackville (13249)—Sympathy by Duke of Athol
(10150)—Silence by Earl of Derby (10177)—Secret 3d by Duke of
Sutherland (6945)—Secret 2d by Locomotive (4242)—Secret by
Short Tail (2621)—White Rose by Gambier (2046)—White Rose
by Young Wynyard (2859)—by bulls of R. & C. Colling.

Brit.-Am. H. B., vol. 1, p. 433, and E. H. B., vol. 27, p. 362,
under dam.

FLORA 17TH—Roan heifer, calved Dec. 5, 1880, bred by A.
Cruickshank, got by Perfection (37185), out of Flora 7th by Royal
Forth (25022)—Flora 6th by Golden Eagle (26267)—Flora 2d by
Baronet 2d (17363)—Flora by Marquis of Bute (18336)—by Count
Fairfax (8991)—by South Durham (9676)—by Uptaker (5534)—
by Miracle (2320)—by Matchem (2280)—by Fitz Remus (2025)—by
Cato (119).

Brit.-Am. H. B., vol. 1, p. 432.

July 31, 1882. JOHN DRYDEN.

From Glasgow to Quebec, by Steamer Buenos Ayrean, of the Allan Line.

LORD GLAMIS (48192)—Red, calved Sept. 20, 1881, bred
by A. Cruickshank, got by Barmpton (37763), out of Garnish by
Royal Duke of Gloster (29864)—Garnet by Prince Alfred (27107)
—Guineas by Prince Imperial (22595)—Golden Chain by Lord
Raglan (13244)—Gold Mint by The Baron (13833)—Pure Gold by
Young Fourth Duke (9037)—Star Pagoda by Duplicate Duke
(6952)—The Mint by Robin-o'-Day (4973)—Brawith Bud by Sir
Walter (2639)—by Young Jerry (8177)—by Roseberry (567)—by
Roseberry (567)—by Constellation (163)—by Hastings (293)—by
Hastings (293)—by Leopold (372).

LAVENDER 30TH—Roan, calved Oct. 17, 1880, bred by A. Cruick-
shank, got by Pride of the Isles (35072), out of Lavender 18th by
Lord Lancaster (26666)—Lavender 14th by Senator (27441)—
Lavender 10th by Count Bickerstaffe 2d (25838)—Lavender 7th

by Friar Tuck (14578)—Lavender 3d by Eclipse (10186)—Lavender 2d by The Queen's Roan (7389)—Lavender by Will Honeycomb (5660)—by Spectator (2688)—by Albion (1619)—by Lancaster (360)—by son of Windsor (698)—by Comet (155).

Recorded in Brit.-Am. H. B., vol. 1, p. 432.

LAVENDER PRIDE—Red, calved April 10, 1883, bred by Amos Cruickshank, and calved in Canada, got by Cumberland (46144), out of Lavender 30th by Pride of the Isles (35072), &c. See dam, next above.

VICTORIA 72d—Roan, calved Oct. 26, 1881, bred by A. Cruickshank (Recorded in Brit. Am. H. B., vol. 1, p. 434), got by Barmpton (37763), out of Victoria 56th by Bridesman (30586)—Victoria 47th by Lord Lansdowne (29128), &c. See Victoria 69th, above.

ABERDEEN CHAMPION (47313)—Red, calved April 1, 1882, bred by Amos Cruickshank, got by Barmpton (37763), out of Costume by Bridesman (30586)—Cactus by Champion of England (17526)—Cicely by Lancaster Royal (18167)—Crocus by Jemmy (11611)—Kitty by Somerset (10858)—Kate by Hawthorn (7071)—Kilmeny by The Peer (5455)—by George (2057)—by Togston (5487)—bred by Mr. Laing, of Longhoughton.

1883. JOHN DRYDEN.

ARBUTUS—Roan, calved Dec. 18, 1881, bred by A. Cruickshank, got by Roan Gauntlet (35284), out of Abarilla by Barmpton Prince (32995)—Abarilla by Earl of Windsor (15968)—Lady by Bushranger (12516)—Mary by Mosstrooper (11827)—Geraldine by Robin-o'-Day (4973)—Angelica by Holkar (4041)—Glendronach by Jopp's Bull (9256)—Kate of Darlington.

LOVELY 37th—Roan, calved Oct. 7, 1882, bred by A. Cruickshank, got by Perfection (37185), out of Lovely 35th by Roan Gauntlet (35284)—Lovely 20th by Lord Lancaster (26666)—Lovely 11th by Allan (21172)—Lovely 10th by Duke of Bedford (23722) —Lovely 6th by Bosquet (14183)—Lovely 3d by The Hero (19934)

—Lovely by Kelly 2d (9265)—Lady Ythan by Robin-o'-Day (4973)—Lady by Favorite (9116)—Marion by Anthony (1640)—Miranda by Anthony (1640)—Merino by Edgcott (1953)—Matilda by Son of Merlin (6522)—White Cow by Acton (1607).

Vol. 29, p. 401, E. H. B., under dam.

1856. R. H. DULANEY.
Welbourne, Va.

SIR EDMUND LYONS (15285)—Roan, calved Feb. 28, 1856, bred by Jonas Webb, Babraham, got by Cheltenham (12588), out of Lady Lucy Lennox by Red Roan Kirtling (10691)—Duchess of Richmond by Pam (6272)—Victoria by Chancellor (3335)—Medusa by Helmsman (2109)—by Speedwell (1473)—by Columella (904).

MISS EMMA—Roan, calved April 17, 1855, bred by Jonas Webb, Babraham, got by Cheltenham (12588), out of Miss Emmaline by Lord of the North (11743)—Lady Godolphin by Paris (7314)—Victoria by Chancellor (3335)—Medusa by Helmsman (2109)—by Speedwell (1473)—by Columella (904).

E. H. B., vol. 12, p. 504, under dam.

MISS ISABEL—Roan, calved March 31, 1855, bred by Jonas Webb, got by Cheltenham (12588), out of Miss Mabel by Lord of the North (11743)—Lady Godolphin by Paris (7314)—Victoria by Chancellor (3335)—Medusa by Helmsman (2109)—by Speedwell (1473)—by Columella (904).

E. H. B., vol. 12, p. 506, under dam.

ROSINA—Red, calved March 10, 1855, bred by R. Stratton, Broad Hinton, got by Chinaman (12599), out of Aileen by Midsummer (8314)—Aileen by Duke of St. Albans (6944)—Strawberry by Colling (902)—by a son of Alexander (1624)—by grandson of Favorite (252).

E. H. B., vol. 12, p. 262, under dam.

ISABELLA—Red, calved May 10, 1855, bred by R. Stratton, got by Ilion (13059), out of Esther by Hero of the West (8150)—Modish by Kenilworth (7118)—Madam by Lottery (4280)—Premium by Phœnix (6290).

E. H. B., vol. 13, p. 502.

ROSE—White, calved in Feb., 1856, bred by Messrs. Atkinson, got by Abraham Parker (9856), out of Rosebud by Expectation (1454)—Wood Rose by Albert (5729)—White Rose by Baronet (1686)—by Duke (1933).

PANSY—Roan, calved in March, 1856, bred by Messrs. Atkinson, got by Colonel (14297), out of Gay Lass by Ethelred (5990)—Gaity by Albert (5729)—Winifred by Fitz Maurice (3807)—Wealthy by Brougham (1746)—Lucky by Marlish (4375)—Dandy by a son of Sir Harry (1444).

ALBA—White, calved March 21, 1857, bred by Mr. R. Stratton, got by Merry Lad (14947), out of Miss Isabella by Cheltenham (12588), &c., as in dam, above. Begotten in England, calved in Virginia.

These cattle were all lost to Col. Dulany by being driven from his farm during the civil war, and there is no trace of them in the A. H. Books.

1833. WALTER DUN.
Near Lexington, Ky.

CAROLINE—Red, calved Jan. 9, 1831, bred by John Stocks, Delph Hill, Yorkshire, England, got by Dashwood 9731, out of Fanny by Stockstry 11035—by Senator (1427)—by Whitefoot 11130—Rosina by Aid-de-Camp (722)—Louisa by Charles (127)—Lily by R. Colling's Son of Favorite (1033)—by R. Colling's Son of Favorite (1033)—by Chapman's Bull (122)—by R. Grimston's Bull (282).

A. H. B., vol. 14, p. 12.

DAISY—Red and white, calved in fall of 1830, bred by John Johnson, Yorkshire, England, got by Wild 11134, out of Lassie by David 9735—Modesty by Old David 10550—bred by Mr. White, Seapool, England.

A. H. B., vol. 14, p. 13.

MULTIFLORA—Roan, calved January 7, 1831, bred by Mr. Wright, Yorkshire, England, got by Walter 11118, out of White Face (bred by John Charge, near Darlington) by Roland (563)—by Constellation (163).

A. H. B., vol. 14, p. 13.

RED ROSE—Red and white, calved in Nov., 1830, bred by Edward Payne, Yorkshire, England, got by Ernesty 10017, out of Rosney by Eryholme (1018)—by Barmpton (54)—by Lancaster (360)—by Wellington (680)—by George (275)—by Favorite (252) —by Punch (531).

A. H. B., vol. 14, p. 12.

WHITE ROSE—Roan, calved in fall of 1830, bred by Mr. John Milwood, Plainstone, Holderness, Yorkshire, England, got by Publicola (1348), out of Fanny by Premier (1331).

A. H. B., vol. 14, p. 13.

PREMIUM—Roan, calved April 8, 1833, bred by Mr. Thos. Sowerby, Jr., got by Maximus (2284), out of Daisy by Lawnsleeves (365)—Rosebud by Lady Bull (354)—by Favorite (252).

Vol. 14, p. 12, and vol. 2, p. 514.

YOUNG CHARLOTTE—Red roan, calved in Dec., 1834, bred by Thomas Smurthwaite, Holme House, England, got by Thorp (2757), dam by Rob Roy (557)—by Sir Charles (5146)—by Surprise (2717)—by Wellington (678)—descended from the Denton and Barmpton stocks.

Vol. 14, p. 13, and vol. 2, p. 601.

SYMMETRY (5382)—Calved March 12, 1832, bred by Henry Peacock, Haddockstone, near Ripon, got by Red Simon (2499), out of Red Rose by Haddock, also called Governor (1077)—Roaney by Wellington (680)—by Blaize (75).

1838. WALTER DUN.

OTHO 794—Light roan, calved in April, 1838, gotten in England by Sol (2655), out of Premium by Maximus (2284)—Daisy by Lawnsleeves (365)—Rosebud by Lady Bull (354)—by Favorite (252).

TARIK 1022—Red and white, calved in April, 1838, got by Sol (2655), out of Young Charlotte by Thorp (2757)—by Rob Roy (557)—by Sir Charles (5146)—by Surprise (2717)—by Wellington (678)—descended from the Denton and Barmpton stocks.

1836. WALTER DUN and SAMUEL SMITH.
Near Lexington. Ky.

MARY ANN—Roan, calved in 1831, bred by Mr. Payne, got by Middlesbro' (1234), out of Prudence by Edward (1002)—Modesty (bred by Mr. Melton) by President (517)—by George (274).
E. H. B., vol. 3, p. 494.

BEAUTY OF WHARFDALE—Got in England by Brutus (1752), out of imp. Adelaide by Magnum Bonum (2243)—Beauty by George (1066)—Beauty by Lancaster (360)—Clara by Lancaster (360)—Venus by Wellington (680)—by George (275)—by Favorite (252)—by Punch (531)—from a sister to the dam of the White Heifer that Traveled.

ADELAIDE—Roan, got by Magnum Bonum (2243), out of Beauty by George (1066)—by Lancaster (360)—by Lancaster (360)—by Wellington (680)—by George (275)—by Favorite (252)—by Punch (531).

JEWESS—Barren.

GEORGE (2059)—Red and white, calved in 1834, bred by Mr. Wood, got by Young Magog (2247), out of Georgiana by Fitz Remus (2025)—by Whitworth (695)—by Charles (127).

OTLEY (4632)—Roan, calved in America, bred (supposed) by Mr. Fawkes, got by Norfolk (2377), out of Mary Ann by Middlesbro' (1234)—Prudence by Edward (1002)—Modesty by President (517)—by George (274).

COMET (1854)—Roan, calved in 1834, bred by Mr. Crofton, got by Emperor (1974), out of Gaudy by Monarch (2326)—by Cato (119)—by Whitworth (695).

1837. RECEIVER-GENERAL DUNN,
Of Canada.

QUEEN—See 21575.

NORTH STAR—See 11265, S. H. R.

LADY JANE—See 2057, A. H. B.

BRITON—In No. 285, A. H. B.

1882. F. A. ELLIS,
Of Boston, Mass. By Ship Bulgarian, from London, June 7, 1882.

ANNA REGINA—Roan, calved Aug. 19, 1878, bred by E. Heinemann, Rattan Park, Willingdon, Sussex, Eng., got by Royal Saxon (39057), out of Anna 9th by Crown of the Realm (30824)—Anna 7th by Rosedale Favorite (29831)—Anna 3d by Mantalini Prince (22276)—Anna 2d by Brilliant Star (17450)—Princess Julia by Lord of Rainsber (13211)—Dowager Queen by Sir Frederick Fairfax (5152)—Adelaide by Albert (727)—Anna by Pilot (496)—by Albion (14)—by Lame Bull (359)—by Shipton (587)—by son of Suwarrow (636)—by Son of Twin Brother to Ben (88)—by Twin Brother to Ben (660).

A. H. B., vol. 26, p. 1181, and E. H. B., vol. 28, p. 428.

ANNA VICTORIA—Roan, calved Sept. 30, 1882, bred by E. Heinemann, got by Royal Sceptre (43967), out of Anna Regina, above.

Vol. 26, p. 1182.

ROSETTA BOOTH—Red, calved May 20, 1874, bred by A. T. Matthews, Church Hanborough, Eynsham Oxon., got by Lieutenant General (31600), out of Rosanna by Sir James (16980)—Rosalie by Uncle Ned (19026)—Roa by Crusade (7938)—Rosamond by Cotherstone (6903)—Red Rose by Young Matchem (4425)—by Fairfax (1023)—by Marske (418)—by Palmsun (7311)—by a bull of Mr. Maynard's—by son of Favorite (252).

A. H. B., vol. 26, p. 1182, and E. H. B., vol. 28, p. 429.

LORD HOMEWORTH 56858 (45099)—White, calved Aug. 24, 1880, bred by Executors of T. C. Booth, Warlaby, got by Royal Stuart (40646), out of Homespun by Royal Benedict (27348)—by King James (28971)—by Lord Blithe (22126)—by Elfin King (17796).

ROYAL HEART 57957 (45528)—Roan, calved Dec. 22, 1880, bred by E. Heinemann, got by Royal Stuart (40646), out of Homely by Knight of St. Patrick (38520)—Homespun by Royal Benedict (27348)—by King James (28971)—by Lord Blithe (22126) —by Elfin King (17796).

ROYAL ROAN 57969 (47049)—Roan, calved Dec. 17, 1881, bred by E. Heinemann, got by Royal Sceptre (43967), out of Rosetta Booth by Lieutenant General (31600). See dam, in this list.

ROYAL GEORGE 57955—Roan, calved January 19, 1882, bred by E. Heinemann, got by Sir Wilfrid (37484), out of Generous (vol. 28, p. 428, E.) by Merry Monarch (22349)—Brevity by Brigade Major (21312)—Georgia by Prince George (13510)—Hopeful by Hopewell (10332)—by Warrior (12287).

1834. THOMAS A. EMMET,
Of New York City.

RED ROSE—Got by a son of Young Albion (2968)—by Sir Marton (1453).

WHITE LILY—White, calved in 1835, got in England by Magnum Bonum (2244), out of Red Rose by Young Albion (2968) —by Sir Marton (1453).

A. H. B., vol. 1, p. 237.

1853. ELIAS FASSETT,

Of Granville, Ohio.

PAGANINI 800—Calved Dec. 6, 1853, gotten in England by Duke of Beaufort (11377), out of imp. Jenny Lind by Minstrel (8687)—Lady Louise by Paris (7314)—Countess by Vanguard (5545)—Dodona by Alabaster (1616)—No. 6 by Dr. Syntax (220) —by Charles (127)—by Henry (301)—Lydia by Favorite (252)— Nell by Mason's White Bull (421)—Fortune by Bolingbroke (86)— by Foljambe (263)—by Hubback (319)—bred by Mr. Maynard.

JENNY LIND—Red, bred by Jonas Webb, Babraham, England, calved Jan. 24, 1849, got by Minstrel (8687), out of Lady Louise by Paris (7314)—Countess by Vanguard (5545)—Dodona by Alabaster (1616)—No. 6 by Dr. Syntax (220)—by Charles (127)—by Henry (301)—Lydia by Favorite (252)—Nell by White Bull (421) —Fortune by Bolingbroke (86)—by Foljambe (263)—by Hubback (319)—bred by Mr. Maynard.

BUTE 303—Bred by Jonas Webb, Babraham, got by Marquis of Bute (11788), out of Jenny Lind by Minstrel (8687), &c., as in Jenny Lind, above.

1839. FAYETTE COUNTY (KY.) CATTLE IMPORT-ING ASSOCIATION.

(REV. R. T. DILLARD AND NELSON DUDLEY, ESQ., Agents.)

Sold July 15, 1840, at the farm of David Sutton, in Fayette County, Ky.

COWS.

VICTORIA—Roan, calved in Aug., 1835, bred by J. E. Maynard, got by Plenipo (4724), out of White Rose by Francisco (2032)—by

Leopold (372)—by Major (397)—by own brother to R. Colling's White Heifer—descended from the stock of Sir James Pennyman.

A. H. B., vol. 2, p. 584. Sold to R. Fisher, for $1,750.

ELIZABETH—Roan, calved in Oct., 1832, bred by J. E. Maynard, got by Plenipo (4724), dam by Alexander (736)—by Marske (418)—by Comet (155).

Sold to A. McClure, for $505.

FASHION—Roan, calved in April, 1832, bred by W. Cooper, got by Young Don Juan (3610), out of Grizzle by Studley Grange (1483)—Young Fanny by Young Dimple (971)—Old Fanny by Layton (2190).

Recorded, E. H. B., vol. 5, p. 359. Sold to G. W. Williams, Paris, for $440.

ELIZABETH 2D—Calved in 1840, bred by J. E. Maynard, calved the property of A. McClure, Lexington, Ky., got by Velocipede (5552), out of Elizabeth by Plenipo (4724), &c., as in dam above.

SPLENDOR—Roan, calved in March, 1834, in Imp. Catalogue, and in E. H. B., vol. 3, p. 635, calved Nov. 22, 1833, bred by Mr. Cattley, got by Bedford Jr. (1701), out of Stately by Baron (1681)— White Rosette by Juniper (1144)—Rosette by White Comet (1582)— Young Rose by Wright's grandson of Favorite (252)—by son of Charge's Gray Bull (872).

The E. H. B., vol. 5, pp. 635 and 639, give Stately, above, by Romulus (1403) instead of Baron (1681).

E. H. B., vol. 3, p. 635, gives date of birth as Nov. 22, 1833. The Company's Catalogue gives it as March, 1834. Sold to B. Gratz, for $650.

MISS HOPPER—Roan, calved in 1835, bred by Thomas Crofton, got by Duke (1933), dam by Waverley (2819)—by a son of St. Albans (2584)—bred by Mr. Salvin, of Burnhall, near Durham, England.

Recorded, E. H. B., vol. 5, p. 671. Sold to W. T. Calmes, for $270.

TULIP—Roan, calved in 1836, bred by Mr. Crofton, Holywell, got by Bachelor (1666), dam by Duke (1933)—by Waverley (2819) —by a son of St. Albans (2584)—bred by Mr. Salvin, of Burnhall, near Durham.

Recorded, E. H. B., vol. 5, p. 1025. Sold to A. McClure, for $700.

BRITANNIA—Roan, calved in Feb., 1838, bred by Mr. Crofton, got by Emperor (1974), out of Miss Hopper by Duke (1933)—by Waverley (2819)—by a son of St. Albans (2584)—bred by Mr. Salvin, of Burnhall, near Durham.

Recorded, E. H. B., vol. 5, p. 671, under dam. Sold to H. Duncan, Bourbon Co., Ky., for $375.

ISABELLA—White, calved Oct. 14, 1839, bred by Thomas Crofton, Holywell, England, got by Melmoth (2291), out of Tulip by Bachelor (1666)—by Duke (1933)—by Waverley (2819)—by a son of St. Albans (2584)—bred by Mr. Salvin, Burnhall, near Durham.

Recorded, E. H. B., vol. 5, p. 1025, under dam. Sold to R. Fisher, Boyle Co., Ky., for $355.

LADY ELIZABETH—Roan, calved Feb. 4, 1838, bred by Mr. Crofton, Holywell, got by Emperor (1974), out of Elvira by Duke (1933)—by Wellington (2824)—by Young Remus (2522)—by Midas (435)—by Traveler (655)—by Bolingbroke (86).

Recorded, E. H. B., vol. 5, p. 316, under dam. Sold to H. Clay, Jr., Bourbon Co., Ky., for $660.

LILY—White, calved in 1834, bred by Leo Severs, got by Count (3506), dam by Linton (2207)—by Jupiter (343)—by Easby (232). Sold to T. Calmes, for $390.

NANCY—White, calved Jan. 1, 1837, bred by L. Severs, got by Reformer (2510), dam by Linton (2207)—by Jupiter (343)—by Easby (232)—by Mr. Parkinson's Red Bull. Sold to C. I. Rogers, for $730.

AVARILDA—White, calved April 8, 1836, bred by W. F. Paley, got by Norfolk (2377), out of Alice by Anson (1639)—Lady Jane by Comet (155)—Cleasby Lady by a son of Favorite (252)—Lucinda by Hutton's Bull (323)—Lucy by Barningham (56).

Recorded, E. H. B., vol. 3, p. 262.   Sold to John Allen, for $920.

BEAUTY—Roan, calved in March, 1834, bought of A. L. Maynard, got by Belvedere (1706), dam by Matchem (2281)—by Alexander (736)—by Marske (418).

Sold to H. Clay, Jr., for $700.

MISS MAYNARD—Roan, calved in 1837, bought of A. L. Maynard, got by Chorister (3378), dam by Matchem (2281)—by Sir Thomas (2636)—by Favorite (2037)—by Harold (291)—by North Star (1158)—by Favorite (252).

Sold to Andrew McClure, for $1,005.

JESSICA—Roan, calved Feb. 22, 1839, bought of Mr. Maynard, got by Velocipede (5552), out of Beauty by Belvedere (1706)—by Matchem (2281)—by Alexander (736)—by Marske (418).

Bought by Joel Higgins, for $330.

MISS LUCK—Roan, calved May 25, 1834, bought of Mr. Whitaker, got by Alison's Roan Bull (2999), out of White Daisy by Jerry (4097)—Old Daisy by Scroope's Danby (3550)—Ash by Easby (232).

Recorded, E. H. B., vol. 5, p. 673.   Sold to H. Clay, Jr., Bourbon Co., Ky., for $800.

ROSABELLA 2D—Roan, calved in Jan., 1839, bought of Mr. Whitaker, got by Velocipede (5552), out of Rosabella by Alexander (736)—Nonsuch by Harold (291)—Golden Pippin by North Star (459)—Beauty by Favorite (252)—Beauty by Favorite (252)—Beauty by Favorite (252)—the dam of Punch by Broken Horn (95)—bred by Mr. Best, of Mansfield.

Bought by W. A. Warner, for $465.

NOTE.—In the pedigree of Rosabella 2d. (5552) is correct. The catalogue of this sale in 1840 refers to Velocipede on page 733. vol. 3, E. H. B., old edition, which is (5552).

ZELIA—Red and white, calved Feb. 10, 1840, bred by W. Cooper, got by Norfolk (2377), out of Fashion by Young Don Juan (3610). See Fashion, in this list.

FLORIA—Roan, calved Jan. 14, 1840, bred by A. L. Maynard, got by Eclipse (9069), out of Beauty by Belvedere (1706), &c., as in dam.

WASHINGTON—Red and white, calved Feb. 22, 1840, bred by Mr. Crofton, got by Carcase (3285), out of Miss Hopper by Duke (1933), &c., as in Miss Hopper.

BULLS.

CARCASE (3285)—Red and white, calved in July, 1837, bred by Mr. Wiley, got by Belshazzar (1704), out of Sultana by Sultan (1485)—Milta by Tarrare (1501)—Mida by Midas (435)—Wildair by grandson of Phenomonon (491)—by Punch (531).

Sold to Benj. Gratz, Lexington, Ky., for $725.

ÆOLUS (2938)—Roan, calved in April, 1836, bred by Mr. Rowlandson, got by Harlsey (2091), out of Tulip (bred by Mr. Charge) by Scipio (1421)—by Sir Stephen (1456).

Sold to R. Fisher, Boyle Co., Ky., for $610.

ECLIPSE (9069)—Calved April 26, 1837, bred by Mr. Arrowsmith, got by Velocipede (5552), out of Strawberry by Young Monarch (2324)—by Matchem (2281)—by Cato (119)—by Sir Oliver (605)—by Trunnell (659)—by Bolingbroke (86).

Sold to R. Fisher, for $1,050.

CROFTON (3523)—Roan, calved Feb. 26, 1839, bred by T. Crofton, got by Melmoth (2291), out of Miss Hopper by Duke (1933)—by Waverley (2819)—by Mr. Salvin's Son of St. Albans (2584)—bred by Mr. Salvin, of Burnhall, near Durham.

Sold to J. Downing, for $155.

TRAJAN 1042—White, calved Dec. 21, 1839, bred by Leo Severs, England, begotten in England by Lord Lieutenant (4260), out of Lily by Count (3506)—by Linton (2207)—by Jupiter (343).

Sold to J. Wheland & Co., Boyle Co., Ky., for $150.

NELSON 741—White, calved Dec. 4, 1839, bred by Mr. Whitaker, gotten in England by Sir Thomas Fairfax (5196), out of Miss Luck by Allison's Roan Bull (2999)—White Daisy by Jerry (4097)—Old Daisy by Scoopes' Danby (3550)—Ash by Easby (232).

This pedigree is entirely different after White Daisy by Jerry (4097) from the A. H. B. See E. H. B., vol. 5, p. 673. Sold to P. Todhunter, for *610.

PRINCE ALBERT 2065—Roan, calved May 25, 1840, bred by J. E. Maynard, got by Carcass (3285), out of Victoria by Plenipo (4724), &c., as in Victoria, above.

ADAM FERGUSSON.

Woodhill, East Flamboro, Wentworth Co., Canada.

BEAUTY—Roan, calved in May, 1833, bred by James Crisp, Northumberland, got by Snowball (2647), dam by Lawnsleeves (365)—by Mason's Charles (127).

C. H. B., vol. 1, p. 210.

AGRICOLA (1614) ALIAS SIR WALTER—Roan, calved in 1833, bred by Mr. Crofton, got by Batchelor (1666), out of Sprightly by Barmpton (1677)—by Waverly (2819)—by son of Washington (674).

Agricola was imported in 1836. Sold to Mr. McKnight, New York. See Albany Cultivator, vol. 1, p. 44.

CHERRY—Roan, calved in 1833, bred by Mr. James Crisp, Doddington, Northumberland, got by Dunston Castle, dam by a son of St. Albans (2584)—by St. Albans (2584)—by Lawnsleeves (365).

C. H. B., vol. 1, p. 239.

MR. FERRIS.

Westchester County, N. Y.

OLD ELEANOR—See [37].

1821. FISH & GRINNELL.

ALBION—See 2318.

MOSS ROSE.

LOTHARIO—Begotten in England by Mr. Ashcroft's George, dam Moss Rose.

FLORA—Sold to C. H. Hall, Oct. 8, 1823.

July, 1871. E. P. P. FOWLER.

From Liverpool to Philadelphia.

BEAUTIFUL STAR—Roan heifer, calved April 11, 1869, bred by Mr. T. Stamper, Highfield House, Oswaldkirk, got by Stonegrave (27575), out of Bride in White by Wiske Boy (23234)—Beauty's Bride by Earl of Oxford (15966)—Bridesmaid by Star 2d (13786)—Young Melody by Lord John (13199)—Melody by Roan Duke (8486)—by Ganthorpe (2049)—Primrose by Herdsman (6076) —by Young Durham (8013)—by Young Warrior (8755)—by Waddingworth (668).

Vol. 17, p. 12788, A. H. B.

CLARET—Red heifer, calved April 18, 1869, bred by T. Stamper, Highfield House, Oswaldkirk, got by Grand Duke of Vladimir (28769), out of Caroline by Wathstone's Hero (25417)— Cygnet by Captain (14229)—Citron by Captain (14229)—Caroline by Chevy Chase (7897)—by Monarch (9407)—by Gainford (2044) —by Samson (5080)—by Young Sovereign (5286)—by White Walton.

Recorded, E. H. B., vol. 19, p. 430, under dam.

CHILTON 2D 22394—Red roan, calved Jan. 3, 1872, bred by T. Stamper, got by Chilton (25774), out of Claret by Grand Duke Vladimir (28769), &c., as in Claret, above.

30

Sept. 16, 1871. E. P. P. FOWLER.

By Ship Hauza, to Baltimore. Sold at Public Sale, at Kearney's Stables, Centre Street, Nov. 21, 1871. They were consigned to Richards, Leftwich & Co.

JANUARY ROSE—Roan heifer, calved Jan. 1, 1869, bred by A. Stubbs, Kirkbank, Richmond, Yorkshire, got by Lord Albert (20143), out of Rose of Christmas by Royal Alfred (18748)—Rose Clarence by Fitz Clarence (14552)—Moss Rose by Gavazzi (11508) —Moss Rosebud by Wilberforce (9830)—Rosebud by Swintonian (9702)—Ervinton Rose by Young Symmetry (7577)—by Shipton (2620)—by Champion (7007)—by Snowball (7521).

A. H. B., vol. 14, p. 581. Sold to I. S. Tanner, Shepherdstown, W. Va., for *205.

FLEDA'S FAREWELL—Roan heifer, calved April 21, 1869, bred by W. R. Bromet, Cocksford, Tadcaster, got by Third Duke of Flanders (23750), out of Fleda by Cannonier (30649)—Duchess by Trumpeter (10977)—by Chevalier (10050)—by The Duke (9030) —by Villager (7683)—by Charles 12th (5853).

Under dam, E. H. B., vol. 20, p. 521. Sold to Jno. W. Fairfax, Leesburg, Va., for *140.

LADY FERN—Roan heifer, calved Jan. 9, 1869, bred by J. Cooper, North Deighton, got by Fifth Duke of Wharfdale (26033), out of Maidenhair by Grand Duke of Wetherby (17997)—Mary 2d by Mare Antony (14895)—My Mary by British Yeoman (8906)— Moretrix by Adrian (5714)—Twin Sister by Noble (4577)—by Young Don Juan (3610)—by Young Isaac (2154)—by Son of Young Dimple (6518)—by Young Dimple (971)—by Snowball (2648)—by Layton.

Under dam, E. H. B., vol. 19, p. 613, and A. H. B., vol. 12, p. 906. Sold to C. E. Coffin, Muirkirk, Md., for *215, and by him to W. Warfield, Lexington, Ky., May 13, 1874, for *700, and by him to Mr. Pierce, of Ohio, June, 1878.

ROSE BLOOM—Roan heifer, calved March 1, 1870, got by Regal Rose [by Baron Booth (21212), dam Rose Louisa], out of

Rose of Thornhill by Blencow (19267)—Ruby by Gen. Murat (17955)—Mirror by Admiral Lyons (12352)—Profile by Benedict (7828)—Portrait by Patron (7315)—by Young Archibald (9902)—by Gainford (10254).

Sold to H. Zeller, Hagerstown, Md.

ELEANOR 4TH—Roan heifer, calved March 16, 1869, bred by A. Stables, Kirkbank, Richmond, Yorkshire, got by Lord Albert (20143), out of Eleanor First Fruits by First Fruits (16048)—Eleanor by Wilberforce (9830)—Ellen by Symmetry (11120)—by Champion (3323)—by Orpheus (473)—by Ambo (746).

Sold to J. P. Thorn, Baltimore, and by him to C. E. Coffin.

ELEANOR 5TH—Roan, calved Nov. 3, 1871 (aboard ship · Hauza), got by Prince Victor (29685), out of Eleanor 4th by Lord Albert (20143), &c., as above.

Sold to C. E. Coffin.

SYCAMORE 2D—Red heifer, calved Feb. 27, 1870, bred by C. H. Dawson, Weston Hall, Olley, Yorkshire, got by Thorndale Lad (23066), out of Sycamore by Royal Oak (16373)—Solferino by Sir Edmund Lyons (15284)—Clairvoyant by Lamartine (11660)—by Matchless (4428)—by True Blue (5522)—by Baronet (1688)—by Alamode (725)—by Mameluke (2257).

Under dam, E. H. B., vol. 21, p. 666. Sold to J. W. Fairfax, Leesburg, Va., for $140.

PRINCE VICTOR 20714 (29685)—Roan, calved April 3, 1868, bred by S. Wiley, Brandsby, got by Earl of Derby (21638), out of Princess Alexandra by Spearman (17025)—Miss Dundas by John Bull (11618)—Lady Zetland by Zetland (14047)—Hermia by Buckingham (3239)—by Belshazzar (1704)—by a bull of Mr. Hunter's, of Gilling.

NOTE.—Observe in A. H. B. the first sire and dam are left out. I am almost certain this bull was never imported. Mr. Fowler brought over one of his calves, and the cow January Rose was in calf by him.

VICTOR FIRST OF SPRINGDALE 21268—White, calved March 13, 1872, bred by A. Stables, got by Prince Victor (29685), out of January Rose by Lord Albert (20143), &c., as in January Rose, above.

LIVERPOOL 12304—Roan, bred by J. Birch, Ormskirk, Lancashire, got by Frederick (26193), out of Silent Thought by Hailstone (21891)—Happiness by Yorkshire Boy (17261)—Joy by Man Friday (13290)—Marjoram by Wharfdale Burley (7713)—Lavender by Warrior (6660)—Lavinia by Emperor (1974)—by Wharton (2833)—by Sir William (2640)—by Young Rockingham (2549)—by Major (2255)—by Northumberland (461)—by Lame Bull (358).

Dec. 1871. E. P. P. FOWLER.

By Ship Massachusetts, from Liverpool to New Orleans.

RIVENHALL ROSE—Red heifer, calved Dec. 29, 1869, bred by J. Upson, Rivenhall, Witham, got by Heir of Wetherby (26362), out of Lilac by Warwick (19120)—Lily of Windsor by Young Windsor (17241)—Mary Ann by His Highness (14708)—Martha by King John (14763)—Minstrel by Duke of Gloster (10153)—White Cow by Wroughton (8789).

CHERRY BUD—Roan heifer, calved in May, 1869, bred by G. Yeats, Studley, Ripon, got by Emperor (23875), out of Cherry by Perseus (18533)—Cherry Cheeks by Cotherstone (6903)—Moss Rose by Lofty (2217)—by son of Lindrick (1170)—by Shylock (2622).

Bought by G. W. Polk, of Tenn.

FANNY 8TH—Roan heifer, calved July 21, 1869, bred by J. Dickinson, Upholland, Wigan, got by Buxton (23497), out of Fanny 3d by Hyde Park (19994). Barren.

Mr. G. W. Polk bought her of Mr. Fowler.

RED BLOOM—Red heifer, calved Dec. 18, 1869, bred by J. Renton, Farnly Otley, got by Lord St. Leonards (29202), out of Bess by Doctor Spence (19578)—Betsy by Eclipse (15976)—Strawberry by Captain Riffa (12551)—Grizzle by Laudable (9282)—by The Stuart (7623)—by Conqueror (3463)—by Norfolk (2377)—

by Atlas (1660)—by Regent (2518)—by Meteor (432)—by Comet (155).

Under dam, vol. 19, p. 409, E. H. B. Bought by G. W. Polk, of Tenn.

OCEANA—Red, bred by Geo. Yates, Studley, Ripon, England, calved in April, 1872, got by Knight of Studley (29005), out of imp. Cherry Bud by Emperor (23875), &c., as in dam above, this volume.

Recorded, S. H. R., vol. 6, p. 524.

JAMES GARDHOUSE.

VERBENA—Roan, calved Feb. 15, 1874, bred by John Outhwaite, Yorkshire, got by Royal Windsor (29890), out of Red Rose by Marquis (16829)—Grace 5th by Richard the Beau (24951)—Grace 3d by Knight of the Border (18161)—Baroness by Baron of Kidsdale (11156)—Duchess by Whitaker's Comet (8771)—Meg by Red Kirk (11975)—Grace by Playfellow (6297)—Meg by Tryo (6620)—by Memnon (4451)—by Duke of Wellington (231)—by a bull of Mr. Mason's.

C. H. B., vol. 4, p. 575.

KNIGHT OF THE BORDER [5496]—Roan, calved June 1, 1877, bred by John Outhwaite, got by Count Grindelwand [2919], out of Verbena, above.

July, 1868. J. L. GIBB.

FLORA 4TH—Roan, calved March 19, 1867, bred by W. S. Marr, Upper Mill, got by Rory O'More (24991), out of Flora 3d by Prince Arthur 2d (22571)—Flora by Vladimir (21043)—Matilda 5th by Cunningham (11323)—Matilda by Robin Hood (8494)—Ruby by Mahomed 2d (10492)—Daisy by Billy (3151)—Maria by Belshazzar (1703)—by Abraham (2905)—by Simon (5134).

C. H. B., vol. 2, p. 481.

MARIGOLD 5TH—Roan, calved April 13, 1867, bred by W. S. Marr, got by Young Pacha (20457), out of Marigold 2d by Lord

Lorne (18258)—Marigold by Rubens (13641)—Rose by Jemmy (11611)—Rosamond by Captain (8925)—Ruth by Duke of St. Albans (8001)—Ruby by Guy Fawkes (7062)—Little Red Rose by Pedestrian (7321)—by Lucian (2228)—by Raby (2473).

C. H. B., vol. 2, p. 636.

1874. J. L. GIBB.

Quebec, Can. By Ship Vicksburg from Liverpool, Aug. 19, 1874.

TOPSY—Roan heifer, calved Nov. 24, 1871, bred by Mr. J. Angus, Bearl, Northumberland, got by Roan Chief (27294), out of Cowslip 2d by Merry Monarch (22349)—Cowslip by Gen. Havelock (16108)—Cherry by Brilliant (7851)—by Traveler (7646)—by Fitz Maurice (3807)—by Reformer (2502)—by Leopold (2199)—by Sir Harry (5155)—by Traveler (655)—by Colonel (152)—by R. Colling's son of Hubback (319).

Under dam, E. H. B., vol. 20, p. 463.

BLOOMER 3D—Red and white heifer, calved May 12, 1873, bred by Mr. Browell, Apperby, Stocksfield, got by Whiff (30299), out of Bloomer by Knight of Broomley (24271)—Cherry Ripe by Knight of Richard Cœur-de-Lion (20080)—Cherry Blossom by Professor Miller (18649)—Cherry Blossom by The Templar 2d (13879)—Laurel Leaf by South Durham (9676)—Laurel Leaf by Pilot (7333)—by Young Sir Harry Liddle (7503)—by Mickley (7234)—by Adonis (2933)—by Lord Prudhoe (8251)—by Hollon's Bull (313).

Under dam, E. H. B., vol. 21, p. 601.

RUBINA—Red and white heifer, calved Nov. 10, 1872, bred by Mr. J. Whyte, Clinterty, Aberdeen, got by K. C. B. (26492), out of Violet by Prince of Warlaby (20593)—Ruby by Prince George (13510)—Garnet by King Arthur (13110)—Georgiana by Silky Laddie (10947)—Rosy by Leonardo (7137)—Pansy by Musician (6234)—by Priam (2452).

Under dam, E. H. B., vol. 20, p. 809.

ROYAL PRINCESS—Roan heifer, calved May 5, 1871, bred by Messrs. Dudding, Panton, Wragby, Lincolnshire, got by The Stuart

(27650), out of Royal Butterfly by Royal Buckingham (20718)—Butterfly by Vanguard (10994)—Beautiful by The Squire (12217)—The Belle by Sugar Plum (10894)—Belladonna by Nero (4557)—Belvorinia by Thorpe (2757)—Belvideria by Beaufort (1696)—Duchess by Czar (945)—by Meteor (1226)—by Marske (418)—by Freeman (269)—by Danby (190)—by son of Danby (190)—by Danby (190).

Under dam E. H. B., vol. 20, p. 749.

RENNIE GWYNNE 2D—Roan heifer, calved April 10, 1872, bred by Mr. H. Caddy, Roughholm, Bootle, Cumberland, got by Waterloo Cherry (27763), out of Rennie Gwynne by Sir Windsor (22927)—Rebecca Gwynne by Knight of Distington (18158)—Ruth Gwynne by Exquisite (14524)—Young Dowager Gwynne by St. Thomas (10777)—Dowager Gwynne by Prime Minister (2456)—by Wallace (5586)—by Marmion (406)—by Merlin (430)—by Layton (366)—by Phenomenon (491)—by Favorite (252)—by Favorite (252)—by Hubback (319)—by Snowdon's Bull (612)—by Waistell's Bull (669)—by Masterman's Bull (422)—by Studley Bull 626).

Under dam, E. H. B., vol. 21, p. 613.

FLOWER GIRL—Roan, calved May 1, 1869, bred by James How, Huntingdon, got by Victorius (25378), out of British Girl by British Hope (21324)—Lenton Lass by Black Knight (19216)—Lenton Girl by Earl De Grey (15956)—Lancaster 18th by Prince Royal (7371)—Lancaster 12th by Will Honeycomb (5660)—by George 3d (7038)—by Spectator (2688)—by Albion (1619)—by Lancaster (360)—by Windsor (698)—by Comet (155).

A. H. B., vol. 13, p. 602.

PRINCE VICTOR 30698—White, calved in August, 1874, bred by Dudding & Son, got by Robert Stephenson (32313), out of Royal Princess by The Stuart (27650), &c., as in dam.

WATERCRESS GIRL—Light roan, calved Aug. 5, 1873, bred by James How, got by Prince of the Realm (22627), out of Flower Girl by Victorious (25378), &c., as in Flower Girl.

A. H. B., vol. 13, p. 972.

1837 or 1838. WILLIAM GIBBONS.
Madison, New Jersey.

MAJESTIC (2249)—Roan, calved in March, 1835, bred by Mr. Crofton, got by Bachelor (1666), out of Lady by Young Wellington (2826)—by Wellington (2824)—Countess by Sir Charles (593)—Princess by St. Albans (1412)—by Cupid (177)—by Simon (590) —by Punch (531)—by Bolingbroke (86).

1833 or 1834. WILLIAM GIBBONS.

VOLAGE—White, calved March 17, 1828, bred by Mr. Whitaker, got by Charles (878), out of Seraphina by Wharfdale (1578) —Surprise (bred by Sir H. Carr Ibbetson) by Palemon (479)— Prodigy by Meteor (432)—Princess by Western Comet (689)—Selina by Favorite (252)—Countess by Cupid (177)—Lady by Grandson of Bolingbroke (280)—Phœnix by Foljambe (263)—Favorite by Alcock's Bull (19)—by Jacob Smith's Bull (608)—by Jolly's Bull (337).

Vol. 3, p. 685, E. H. B.

ARTHUR (3040)—Roan, calved July 10, 1838, bred by Mr. Whitaker, got by Charles (1815), out of Voluna by Danby (3550) —Lingcropper by Remus (550)—Pink by Sedbury (1424)—Beauty by Hollings (2131) —Lingcropper by Marske (418)—descended from R. Alcock's Bull (19).

May 28, 1881. R. GIBSON.
Ilderton, Ont., Can. By Steamship Quebec, from Liverpool.

ROWFANT KIRKLEVINGTON 3D—Red heifer, calved Sept. 11, 1877, bred by Sir C. M. Lampson, Bart., Rowfant, Crawley, Sussex, got by 3d Duke of Hillhurst (30975), out of Kirklevington Duchess 5th by 2d Duke of Claro (21576)—Duchess of Kent by Lord Liverpool (22168)—Kirklevington 14th by 4th Duke of Oxford (11387)—Kirklevington 7th by Earl of Derby (10177)—Kirklevington 4th by Earl of Liverpool (9061)—Kirklevington by Duke of Northumerland (1940)—by Belvedere (1706)—by Son of 2d Hubback (2683)—a cow of Mr. Bates', descended from the herd of Mr. Maynard.

A. H. B., vol. 21, p. 16669, and E. H. B., vol. 26, p. 514, under dam.

VISCOUNTESS BARRINGTON 2D—Roan heifer, calved Nov. 15, 1877, bred by Earl of Lathom, Lathom, Lancashire, got by Baron Oxford 4th (25580), out of Grand Duchess of Barrington 2d by Duke of Brailes (23724)—Grand Duchess of Barrington by Grand Duke 7th (19877)—Countess of Barrington 2d by 9th Duke of Oxford (17738)—Countess of Barrington by 3d Grand Duke (16182)—by Grand Turk (12767)—by Earl of Derby (10177)—by Earl of Liverpool (9061)—by 2d Duke of Cambridge (3638)—by Belvedere (1706)—by son of Herdsman (304)—by Wonderful (700) —by Alfred (23)—by Young Favorite (9946).

A. H. B., vol. 21, p. 16691, and E. H. B., vol. 27, p. 283, under dam.

HAVERING WATERLOO 3D—Red heifer, calved March 31, 1879, bred by D. McIntosh, Havering Park, got by Duke of Havering (33664), out of Waterloo 38th by Grand Duke 11th (21849)— Waterloo 27th by Duke of Geneva (19614)—Waterloo 17th by Red Knight (11976)—Waterloo 14th by Grand Duke (10284)—Waterloo 13th by 3d Duke of Oxford (9047)—Waterloo 9th by 2d Cleveland Lad (3408)—Waterloo 6th by Duke of Northumberland (1940)— Waterloo 3d by Norfolk (2377)—by Waterloo (2816)—by Waterloo (2816).

A. H. B., vol. 22, p. 17178, and E. H. B., vol. 26, p. 540, under dam.

BARON KIRKLEVINGTON 42801—Red and white, calved Nov. 21, 1881, bred by Sir C. Lampson, calved the property of C. M. Lansing, Niagara, Ont., got by 2d Duke of Sussex (43130), out of Rowfant Kirklevington 3d by 3d Duke of Hillhurst (30975), &c., as in dam, above.

OXFORD DUKE 53045 (45297)—Roan bull, calved Nov. 2. 1879, bred by E. Holden, Laurel Mount, Shepley, Yorkshire, got by Duke of Tregunter 5th (33743), out of Lady of Oxford 17th by 9th Duke of Geneva (28391)—Lady of Oxford 13th by Baron of Oxford (23371)—7th Lady of Oxford by 6th Duke of Thorndale (23794)—2d Lady of Oxford by 2d Grand Duke (12961)—Oxford 13th by 3d Duke of York (10166)—Oxford 5th by Duke of Northumberland (1940)—Oxford 2d by Short Tail (2621)—Matchem Cow by Matchem (2281)—by Young Wynyard (2859).

ROWFANT DUKE OF OXFORD 3d (47010)—Red and white, calved April 29, 1881, bred by Sir C. M. Lampson, Bart., got by Grand Duke 36th (43306), out of 12th Maid of Oxford by Duke of Underley (33745)—10th Maid of Oxford by 4th Duke of Geneva (30958)—3d Maid of Oxford by Grand Duke of Oxford (16184)—Oxford 20th by Marquis of Carrabas (11789)—Oxford 5th by Duke of Northumberland (1940)—Oxford 2d by Short Tail (2621)—Matchem Cow by Matchem (2281)—by Young Wynyard (2859).

VISCOUNTESS BARRINGTON THIRD ALIAS GRAND DUCHESS OF BARRINGTON FIFTH—White, calved Sept. 9, 1881, bred by Wm. Ashburner, calved in Canada, got by Duke of Oxford 49th (41416), out of Viscountess Barrington 2d by Baron Oxford 4th (25580)—Grand Duchess of Barrington 2d by Duke of Brailes (23724), &c., as in Viscountess of Barrington 2d, in this list.

A. H. B., vol. 23, p. 17860.

GRAND DUCHESS BARRINGTON 4TH—White heifer, calved Oct. 12, 1880, bred by W. Ashburner, Coneshead Grange, Ulvorston, Lancashire, got by Duke of Oxford 49th (41416), out of Viscountess Barrington 2d, by Baron Oxford 4th (25580), &c., as above.

A. H. B., vol. 25, p. 770, and E. H. B., vol. 27, p. 283, under dam.

DUCHESS WILD EYES—Roan heifer, calved Oct. 23, 1880, bred by Sir W. H. Salt, Bart., New Parks, Leicester, got by Duke of Oxford 47th (41414), out of Wild Duchess of Gloster by 3d Duke of Gloster (33653)—Wild Oxford by 2d Lord Oxford (20215) —Wild Eyes 24th by Lord Barrington 3d (16382)—Wild Eyes 16th by 2d Duke of Oxford (9046)—Wild Eyes 15th by 4th Duke of Northumberland (3649)—by Duke of Northumberland (1940)— by Belvedere (1706)—by Emperor (1975)—by Wonderful (700)—by Cleveland (145)—by Butterfly (104)—by Hollon's Bull (313)— by Mowbray's Bull (2342)—by Masterman's Bull (422)—descended from the stock of M. Dobison.

E. H. B., vol. 27, p. 564, under dam.

ROWFANT DUKE OF OXFORD 2d 42489 (43927)—Red bull, calved May 25, 1879, bred by Sir C. M. Lampson, Bart., got by Duke of Underley 2d (36551), out of Rowfant Oxford by 3d Duke of Hillhurst (30975)—Grand Duchess of Oxford 11th by Grand Duke 10th (21848)—Grand Duchess of Oxford 5th by Priam (18567)—Countess of Oxford by Earl of Warwick (11412)—Oxford 15th by 4th Duke of York (10167)—by 2d Duke of Northumberland (3646)—by Short Tail (2621)—by Matchem (2281)—by Young Wynyard (2859).

Nov., 1881. R. GIBSON,
Of Ilderton, Ont., Canada.

WILD EYES LASSIE 3d—Roan cow, calved Oct. 16, 1877, bred by S. P. Foster, Killhow, Carlisle, got by Duke of Ormskirk (36526), out of Wild Eyes Lassie 2d by 22d Duke of Oxford (31000) —Wild Eyes Lassie by 3d Duke of Claro (23729)—Wild Eyes 24th by 4th Duke of Oxford (11387)—Wild Eyes 22d by Wild Duke (19148)—Wild Eyes 20th by Lord Barrington 5th (13170)—Wild Eyes 16th by 2d Duke of Oxford (9046)—Wild Eyes 15th by 4th Duke of Northumberland (3649)—Wild Eyes 8th by Duke of North-umberland (1940)—Wild Eyes 2d by Belvedere (1706)—Wild Eyes by Emperor (1975)—by Wonderful (700)—by Cleveland (145)—by Butterfly (104)—by Hollon's Bull (313)—by Mowbray's Bull (2342) —by Masterman's Bull (422)—descended from M. Dobison's stock.

A. H. B., vol. 25, p. 771, and E. H. B., vol. 27, p. 398.

WILD EYES WINSOME 3D—Roan heifer, calved Oct. 23, 1880, bred by J. Rigg, Wrotham Hill Park, Sevenoaks, Kent, got by Duke of Oxford 26th (33708), out of Lightburne Winsome by Duke of Oxford 23d (31001)—Winsome 7th by Grand Duke 10th (21848)—Winsome by Oxford 2d (18507)—Beauty by Crusade (7938)—Bright Eyes by 3d Duke of York (10166)—Wild Eyes 23d by 2d Cleveland Lad (3408)—Wild Eyes 9th by Duke of Northumberland (1940)—Wild Eyes 3d by Belvedere (1706)—Wild Eyes by Emperor (1975), &c., as above.

A. H. B., vol. 25, p. 848, and E. H. B., vol. 27, p. 557, under dam.

WILD EYES WINSOME 4TH—Roan, calved Jan. 26, 1881, bred by J. Rigg, got by Duke of Oxford 26th (33708), out of Wild Eyes Winsome by Duke of Underly 2d (36551)—Lightburne Winsome by Duke of Oxford 23d (31001)—Winsome 7th by Grand Duke 10th (21848)—Winsome by Oxford 2d (18507)—Beauty by Crusade (7938), &c., as above.

A. H. B., vol. 26, p. 560, and E. H. B., vol. 28, p. 561, under dam.

LADY YORK AND OXFORD BATES—Roan heifer, calved Feb. 23, 1879, bred by Mrs. Fawcett, Scaleby Castle, Carlisle, got by Baron Turncroft Oxford 4th (37822), out of Lady York and Thorndale Bates 2d by 8th Duke of York (28480)—Lady Tregunter Bates by 2d Duke of Tregunter (26022)—Lady Thorndale Bates by 4th Duke of Thorndale (17750)—Lady Bates 3d by 4th Duke of Oxford (11387)—Lady Bates 2d by The Buck (13836)—Lady Bates by Duke of Gloster (11382)—Lady Blanche by 4th Duke of York (10167)—Lady Barrington 8th by 2d Duke of Oxford (9046)—Lady Barrington 5th by 4th Duke of Northumberland (3649)—Lady Barrington 3d by Cleveland Lad (3407)—Lady Barrington 2d by Belvedere (1706)—Lady Barrington by a son of Herdsman (304)—Young Alicia by Wonderful (700)—by Alfred (23)—by Young Favorite (6994).

A. H. B., vol. 26, p. 1104, and E. H. B., vol. 28, p. 381.

LADY YORK AND THORNDALE BATES 6TH—Roan heifer, calved Jan. 29, 1881, bred by Mrs. Fawcett, got by Grand Duke of Kirklevington 4th (43317), out of Lady York and Thorndale Bates 3d by Duke of York 8th (28480)—Lady Tregunter Bates by Duke of Tregunter 2d (26022)—Lady Thorndale Bates by Duke of Thorndale 4th (17750)—Lady Bates 3d by Duke of Oxford 4th (11387)—Lady Bates 2d by The Buck (13836), &c., as above.

A. H. B., vol. 24, p. 18618.

LADY YORK AND THORNDALE BATES 7TH—Roan heifer, calved Feb. 25, 1881, bred by Mrs. Fawcett, got by Grand Duke of Kirklevington 4th (43317), out of Lady York and Thorndale Bates 2d by Duke of York 8th (28480)—Lady Tregunter Bates by Duke of Tregunter 2d (26022)—Lady Thorndale Bates by Duke of Thorndale 4th (17750)—Lady Bates 3d by Duke of Oxford 4th (11387)—Lady Bates by The Buck (13836), &c., as above.

LALLY BARRINGTON 6TH—Roan, calved April 10, 1881, bred by the Earl of Lathom, Lathom, Lancashire, got by 51st Duke of Oxford (43122), out of Lally Barrington 3d by Duke of Ormskirk (36526)—Lally 18th by 8th Duke of Geneva (28390)—Lally 13th by 3d Duke of Claro (23729)—Lally 3d by 4th Duke of Oxford (11387) —Lally by Earl of Derby (10177)—Olive Leaf 3d by Earl of Liverpool (9061)—Olive Leaf 2d by 2d Duke of Cambridge (3638)—Olive Leaf by Belvedere (1706)—Lady Barrington by son of Herdsman (304)— Young Alicia by Wonderful (700)—Old Alicia by Alfred (23)—by Young Favorite (6994).

E. H. B., vol. 28, p. 469, under dam.

ROWFANT KIRKLEVINGTON 5TH—Red, calved Nov. 23, 1879, bred by Sir C. M. Lampson, Bart., got by 2d Duke of Underley (36551), out of Kirklevington Duchess 5th by 2d Duke of Claro (21576)—Duchess of Kent by Lord Liverpool (22168)—Kirklevington 14th by 4th Duke of Oxford (11387)—Kirklevington 7th by Earl of Derby (10177)—Kirklevington 4th by Earl of Liverpool

(9061)—Kirklevington 1st by Duke of Northumberland (1940)—
Nell Gwynne by Belvedere (1706)—Northallerton by Son of 2d Hub-
back (2683)—a cow of Mr. Bates', descended from the stock of Mr.
Maynard, of Eryholme.

E. H. B., vol. 26, p. 514, under dam.

KIRKLEVINGTON DUCHESS 27TH—Red, calved April 22,
1880, bred by Mrs. and Mr. M. Fawcett, got by Lally's Grand Duke
(43449), out of Kirklevington Duchess 11th by 2d Duke of Gloster
(28392)—Kirklevington Duchess 5th by 2d Duke of Claro (21576)—
Duchess of Kent by Lord Liverpool (22168), &c., as in dam, above.

E. H. B., vol. 27, p. 388, under dam.

LADY ROTHESAY BATES—Roan cow, calved April 6, 1877,
bred by Colonel Kingscote, Kingscote, got by Duke of Rothesay
(36534)—Lady Bates 2d by 3d Duke of Claro (23729)—Lady Bates
by Grand Duke 15th (21852)—Lily by Lord Liverpool (22168)—
Levity by Lord Thoresby (14856)—Lily Bell by Horrox (11591)—
Lily by 2d Duke of Oxford (9044)—Harmless by Cleveland Lad
(3407)—Hawkey by Red Rose Bull (2493)—Hart by Rex (1375)—
bought of Mr. Richardson, of Hart.

A. H. B., vol. 24, p. 18612, and E. H. B., vol. 25, p. 528, under
dam.

LADY HILLHURST BATES 2D—Roan, calved July 29, 1878,
bred by Colonel Kingscote, got by Duke of Hillhurst (28401), out
of Lady Bates 2d by 3d Duke of Claro (23729), &c., as in dam,
above.

E. H. B., vol. 27, p. 469.

GRAND DUCHESS CAROLINA 3D—Roan, calved Aug. 10,
1879, bred by J. Rigg, Wrotham Hill Park, got by Duke of Oxford
26th (33708), out of Grand Duchess Carolina 2d by Duke of Geneva
8th (28390)—Grand Duchess Carolina by Grand Duke 10th
(21848)—Carolina 5th by Duke of York 7th (17754)—Carolina 2d
by Douglas (12714)—Caroline by 2d Cleveland Lad (3408)—Careless
by Short Tail (2621)—Craggs by Son of 2d Hubback (2683)—

Craggs, from Mr. Bates' herd, descended from the stock of Mr. Maynard, of Eryholme.

A. H. B., vol. 25, p. 647, and E. H. B., vol. 26, p. 611, under dam.

SURMISE DUCHESS 19TH—Roan, calved Sept. 8, 1879, bred by Sir C. M. Lampson, Bart., got by 2d Duke of Underley (36551), out of Fancy Duchess by 9th Duke of Geneva (28391)—Fancy 2d by Rowfant 1st (22767)—Fancy by 4th Duke of Thorndale (17750) —Surmise by Duke of Gloster (11382)—Silence by Earl of Derby (10177)—Secret 3d by Duke of Sutherland (6945)—Secret 2d by Locomotive (4242)—Secret by Short Tail (2621)—White Rose 1st by Gambier (2046)—White Rose by Young Wynyard (2859)—by a bull owned by Messrs. Colling.

A. H. B., vol. 24, p. 18799, and E. H. B., vol. 26, p. 514, under dam.

SURMISE DUCHESS 25TH—Roan, calved July 31, 1880, bred by Sir C. M. Lampson, Bart., got by 2d Marquis of Oxford (37055), out of Fancy Duchess by 9th Duke of Geneva (28391), &c., as in dam, above.

A. H. B., vol. 24, p. 18799, and E. H. B., vol. 27, p. 474, under dam.

ROWFANT FANTAIL—Red cow, calved Nov. 18, 1878, bred by Sir C. M. Lampson, Baronet, got by 2d Duke of Underley (36551), out of Fantail Duchess 2d by 9th Duke of Geneva (28391) —Fantail by Barleycorn (17348)—Fair Helen by Gen. Canrobert (12927)—Florella by 5th Duke of York (10168)—Flirt by 4th Duke of Northumberland (3649)—Flirt by Short Tail (2621)—Fletcher 2d by Belvedere (1706)—Fletcher by Young Wynyard (2859)—descended from J. Brown's Red Bull (97).

E. H. B., vol. 26, p. 514, under dam.

ROWFANT PEACH 3D—Roan, calved June 15, 1879, bred by Sir C. M. Lampson, Bart., got by 2d Duke of Underley (36551), out of Peach Blossom 12th by 8th Duke of York (28480)—Peach Blossom 6th by Earl of Gloster (21644)—Peach Blossom 2d by

Baron Westbury (19287)—Peach by Gen. Canrobert (12927)—Poppy by 2d Cleveland Lad (3408)—Place 2d by 2d Earl of Darlington (1945)—Place by Son of 2d Hubback (2683)—a cow of Mr. Bates'.

E. H. B., vol. 26, p. 515, under dam.

FAWSLEY DUCHESS—Roan, calved Nov. 4, 1877, bred by Sir C. M. Lampson, Bart., got by 2d Duke of Underley (36551), out of Oxford Fawsley 4th by Grand Duke of Kent (26289)—Oxford Fawsley 2d by Lord Oxford 2d (20215)—Fawsley 3d by 4th Grand Duke (19874)—Coquilicot by Duke of Cambridge (12742)—Blouzelind by Earl of Dublin (10178)—Candytuft by Janizary (8175)—Cathleen by Philip (1774)—Rosy by Rob Roy (557)—by Satellite (1420)—by Sir Dimple (594)—by Styford (629).

E. H. B., vol. 26, p. 514, under dam.

WILD PRINCE 14TH 49525 (45806)—Roan, calved March 13, 1880, bred by Earl of Lathom, got by 4th Baron of Oxford (25580), out of Lady Wild Eyes by 8th Duke of Geneva (28390)—Wild Eyes 24th by 4th Duke of Oxford (11387)—Wild Eyes 22d by Wild Duke (19148)—Wild Eyes 20th by Lord Barrington 1st (13170)—Wild Eyes 16th by 2d Duke of Oxford (9046)—Wild Eyes 15th by 4th Duke of Northumberland (3649)—Wild Eyes 8th by Duke of Northumberland (1940)—Wild Eyes 2d by Belvedere (1706)—Wild Eyes by Emperor (1975)—by Wonderful (700)—by Cleveland (145)—by Butterfly (104)—by Hollon's Bull (313)—by Mowbray's Bull (2342)—by Masterman's Bull (422)—descended from M. Dobison's stock.

July 20, 1882. RICHARD GIBSON.

From Liverpool to Quebec.

GRAND DUCHESS OF WATERLOO—Roan, calved Sept. 2, 1877, bred by H. Sharpley, Esq., got by Grand Duke 27th (34067), out of Oxford's Waterloo 4th by Oxford's Baronet (29499)—Oxford's Waterloo 3d by Imperial Oxford (18084)—Oxford's Waterloo by Lord Oxford 2d (20215)—Waterloo 19th by 2d Grand Duke (12961)—Waterloo 15th by Matadore (11800)—Waterloo 12th by 3d

Duke of York (10166)—Waterloo 4th by Cleveland Lad (3407)—
Waterloo 3d by Norfolk (2377)—Waterloo Cow by Waterloo (2816)
—by Waterloo (2816).

E. H. B., vol. 24, p. 644, under dam.

GRAND DUCHESS OF WATERLOO 2d—Red, calved May 3,
1881, bred by H. Sharpley, Esq., got by Grand Duke 33d (39946),
out of Grand Duchess of Waterloo by Grand Duke 27th (34067)—
Oxford's Waterloo 4th by Oxford's Baronet (29499), &c., as in dam,
above.

E. H. B., vol. 28, p. 577, under dam.

LADY YORK AND UNDERLEY BATES—Roan, calved Dec.
21, 1877, bred by Mrs. Fawcett, Scaleby Castle, got by 2d Duke of
Underley (36551), out of Lady York and Thorndale Bates 2d by 8th
Duke of York (28480)—Lady Tregunter Bates by 2d Duke of Tregunter (26022)—Lady Thorndale Bates by 4th Duke of Thorndale
(17750)—Lady Bates 3d by 4th Duke of Oxford (11387)—Lady
Bates 2d by The Buck (13836)—Lady Bates by Duke of Gloster
(11382)—Lady Blanche by 4th Duke of York (10167)—Lady Barrington 8th by 2d Duke of Oxford (9046)—Lady Barrington 5th by
4th Duke of Northumberland (3649)—Lady Barrington 3d by
Cleveland Lad (3407)—Lady Barrington 2d by Belvedere (1706)—
Lady Barrington by son of Herdsman (304)—Young Alicia by
Wonderful (700)—Old Alicia by Alfred (23)—by Young Favorite
(6994).

E. H. B., vol. 28, p. 410.

WILD DUCHESS OF GENEVA 3d—Red, calved Dec. 26,
1873, bred by E. H. Cheney, got by 9th Duke of Geneva (28391),
out of Wild Oxford by 2d Lord Oxford (20215)—Wild Eyes 24th
by Lord Barrington 3d (16382)—Wild Eyes 16th by 2d Duke of
Oxford (9046)—Wild Eyes 15th by 4th Duke of Northumberland
(3649)—Wild Eyes 8th by Duke of Northumberland (1940)—Wild
Eyes 2d by Belvedere (1706)—Wild Eyes by Emperor (1975)—by
Wonderful (700)—by Cleveland (145)—by Butterfly (104)—by

32

Hollon's Bull (313)—by Mowbray's Bull (2342)—by Masterman's Bull (422)—descended from M. Dobison's stock.

E. H. B., vol. 24, p. 304.

LADY TURNCROFT WILD EYES 2D—Red, calved Nov. 29, 1879, bred by Rev. P. Graham, got by 7th Duke of Tregunter (38194), out of Lady Ashton Wild Eyes 2d by Grand Duke of Thorndale 2d (31298)—Winsome 10th by 18th Duke of Oxford (25995)—Winsome 4th by Grand Duke 10th (21848)—Winsome by Oxford 2d (18507)—Beauty by Crusade (7938)—Bright Eyes by 3d Duke of York (10166)—Wild Eyes 23d by 2d Cleveland Lad (3408)—Wild Eyes 9th by Duke of Northumberland (1940)—Wild Eyes 3d by Belvedere (1706)—Wild Eyes by Emperor (1975), &c., as above.

A. H. B., vol. 25, p. 770, and E. H. B., vol. 26, p. 455, under dam.

LADY TURNCROFT WILD EYES 3D—Red, calved Nov. 14, 1880, bred by R. P. Graham, got by 7th Duke of Tregunter (38194), out of Lady Ashton Wild Eyes 2d by Grand Duke of Thorndale 2d (31298)—Winsome 10th by 18th Duke of Oxford (25995), &c., as above.

A. H. B., vol. 26, p. 1104, and E. H. B., vol. 28, p. 409, under dam.

WATERLOO 42D—Red, calved July 17, 1875, bred by Lord Penrhyn, got by Grand Duke 20th (31281), out of Waterloo 31st by Grand Duke 11th (21849)—Waterloo 17th by Red Knight (11976)—Waterloo 14th by Grand Duke (10284)—Waterloo 13th by 3d Duke of Oxford (9047)—Waterloo 9th by 2d Cleveland Lad (3408)—Waterloo 6th by Duke of Northumberland (1940)—Waterloo 3d by Norfolk (2377)—Waterloo Cow by Waterloo (2816)—by Waterloo (2816).

E. H. B., vol. 26, p. 371.

MARCHIONESS OF TURNCROFT—Red and white, calved July 19, 1876, bred by Rev. P. Graham, got by Baron Turncroft Oxford 2d (33087), out of Marchioness 5th by 2d Duke of Collingham (23730)—Siddington by 4th Duke of Oxford (11387)—Kirklevington 7th by Earl of Derby (10177)—Kirklevington 4th by

Earl of Liverpool (9061)—Kirklevington 1st by Duke of Northumberland (1940)—Nell Gwynne by Belvedere (1706)—Northallerton by Son of 2d Hubback (2683)—a cow of Mr. Bates', descended from the stock of Mr. Maynard, of Eryholme.

A. H. B., vol. 25, p. 845, and E. H. B., vol. 23, p. 468, under dam.

MARCHIONESS OF TURNCROFT 2D—Red and white, calved May 13, 1878, bred by Rev. P. Graham, got by Baron Turncroft Oxford 2d (33087), out of Marchioness 5th by 2d Duke of Collingham (23730), &c., as above.

A. H. B., vol. 25, p. 770, and E. H. B., vol. 28, p. 410.

SURMISE DUCHESS 16TH—Roan, calved April 24, 1878, bred by Sir Curtis Lampson, got by 2d Duke of Underley (36551), out of Fancy Duchess by 9th Duke of Geneva (28391)—Fancy 2d by Rowfant 1st (22767)—Fancy by 4th Duke of Thorndale (17750)—Surmise by Duke of Gloster (11382)—Silence by Earl of Derby (10177)—Secret 3d by Duke of Sutherland (6945)—Secret 2d by Locomotive (4242)—Secret by Short Tail (2621)—White Rose 1st by Gambier (2046)—White Rose by Young Wynyard (2859)—bulls owned by Messrs. Colling.

A. H. B., vol. 25, p. 771, and E. H. B., vol. 26, p. 514, under dam.

WILD DUKE OF GENEVA 14TH (48946)—Red, calved April 20, 1882, bred by Sir W. G. Armstrong, Bart., Cragside, got by Duke of Oxford 48th (41415), out of Wild Duchess of Geneva 3d by 9th Duke of Geneva (28391), &c., as in dam, above.

VISCOUNT OXFORD 6TH (47216)—Roan, calved Sept. 18, 1880, bred by T. Holford, got by 4th Baron of Oxford (25580), out of Viscountess Oxford by 23d Grand Duke (34063)—Baroness of Oxford 3d by Duke of Hillhurst (28401)—Baroness of Oxford by 2d Duke of Claro (21576)—Lady Oxford 5th by 3d Duke of Thorndale (17749)—Lady Oxford 4th by 2d Grand Duke (12961)—Maid of Oxford by Lord of Eryholme (12205)—Oxford 13th by

3d Duke of York (10166)—Oxford 5th by Duke of Northumberland (1940)—Oxford 2d by Short Tail (2621)—Matchem Cow by Matchem (2281)—by Young Wynyard (2859).

GRAND DUKE OF WATERLOO 51878 ALIAS BLUCHER—Red and white, calved Aug. 10, 1882, bred by J. P. Clark, got by Lord Wildrake (45175), out of Waterloo 42d by Grand Duke 20th (31281)—Waterloo 31st by Grand Duke 11th (21849)—Waterloo 17th by Red Knight (11976)—Waterloo 14th by Grand Duke (10284)—Waterloo 13th by 3d Duke of Oxford (9047)—Waterloo 9th by 2d Cleveland Lad (3408)—Waterloo 6th by Duke of Northumberland (1940)—Waterloo 3d by Norfolk (2377)—Waterloo Cow by Waterloo (2816)—by Waterloo (2816).

LADY BARRINGTON BATES—Red, calved Oct. 17, 1882, bred by Rev. P. Graham, got by Grand Duke 39th (43308), out of Lady York and Underley Bates by 2d Duke of Underley (36551), &c., as in dam, above.

A. H. B., vol. 25, p. 770.

MARQUIS OF KIRKLEVINGTON 52664—Red and white, calved Nov. 28, 1882, bred by Rev. P. Graham, got by Grand Duke 39th (43308), out of Marchioness of Turncroft by Baron Turncroft Oxford 2d (33087), &c., as in dam, above.

SURMISE DUKE—White, calved Aug. 5, 1882, bred by Sir Curtis Lampson and calved R. Gibson's, got by Grand Duke 37th (43307), out of Surmise Duchess 16th by 2d Duke of Underley (36551), &c., as in dam, above.

A. H. B., vol. 25, p. 771, under dam.

WILD EYES LE GRAND—Red, calved Sept. 17, 1882, bred by Rev. P. Graham, got by Grand Duke 39th (43308), out of Lady Turncroft Wild Eyes 2d by 7th Duke of Tregunter (38195), &c., as in dam, above.

A. H. B., vol. 25, p. 770, under dam.

GRAND DUKE OF WATERLOO 51879—Red, calved April 8, 1883, bred by Henry Sharpley, Timber, Eng., got by Grand

Duke 33d (39946), out of Grand Duchess of Waterloo by Grand Duke 27th (34068), &c., as in dam, above.

Oct. II, 1883. RICHARD GIBSON,
Of Delaware, Ont., Can. By Steamship Ontario, to Quebec.

DUKE OF OXFORD 60TH 55734 (46265)—Red, calved Oct. 5, 1881, bred by the Duke of Devonshire, got by Duke of Gloster 7th (39735), out of Grand Duchess of Oxford 27th by Baron Oxford 4th (25580)—Grand Duchess of Oxford 6th by Imperial Oxford (18084)—Grand Duchess of Oxford 4th by Grand Duke of Wetherby (17997)—Oxford 15th by 4th Duke of York (10167)—by 2d Duke of Northumberland (3646)—by Short Tail (2621)—by Matchem (2281)—by Young Wynyard (2859).

WILD WINSOME 4TH—Red, calved Dec. 14, 1878, bred by Earl of Feversham, Duncombe Park, got by 5th Duke of Tregunter (33743), out of Wild Winsome by Duke of Oxford 20th (28432) —Winsome 5th by Grand Duke 10th (21848)—Winsome 2d by Lord Oxford (20214)—by Oxford 2d (18507)—by Crusade (7938)— by 3d Duke of York (10166)—by 2d Cleveland Lad (3408)—by Duke of Northumberland (1940)—by Belvedere (1706)—by Emperor (1975)—by Wonderful (700)—by Cleveland (145)—by Butterfly (104)—by Hollon's Bull (313)—by Mowbray's Bull (2342)—by Masterman's Bull (422)—descended from the stock of M. Dobison.
E. H. B., vol. 28, p. 385, and A. H. B., vol. 26, p. 744.

LILY 3D—Roan, calved April 9, 1880, bred by the Duke of Devonshire, got by Duke of Gloster 7th (39735), out of Lily 2d by 5th Duke of Wetherby (31033)—Lily by Lord Liverpool (22168)— Levity by Lord Thoresby (14856)—by Horrox (11591)—by 2d Duke of Oxford (9046)—by Cleveland Lad (3407)—by Red Rose Bull (2493)—by Rex (1375)—bred by Mr. Richardson, of Hart.
A. H. B., vol. 26, p. 744, and E. H. B., vol. 29, p. 413.

BARON HOLKER—Roan, calved Nov. 22, 1883, bred by Duke of Devonshire (calved the property of R. Gibson), got by Baron Oxford 8th (41057), out of Lily 3d, above.

Oct. 25, 1883. R. GIBSON,

Of Delaware, Ont., Can. By Steamship Dominion, to Quebec.

LILY 4TH—Red and little white, calved April 12, 1881, bred by The Duke of Devonshire, got by Duke of Gloster 7th (39735), out of Lily 2d by Duke of Wetherby 5th (31033)—Lily by Lord Liverpool (22168)—Levity by Lord Thoresby (14856)—by Horrox (11591)—by 2d Duke of Oxford (9046)—by Cleveland Lad (3407)—by Red Rose Bull (2493)—by Rex (1375)—bred by Mr. Richardson, of Hart.

Vol. 28, p. 363, E. H. B., under dam, and vol. 26, p. 744, A. H. B.

1858. JOHN GILL.

Toronto Township, Can.

JENNY LIND—Roan, calved March 26, 1852, bred by Francis Jordan, Eastburn, Driffield, Yorkshire, got by Lord Grey (10446), out of Jenny Lind by Lord Marmion (8244)—by Sir Launcelot (5166)—by Mowthorpe (2343).

C. H. B., vol. 1, p. 323.

COBDEN 13670 [136]—Roan, calved, Jan. 15, 1859, bred by F. Jordan, got by Emperor [224], out of Jenny Lind by Lord Gray (10446), &c., as in Jenny Lind, above.

1837. GORDON & BRADFORD.

Nashville, Tenn.

HIBERNIA—Vol. 24, p. 18555.

1872. J. E. GOULD.

E. Whitby, Can.

ROSE—Red and white, calved July 14, 1869, bred by Charles Hardwick, Somerset, got by James 1st (24202), out of Columbine by Young Duke of Cambridge (17708)—Cowslip by Rivers (10716) Raspberry by Mozart (11830)—Cherry by Sterling (5330)—Wide

2d by Frederick (3836)—Old Wide by Favorite (3768)—by Tathwell Studley (5401)—by son of Waddingworth (668)—a Turnell cow.

C. H. B., vol. 3, p. 736.

EMILY—Roan, calved June 15, 1870, bred by Chas. Hardwick, Somerset, got by James 1st (24202), out of Columbine by Young Duke of Cambridge (17708)—Cowslip by Rivers (10716)—Raspberry by Mozart (11830)—Cherry by Sterling (5330)—Wide 2d by Frederick (3836)—Old Wide by Favorite (3768)—by Tathwell Studley (5401)—by a son of Waddingworth (668)—a Turnell cow.

C. H. B., vol. 3, p. 453.

EMILY 2D—Light roan, calved March 29, 1872, bred by Chas. Hardwick, got by Lygon (24494), out of Emily by James 1st (24202), &c., as in Emily, above.

C. H. B., vol. 3, p. 453.

VISCOUNT [4197]—Light roan, calved April 29, 1871, bred by Chas. Hardwick, got by Lygon (24494), out of Airy by Lord Churchill (20161)—Anemone by Young Duke of Cambridge (17708) —Airy by Squire Gwynne (12140)—Agnes by Fanatic (8054)—Almond by Splendid (5298)—Acorn by Cucumber (1891)—by Edmund (1954)—by Grazier (1085)—by Statesman (621)—by Plato (506)—by a grandson of Comet (155).

May 5, 1881. GOVERNMENT AGRICULTURAL EXPERIMENT COLLEGE,
Of Guelph. Ont.

SIR LEONARD (45613)—Roan, calved Jan. 4, 1880, bred by Hugh Aylmer, West Dereham Abbey, Norfolk, got by Sir Wilfrid (37484), out of Countess 3d by High Sheriff (26392)—Countess 2d by British Crown (21322)—Countess by Sir Sam (25171)—Calendula by Majestic (13279)—Calomel by Hamlet (8126)—by Leonard (4210) —by Buckingham (3239)—descended from the stock of Sir M. W. Ridley, Bart.

BETA—Red heifer, calved Sept. 9, 1879, bred by H. Aylmer, got by Sir Wilfrid (37484), out of Beautiful Star by Hyperion (34196)—

Balmful by High Sheriff (26392)—Banter by Cedric (19415)—
Blithsome by Sir James (16980)—Bashful by Prince Imperial (15095)
—Blissful by Grand Duke (10284)—by Baron Warlaby (7813)—by
Leonard (4210)—by Young Matchem (2282)—by Jerry (4097)—by
Young Pilot (4702)—by Pilot (496)—by son of Apollo (36).

Aug. 31, 1876. GOVERNMENT FARM,

Of Guelph, Canada. By Prof. Brown, per Ship Texas, from Liverpool.

ROSALIE—Roan heifer, calved July 25, 1874, bred by Her
Majesty the Queen, got by Prince Albert 2d (29588), out of Rosebud by Rajah (22670)—Rosabella by Goldsmith (10277)—Darling
by Goldsmith (10277)—by Brilliant (8905).

E. H. B., vol. 21, p. 531, under dam.

DUKE OF BEDFORD (36466)—Roan, calved Jan. 6, 1875,
bred by Mrs. Scott, Thorpe, Chestsy, Surrey, got by King Tom
(31521), out of Louise by Prince Louis (22603)—Beauty by Fitz
Clarence (14552)—Miss Folly by Prince Alfred (13494)—Folly by
Paris (7314)—by Vanguard (5545)—by Robin Hood (4970)—by
Anticipation (750)—by Emperor (1014)—by Young Windsor (699)
—by Windsor (698).

MANRICO 2D—Red and white, calved Dec. 8, 1876, bred by
the Queen, got by Manrico (26805), out of Rosalie by Prince Albert
2d (29588), &c., as in dam, above.

No. 1257 in Brit.-Am. H. B.

Aug. 29, 1871. GOVERNMENT STOCK FARM.

Charlottetown, Prince Edward Island. From Liverpool, by Ship Moneynash.

GLENSMAN (28716)—Roan, calved May 19, 1870, bred by Mr.
R. S. Bruere, Braithwaite Hall, got by Booth's Kinsman (25658),
out of Rose Wreath by Windsor (14013)—Rose Garland by Baron
Warlaby (7813)—Garland by The Silky Laddie (10947)—Damsel
by Rouge (5012)—by Shipton (2620)—by Cleveland (3404)—by
Danby (3550)—by son of Studley Grange (1483)—by son of Duke
(225)—by Midas (4470)—by Nailer (4528)—by Ambo (746).

Aug. 13, 1881. GOVERNMENT OF ST. JOHNS,

Of New Brunswick. Can.   Per Steamer Scandinavian. Aug. 13. 1881. from Liverpool. by Simon Beattie.

QUEEN ANNE—Roan, calved Feb. 14, 1880, bred by T. Marshall. Howes. Annan, Scotland, got by Conductor (36389), out of Queen Adelaide by Prince Thomas (35183)—Queen Duchess by Marquis of Lorne (29298)—Queen Beauty by Knight (24267)—Red Beauty by Knight of the Tyne (20087)—Beauty by Kossuth (11646) —by Sir William (8598)—by Roger (13615)—by Studley (628)—by Young Rockingham (2547)—by Wonder (2853)—by Denton (198) —by Ladrone (353)—by Henry (301).

E. H. B., vol. 27. p. 506. under dam.

Roan cow, calved Feb. 15, 1880, bred by T. Marshall. Howes, got by Prince Thomas (35183), out of Flora 3d by Bright Hope (28081)—Caroline by Comet (21449)—Flora 2d by Earl of Derby (12810)—Moss Rose by Prince Edward (6334)—by Traveler (6617) —by Charles (3343)—by Commodore (5874)—by Cupid (5900)—by Parson (1306)—by St. John (572)—by Pope (514)—by Chilton (136).

Red and white calf. calved Jan. 5, 1880. bred by T. Marshall. Howes. got by Prince Thomas (35183), out of Queen Duchess by Marquis of Lorne (29298)—Queen Beauty by Knight (24267), &c., as in Queen Anne, above.

Roan heifer, calved March 24. 1880, bred by T. Marshall. Howes. got by Prince Thomas (35183). out of Cherry by Bright Hope (28081)—Cherry Duchess by Mr. Banting (22366)—Red Duchess 4th by Bumper (10005)—by Young Peer (13462)—by Duke of St. Albans (8001)—by Tyne (7653)—by Wolsington (2852)—by Brandon (3206)—by Young Sovereign—by Lonsdale (379)—by Irishman (329).

STRAWBERRY DUCHESS 13TH—Red and white, calved Jan. 4, 1880, bred by H. Aylmer, West Dereham Abbey, Norfolk, got by Sir Simeon (42412). out of Strawberry Duchess 9th by Sir Wilfrid (37484)—Strawberry Duchess 4th by Royal Killerby (32396)—

33

Strawberry Duchess by Prince Christian (22581)—Windsor Strawberry by Imperial Buckingham (19998)—by Duke of Buckingham (14428)—by Duke of Cambridge (12747)—by St. Thomas (2824) —by Chorister (3378)—by Tom Gwynne (5498)—by Wellington (2824)—by Marmion (406)—by Ossian (476).

E. H. B., vol. 27, p. 287, under dam.

FRANCESCA—Red, calved Feb. 13, 1880, bred by H. Aylmer, got by Sir Simeon (42412), out of Florence by Sir Wilfrid (37484) —Fleurette by Royal Prince (32404)—Fleur-de-lis by Duke of Edinburgh (23741)—Feodorovna by Zaratan (21134)—by Newton (20403)—by The Captain (5422)—by Hero (4021)—by Percy (1314) —by Lord Grantham's son of Comet (155).

E. H. B., vol. 27, p. 286, under dam.

MUSKETEER—Roan bull, calved June 17, 1880, bred by A. & A. Mitchell, Alloa, got by Consort (42967), out of White Rose by Foggathorpe 1st (31179)—Moss Rose by Colin (25795)—Melrose Bud by Southwick (15321)—Roan Moss Rose by White Prince (12298)—Young Moss Rose by Dusty Miller (9055)—by Earl of Durham (5965)—by Newton (2367)—by Emperor (1974)—bred from the Chilton stock.

E. H. B., vol. 28, p. 510, under dam.

BELLMAN (44406)—Roan, calved July 9, 1880, bred by A. & A. Mitchell, got by Aerolite (40955), out of Bellona by Brocklesby (36288)—Bridal Belle by Foggathorpe 1st (31179)—Belle Isle by Island Chief (28898)—Belle by Red Friar (24913)—by Arthur Gwynne (19244)—by Cardigan (12556)—by Lord Fanny (13187) —by Noble (4578)—by Newton (2367)—by Baronet (1686)—by Reformer (2502)—by Margrave (2263)—by Leopold (2199)—by Hector (2103)—by Surly (2715)—by Traveler (655)—by Colonel (152).

June, 1872. W. R. GRACE,
Of 66 Pine Street, N. Y. By Ship Hutton.

EMPEROR 26529—Roan, calved May 20, 1871, bred by W. D. Dunlop, Monasterboice House, Ireland, got by Gallant Knight (26214), out of Miss Gwynne by Mountebank (24626)—Cherry by Downshire (15898)—Old Cherry by son of Eden (3689)—Fanny by Hopewell (10332)—by Albion (7771).

OLGA—Mr. Grace writes me that he has mislaid the pedigree of Olga.

GEO. GRANT.
Kansas.

DIMPLE—Roan, calved April 30, 1874, bred by the Queen, got by Spring Buck (32585), out of Diadem by Prince of Saxe Coburg (20576)—Dignity by Prince Alfred (13494)—Dairymaid by Lord Foppington (10437)—Blossom by Goldsmith (10277)—Strawberry by Brilliant (8905).

Nov. 15, 1871. GEO. GRANT,
Of Madura, Clay Co., Kansas. By Ship Spain.

FAIRY VERULAM—White heifer, calved March 17, 1871, bred by Mrs. Strickland, Cokethorpe Park, Oxon., got by Earl of Verulam (26077), out of Fairy Oxford by 12th Duke of Oxford (19633)—Fairy Princess by Hayman (16245)—Fairy by St. Patrick (12038)—Fancy by The Star (10951)—Flora by The Admiral (8806) —by Bright (1739)—by Young Lancaster (1162)—by Alfred (23) —by Windsor (698)—by Cupid (177)—by Barker's Bull of Layton (53).

E. H. B., vol. 20, p. 511, under dam.

NIOBE 8TH—Red heifer, calved Dec. 15, 1870, bred by Mrs. Strickland, got by Earl of Virulam (26077), out of Niobe 4th by Oxartes (22471)—Niobe by Sultan (15358)—Brownie by Neptune (11847)—Crisis by Essex Hero (9096)—Countess by Malbro

(3645)—Fairy by Harlequin (3975)—by Captain (3273)—by Emperor (1974)—by Snowball (2647)—by a son of St. Albans (2584)—by son of Lawnsleeves (365)—by Barnaby (1678).

E. H. B., vol. 20, p. 673, under dam.

LORD OF THE MANOR 17583 (31727)—Roan, calved April 10, 1871, bred by Mrs. Strickland, got by Earl of Verulam (26077), out of Red Rose 2d by Oxartes (22471)—Star by Walter (21058)—Rosa by Tomboy (19005)—Quickly by Neptune (11847)—by Essex Hero (9096)—by Duke of Malbro (3645)—by Harlequin (3975)—by Captain (3273)—by Emperor (1974)—by Snowball (2647)—by son of St. Albans (2584)—by Col. Trotter's son of Lawnsleeves (365)—by Barnaby (1678).

GOGGLES 17193—Red bull, calved Nov. 28, 1871, bred by Sir G. R. Phillips, Bart., Western Park, Shipston-on-Stour, got by 18th Duke of Oxford (25995), out of Guilia by Duke of Darlington (21586)—Gionetta by Sarawak (15238)—Smock Frock by Earl of Dublin (10178)—London Pride by Janizary (8175)—by Snowball (8602)—by Caliph (1774)—by Norman (2379)—by White Boy (1580)—by Wyville's Bull—a bull bred by Mr. Charge.

Imported 1872.

May 24, 1876. GEO. GRANT.

Of Victoria, Ellis Co., Kansas. From London.

COLD CREAM 10TH—Red and white heifer, calved May 7, 1874, bred by Her Majesty the Queen, got by King Tom (31521), out of Cold Cream 5th by England's Glory (23889)—Comely by Buckingham (17471)—Cold Cream by Earl of Dublin (10178)—Pansy by Gray Friar (9172)—Freckle by Fawsley (6004)—Furbelow by Little John (4232)—Erato by Marcellus (2260)—Beatrice by Caliph (1774)—Quickly by Swing (2721)—Alamode by Argus (759)—Valuable by Defender (194)—by Petrarch (488)—by brother to R. Colling's White Heifer—by Butterfly (104)—by Globe (278).

A. H. B., vol. 16, p. 12019.

ROYAL WINDSOR 27816—Red and white bull, calved January 7, 1877, bred by Her Majesty the Queen, calved G. Grant's, got by Manrico (26805), out of Cold Cream 10th by King Tom (31521), &c., as above.

ROSE LEAF—Red and white heifer, calved June 18, 1874, bred by the Queen, got by King Tom (31521), out of Rosy by England's Glory (23889)—Rosabella by Prince Louis (22603)—Red Rose by Buckingham (17474)—Rosabella by Goldsmith (10277)—Darling by Goldsmith (10277)—Darling by Brilliant (8905).

A. H. B., vol. 16, p. 12331, and E. H. B., vol. 21, p. 531, under dam.

MATILDA—Roan heifer, calved July 10, 1874, bred by the Queen, got by Prince Albert 2d (29588), out of Mary by Prince Louis (22603)—Evening Star by Sir Ronald (13747)—Sunrise by Lord of the Isles (9325)—Tidy by Chain Bull.

A. H. B., vol. 16, p. 12216.

PEERESS—Red and white heifer, calved Aug. 10, 1875, bred by the Queen, got by King Tom (31521), out of Pastime by My Lord of Dublin (29411)—Pink by Prince Louis (22603)—Pretty Lass by Clansman (17569)—Pretty Maid by Prince Albert (13494)—Pineapple by Sir Ronald (13747)—Pink by Lord of the Isles (9325).

A. H. B., vol. 16, p. 12278.

ROYAL GEORGE 27797—Red and white, calved Oct. 25, 1873, bred by Her Majesty the Queen, Prince Consort's Shaw Farm, Windsor, got by Royal Benedict (27348), out of Cold Cream 4th by England's Glory (23889)—Comely by Buckingham (17471)—Cold Cream by Earl of Dublin (10178)—Pansy by Gray Friar (9172)—Freckle by Fawsley (6004)—by Little John (4232)—by Marcellus (2260)—by Caliph (1774)—by Swing (2721)—by Argus (759)—by Defender (194)—by Petrarch (488)—by brother to R. Colling's White Heifer—by Butterfly (104)—by Globe (278).

E. H. B., vol. 21, p. 530, under dam.

ROSA—Roan heifer, calved Jan. 21, 1874, bred by Her Majesty the Queen, Prince Consort's Shaw Farm, Windsor, got by Prince

Albert 2d (29588), out of Emily by Prince Arthur (20535)—Eleanor by Friar John (12905)—Ellanna by Count (10079)—Eleanor by son of Boz (3201)—White Rose by Waverley (5613)—Rosa by Cedric (3311)—by Fisher's son of Favorite (1028)—by Cœlebs (897)—by Rose's Bull (5009)—by Fisher's Bull (3799).

A. H. B., vol. 16, p. 12321, and E. H. B., vol. 21, p. 530, under dam.

MINNETTE—Red and white heifer, calved Feb. 26, 1874, bred by Her Majesty the Queen, got by King Tom (31521), out of Milkmaid by Prince of Saxe Coburg (20576)—Myrtle by Buckingham (17471)—Minnette by Lord Foppington (10437)—Maggy by Goldsmith (10277).

E. H. B., vol. 21, p. 531, under dam, and A. H. B., vol. 16, p. 12234.

COUNTESS 2D—Roan heifer, calved Feb. 13, 1874, bred by Her Majesty the Queen, got by Prince Albert 2d (29588), out of Cecilia by England's Glory (23889)—Countess by Lord Hopewell (18239)—Calceolarie by Prince Alfred (13494)—Cowslip by Brilliant (8905)—Lily by Sam Slick.

A. H. B., vol 16, p. 12021.

MANRICO 2D 30265—Red, calved Jan. 4, 1877, bred by Her Majesty the Queen (calved in America), got by Manrico (26805), out of Countess 2d by Prince Albert 2d (29588), &c., as above.

1828. MR. GREEN.
New York.

BANQUO 1226.

May, 1884. GREEN BROS.
Inniskip, Ontario, Canada. By Ship Mississippi, from Liverpool to Quebec.

ENTERPRISE—Red roan bull, calved March 18, 1883, bred by Mr. Duthie, Collynie, Scotland, got by Ventriloquist (44180), out of Evangeline 2d by Dipthong (17681)—Evangeline by Hotspur

(21960)—Pride of the Dairy by Guy Fawkes (12981)—Bashful by Young Ury (10984)—Likely by The Pasha (7612)—Helen by 2d Duke of Northumberland (3646)—Mary Ann by Sillery (5131)—Miss Gibson by Carleton (843)—Dora by Diamond (205)—Kitty by Diamond (205).

EARL OF ROSEBERRY—Roan, calved May 25, 1883, bred by Mr. Marr, Upper Mill, got by Athabasca (47359)—Emma 2d by Golden Eagle (26267)—Emma by Leopold (24320)—Victoria by Sir Thomas Stanley (25176)—Duchess by Marquis of Bute (18336) —by Count Fairfax (8991)—by South Durham (9676)—by Uptaker (5534)—by Miracle (2320)—by Matchem (2280)—by Fitz Remus (2025)—by Cato (119)—by Whitworth (695)—bred by Mr. Mason, of Chilton.

MONOGRAM 20TH—Red and white, calved March 6, 1882, bred by Alexander Scott, Towie Barclay, Scotland, got by Vienna (45731), out of Monogram 14th by Prince Frederick (42178)—Monogram 5th (vol. 28, p. 571, E.) by Lord Warden (31766)—Monogram 2d by Gen. Grant (36683)—Monogram by Sir Charles 2d (20812)—Lass o' Doune by Inheritor (13065)—Fortuneteller by Dannecker (7949)—Enchantress by Ury (17157)—by Heriot (4017)—by Gray Diomed (2076)—by Juniper (1144).

PROUD DUCHESS—Red, calved Dec. 1, 1881, bred by A. E. Hector, Collyhill, Scotland, got by Norman (45272), out of Venus 2d (imported by Mr. J. J. Hill, of Minn.) by British Champion (36273) —Venus by Chronometer (33385)—Verbena 2d by Rajah (32232)—Verbena by Golden Eagle (26267)—Violet by Sir Thomas Stanley (25176)—Duchess by Marquis of Bute (13336)—Duchess by Count Fairfax (8991)—by South Durham (9676)—by Uptaker (5534)—by Miracle (2320)—by Matchem (2280)—by Fitz Remus (2025)—by Cato (119)—by Whitworth (695)—bred by Mr. Mason, of Chilton.

THE BELLE—Red, calved Dec. 5, 1882, bred by Mr. James Williamson, Pildoulsie, Scotland, got by Forward (46375), out of Menai by Comet (41250)—Countess Cavour (vol. 27, p. 627, E.)

by Oregon (34953)—Eglantine by Nobility (34912)—Bella 2d by Earl Russell (33826)—Missie by Banker (19255)—Bella by Cœur-de-Lion 2d (17581)—Jessica by California (12628)—Kate by Bang (17346)—Rosa by Duplicate Duke (6956)—Lady by Sir Thomas Fairfax (5196)—Miss Ramsden by Duke (3630)—by Reveller (2528)—by Grazier (1085)—by Cato (857)—by Atlas (42)—by Favorite (252)—by Robinson's Bull (4974)—by Badsworth (47).

VAIN MAID—Deep roan, calved Feb. 15, 1881, bred by A. F. Nares, Bracktor, Old Meldrum, Scotland, got by Edgar (41501), out of Vanity by Earl of Granville (28491)—Victoria 3d by Golden Eagle (26267)—Victoria 2d by Leopold (24328)—Victoria by Sir Thomas Stanley (25176)—Duchess by Marquis of Bute (18336)—by Count Fairfax (8991), &c., as in Vain Duchess, of importation 1883.

E. H. B., vol. 28, p. 521, under dam.

April 21, 1883. FRANCIS GREEN,

Of Inniskip, Ont.   From Liverpool, by Steamer Texas, to Quebec.

CLEMATIS—Red heifer, calved Sept. 19, 1881, bred by A. Cruickshank, got by Perfection (37185), out of Custard by Royal Duke of Gloster (29864)—Princess Royal by Champion of England (17526)—Carmine by The Czar (20947)—Cressida by John Bull (11618)—Clipper by Billy (3151)—by Danby (6918)—by Tip Top (7633)—bred by Mr. Mason.

Vol. 29, p. 401, E. H. B., under dam.

ELIZA 9TH—Roan, calved Dec. 24, 1880, bred by W. Duthie, got by Border Chief (37874), out of Eliza 7th by Forth 3d (26185)—Eliza 5th by Dipthong 3d (21547)—Emily by Paddy (22483)—Eliza 3d by Victor Emanuel (15410)—Eliza Groat by John o' Groat (13090)—Eliza 2d by Lord March (10457)—Eliza Jane by Prince Albert (4301)—Eliza by Election (1921)—Miss Ada by Navarino (2352)—by Top Knot (1537)—by Cobourg (1841).

COUNTESS 5TH—Roan calved Feb. 17. 1881. bred by W. Duthie. got by Star of the Border (44093). out of Countess 4th by Gladstone (31253)—Countess 2d by Heir of England (24122)—Countess by Marmaduke (20284)—Fragrant by John Bull (11618) —by Matadore (11800)—by 2d Duke of Northumberland (3646)—by Mahomed (6170)—by Sillery (5131)—by Carleton (843)—by Diamond (205)—by Diamond (205).

JEWEL 8TH—Roan. calved Dec. 10. 1880. bred by W. Duthie. got by Mountain Chief. (38767). out of Jewel 7th by Blucher (33170)—Jewel 4th by Lord Granville (24395)—Jewel 2d by Royal Standard (22803)—Jewel by Lord Ythan (14858)—Wee Thing by Young Van Dunck (19044)—by Burleigh (17474)—by The Pasha (7612)—by Sillery (5131)—by Carleton (843)—by Diamond (205)—by Diamond (205).

FRANCIS GREEN.

CLARA 40TH—Red. calved Feb. 5. 1882. bred by W. S. Marr, Upper Mill. Tarves. Scotland. got by Bentick (42787). out of Clara 35th by Royal Prince (35398)—Clara 24th by Prince Louis (27158) —Clara 14th by Lord of Lorn (18258)—by Speculator (13775)—by Young Grand Duke (12964)—by Punch (16774)—by The Pasha (7612)—by Mahomed (6170)—by Reformer (2509)—by Raby (2474) —by Nesham (1261)—by Marske (418).

E. H. B., vol. 29. p. 580. under dam.

EARL OF MARR (47815)—Roan, calved May 2. 1881. bred by W. S. Marr, got by Bentick (42787). out of Emma 2d by Golden Eagle (26267)—Emma by Leopold (24328)—Victoria by Sir Thomas Stanley (25176)—Duchess by Marquis of Bute (18336)—by Count Fairfax (8991)—by South Durham (9676)—by Uptaker (5534)—by Miracle (2320)—by Matchem (2280)—by Fitz Remus (2025)—by Cato (119).

34

PATCHOULI—Roan, calved Dec. 1, 1883, bred by W. S. Marr, got by Athabasca (47359), out of Princess Royal 23d by Bentick (42787), &c., as in dam, in this list.

JEWESS—Red and white, calved Aug. 29, 1883, bred by Wm. Duthie, got by Cayhurst (47560), out of Jewel 8th, above.

VAIN DUCHESS—Red, calved April 13, 1882, bred by A. F. Nares, Brucktor, Old Meldrum, Scotland, got by Edgar (41501), out of Vanity by Earl Granville (28491)—Victoria 3d by Golden Eagle (26267)—Victoria 2d by Leopold (24328)—Victoria by Sir Thomas Stanley (25176)—Duchess by Marquis of Bute (18336)—by Count Fairfax (8991)—by South Durham (9676)—by Uptaker (5534)—by Miracle (2320)—by Matchem (2280)—by Fitz Remus (2025)—by Cato (119).

PRINCESS ROYAL 23D—Roan, calved April 19, 1881, bred by W. S. Marr, got by Bentick (42787), out of Princess Royal 19th by The Baron (35738)—Princess Royal 15th by Heir of Englishman (24122)—Princess Royal 10th by Humboldt (31398)—Princess Royal 8th by Bosquet (14183)—Princess Royal 3d by Grand Duke (10284)—by Robin-o'-Day (4973)—by Leander (4199)—by son of Matchem (2281)—by Sir Henry (14446)—by Young Neswick (1268) —by son of Rose's Red Bull (5009)—by Southampton—by Prince (521).

MYSIE 34TH—Red, calved April 27, 1881, bred by A. Scott, Towie Barclay, Turniff, Scotland, got by Lord Clemsford (45073), out of Mysie 29th by Roseberry (39016)—Mysie 28th by Royal Heir (35376)—Mysie 24th by Vanguard (30204)—Mysie 23d by Matchless (29345)—Mysie 22d by Cherry Duke 2d (14265)—Mysie 8th by Young Grand Duke (12964)—Mysie 2d by The Hero (10934) —Mysie by Kelly 2d (9265)—Molly by The Pasha (7612)—by Mahomed (6170)—by Sillery (5131)—by Carleton (843)—by Diamond (205)—by Diamond (205).

NOTE—Mysie 34th dropped, Nov. 7, 1883, a roan cow calf, by Athabasca (47359), which died.

April 21, 1875. B. B. GROOM & SON,

Per the Nova Scotia, from Liverpool, April 6, 1875, landing at Baltimore, Md., April 21, 1875; arrived in Kentucky, April 29, 1875.

EIGHTH MAID OF OXFORD—Roan, calved Dec. 9, 1870, bred by Messrs. F. Leney & Sons, Orpines, Wateringbury, got by Lord of Oxford 2d (20215), out of 7th Maid of Oxford by 7th Duke of Airdrie (23718)—Maid of Oxford 2d by Grand Duke of Oxford (16184)—Oxford 20th by Marquis of Carrabas (11789)—Oxford 5th by Duke of Northumberland (1940)—Oxford 2d by Short Tail (2621)—by Matchem (2281)—by Young Wynyard (2859).

Recorded under her dam, vol. 19, p. 615, E. H. B., and vol. 15, p. 732, A. H. B.

FIDGET 7TH—Red and white cow, calved Dec. 24, 1866, bred by Mr. A. Dugdale, Rose Hill, Burnley, Lancashire, got by 3d Lord Oxford (22200), out of Fidget 6th by 7th Duke of York (17754)—Fidget 4th by 4th Duke of Oxford (10167)—Fidget 2d by Duke of Northumberland (1940)—Fidget by Earl of Darlington (1945)—Fletcher by son of Young Wynyard (2859)—descended from J. Brown's Red Bull.

Recorded, A. H. B., vol. 15, p. 553, and E. H. B., vol. 20, p. 520.

FENNEL DUCHESS 5TH—Roan cow, calved Jan. 1, 1870, bred by Mr. T. Atherton, Chapel House, got by 13th Duke of Oxford (21604), out of Fennel 3d by Cherry Duke 2d (14265)—Fennel by 5th Duke of York (10168)—Filbert by 2d Cleveland Lad (3408)—Felicia by 4th Duke of Northumberland (3649)—Fanny by Short Tail (2621)—Fletcher 2d by Belvedere (1706)—Fletcher by son of Young Wynyard (2859)—descended from J. Brown's Red Bull (97).

Recorded, E. H. B., vol. 21, p. 537, and S. H. R., vol. 6, p. 422. Sold Oct. 14, 1875, at public sale, to B. Sumner, of Woodstock, Conn., for $51.00. Again sold by assignees of B. B. Groom, June 19 and 24, 1878, for $60.00.

ANNETTE OF KNIGHTLY HALL—Roan heifer, calved July 10, 1874, bred by Mr. G. Allan, Knightly Hall, got by 2d Duke

of Wetherby (21618), out of Annette 4th by 8th Duke of York
(28480)—Annette 3d by Lord Oxford 2d (20215)—Annette by
Horrax (11591)—Active by 4th Duke of York (10167)—Actress by
Duke of Northumberland (1940)—Anna by Short Tail (2621)—
Acomb by Belvedere (1706)—a cow of Mr. Bates', Kirklevington.

A. H. B., vol. 22, p. 17032. Sold at Vinewood by public sale,
Oct. 14, 1875, to I. Collard, Des Moines, Iowa, for $2,000.00.

GEORGIE HILLHURST 4TH—Red and white, calved Nov. 10,
1875, bred by Col. Kingscote, Walton, Under Edge, got by Duke
of Hillhurst (28401), out of Georgia Walton by 2d Earl of Walton
(19672)—Georgiana 6th by 4th Duke of Oxford (11387)—Georgiana 5th by Gen. Canrobert (12926)—Georgiana by St. Bernard
(15227)—by Lord Geo. Bentick (10444)—by King Pippin (14769)
—by Earl Stanhope (5966).

Recorded, A. H. B., vol. 15, p. 575, and S. H. R., vol. 6, p. 432,
under dam. This heifer was calved at Vinewood after the arrival
of her dam, and was sold to C. C. Chiles, of Independence, Mo.,
by B. B. Groom's assignees, June 19 and 20, 1878, for $525.00.

RUBY DUCHESS—Roan heifer, calved April 2, 1872, bred by
the Rev. E. C. Perry, Leighford Vicarage, Stafford, got by 8th
Duke of York (28480), out of Ruby 3d by Lord Ravenswood
(20222)—Ruby by Baron Bickerstaff (15611)—Raspberry by Gen.
Canrobert (12927)—Nightshade by Earl Derby (10177)—bred from
the stock of Mr. Neason.

Recorded, A. H. B., vol. 20, p. 16179. Sold, Oct. 14, 1875, at
Vinewood, at public sale, to I. R. Craig, Canada, for $1,225.

COSSET—Red heifer, calved Oct. 18, 1872, bred by Col. Kingscote, Kingscote, got by Duke of Hillhurst (28401), out of Comfort
by 3d Duke of Clarence (23727)—Clove by Duke of Wharfdale
(19648)—Cinnamon by Gen. Canrobert (12927)—Cleveland by
Tinsel (13886)—by Earl of Beverly (3674)—by Shipton Bridge
Bull (5203)—by 2d Earl of Darlington (1945).

Recorded, vol. 20, p. 455, E. H. B., under dam. Sold at public
sale, Oct. 14, 1875, by B. B. Groom, at Vinewood, under name of
Duchess of Clarence, to I. H. Spears & Sons, Tallula, Ill., for $4,100.

LADY HILDA 5TH—Roan heifer, calved Feb. 17, 1874, bred by Mr. G. Allen, Knightly Hall, got by 2d Duke of Wetherby (21618), out of Lady Hilda by Lord Liverpool (22167)—Lady Hudson by 4th Duke of Oxford (11387)—Hudson 3d by Gen. Canrobert (12927)—Hudson 2d by Horrox (11591)—Hudson by The Duke (8676).

Recorded, A. H. B., vol. 15, p. 512, as Duchess Hudson. Sold by assignees of B. B. Groom June 19 and 20, 1878, at public auction, to Thos. Grundy, Washington Co., Ky., for $300.

GEORGIE HILLHURST 3D—Red and white heifer, calved March 22, 1873, bred by Col. Kingscote, Kingscote, got by Duke of Hillhurst (28401), out of Georgie Walton by 2d Earl of Walton (19672)—Georgiana 6th by 4th Duke of Oxford (11387)—Georgiana 5th by Gen. Canrobert (12926)—Georgiana by St. Bernard (15227)—by Lord Geo. Bentinck (10444)—by King Pippin (14769)—by Earl Stanhope (5966).

Recorded, A. H. B., vol. 25, p. 1288. Sold at public sale at Vinewood, Oct. 14, 1875, and bought by Avery & Murphy, Detroit, Mich., for $2,800.

DUCHESS OF KENT—Red cow, calved Sept. 29, 1866, bred by Mr. A. Dugdale, Rose Hill, Burnley, Lancashire, got by Lord Liverpool (22168), out of Kirklevington 14th by 4th Duke of Oxford (11387)—Kirklevington 7th by Earl of Derby (10177)—Kirklevington 4th by Earl of Liverpool (9061)—Kirklevington 1st by Duke of Northumberland (1940)—by Belvedere (1706)—by a son of 2d Hubback (2683)—a cow of Mr. Bates', descended from the stock of Mr. Maynard, of Eryholme.

Recorded, E. H. B., vol. 19, p. 485, and A. H. B., vol. 15, p. 628, as Kirklevington Duchess of Kent. Sold to A. L. Hamilton, June 19, 1878.

KIRKLEVINGTON ROSE—Red cow, calved Sept. 23, 1867, bred by Mr. A. Dugdale, Rosehill, Burnley, Lancashire, got by Earl of Gloster (21644), out of Kirklevington 14th by 4th Duke of

Oxford (11387)—Kirklevington 7th by Earl of Derby (10177)—
Kirklevington 4th by Earl of Liverpool (9061)—Kirklevington 1st
by Duke of Northumberland (1940)—by Belvedere (1706)—by son
of 2d Hubback (2683)—a cow of Mr. Bates', descended from the
stock of Mr. Maynard, of Eryholme.

Recorded, E. H. B., vol. 19, p. 567, and A. H. B., vol. 15, p. 629.
Sold, June 19 and 20, 1878, by assignees of B. B. Groom, at Vinewood, at public sale, to C. L. & B. Cannon, Burlington, Vt., for
$750.

GEORGIANA 13TH—Roan cow, calved April 16, 1869, bred by
Mr. T. Bell, of Brockton Hall, got by Prince of Oxford (27172),
out of Georgiana 8th by 2d Baron Westbury (19288)—Georgiana
5th by Gen. Canrobert (12926)—Georgiana by St. Bernard (15227)
—by Lord Geo. Bentick (10444)—by King Pippin (14769)—by
Earl Stanhope (5966).

Recorded, E. H. B., vol. 20, p. 538, and A. H. B., vol. 15, p. 576.
Sold at sale of assignees of B. B. Groom, June 19 and 20, 1878, to
C. D. Chenault, Richmond, Ky., for $300.

GEORGIANA 15TH—Roan cow, calved Aug. 14, 1871, bred by
Mr. T. Bell, of Brockton Hall, got by 8th Duke of York (28480),
out of Georgiana 12th by Earl of Gloster (21644)—Georgiana 7th
by 4th Duke of Oxford (11387)—Georgiana 5th by Gen. Canrobert
(12926)—Georgiana by St. Bernard (15227)—by Lord Geo. Bentinck (10444)—by King Pippin (14769)—by Earl Stanhope (5966).

Recorded, A. H. B., vol. 15, p. 576. See E. H. B., vol. 22, p.
293: Georgiana 15th is there recorded as bred by G. Allen, got by
13th Duke of Oxford, out of Georgiana 12th, and calved May 11,
1873. Also see vol. 21, p. 537. Did Georgiana 12th have two
heifers named Georgiana 15th?

GEORGIE WALTON—Red and white cow, calved March 17,
1868, bred by Col. Kingscote, Kingscote, Gloucestershire, got by
2d Earl of Walton (19672), out of Georgiana 6th by 4th Duke
of Oxford (11387)—Georgiana 5th by Gen. Canrobert (12926)—

Georgiana by St. Bernard (15227)—by Lord Geo. Bentinck (10444) —by King Pippin (14769)—by Earl Stanhope (5966).

Recorded A. H. B., vol. 15, p. 575, E. H. B., vol. 20, p. 538, and S. H. R., vol. 6, p. 431. Sold by B. B. Groom's assignees, June 19 and 20, 1878, at Vinewood, at public sale, and bought by T. C. Anderson, of Mt. Sterling, Ky., for $190.

KIRKLEVINGTON LADY 4TH—Roan heifer, calved April 19, 1874, bred by Mr. G. Allen, Knightly Hall, got by 2d Duke of Wetherby (21618), out of Lady Kirklevington 2d by Earl of Oxford (21651)—Kirklevington Lady by Grand Duke of York (12966) —Kirklevington 9th by Gen. Canrobert (12926)—Kirklevington 7th by Earl of Derby (10177)—Kirklevington 4th by Earl of Liverpool (9061)—Kirklevington by Duke of Northumberland (1940)— Nell Gwynne by Belvedere (1706)—Northallerton by Son of 2d Hubback (2683)—a cow of Mr. Bates', descended from the stock of Mr. Maynard, of Eryholme.

Recorded, A. H. B., vol. 15, p. 628. Sold to Avery & Murphy, Detroit, Mich., Oct. 14, 1875, at public sale, for $2,550.

KIRKLEVINGTON LADY 5TH—Roan heifer, calved April 9, 1874, bred by Mr. G. Allen, Knightly Hall, got by 2d Duke of Wetherby (21618), out of Lady Kirklevington 3d by 8th Duke of York (28480)—Lady Kirklevington 2d by Earl of Oxford (21615) —Kirklevington Lady by Grand Duke of York (12966)—Kirklevington 9th by Gen. Canrobert (12926)—Kirklevington 7th by Earl of Derby (10177)—Kirklevington 4th by Earl of Liverpool (9061)—Kirklevington by Duke of Northumberland (1940)—Nell Gwynne by Belvedere (1706)—Northallerton by Son of 2d Hubback (2683)—a cow of Mr. Bates', descended from the stock of Mr. Maynard, of Eryholme.

Recorded, A. H. B., vol. 15, p. 629. Sold, Oct. 14, 1875, at Vinewood, at public sale, to Avery & Murphy, Detroit, Mich., for $3,900.

DUCHESS OF UNDER-EDGE—White, calved Aug. 3, 1875, bred by Col. Kingscote, Wotton-under-Edge, got by Duke of Wortley (33759), out of Georgiana 15th by 8th Duke of York

(28480)—Georgiana 12th by Earl of Gloster (21644)—Georgiana 7th by 4th Duke of Oxford (11387)—Georgiana 5th by Gen. Canrobert (12926)—Georgiana by St. Bernard (15227)—by Lord George Bentinck (10444)—by King Pippin (14769)—by Earl Stanhope (5966).

Recorded, A. H. B., vol. 15, p. 575, as Georgia Kingscote. This heifer was calved after her dam reached Kentucky, and was sold at public sale at Vinewood, Oct. 14, 1875, to Mr. Emory Cobb, of Kankakee, Ill., for $700.

KIRKLEVINGTON DUCHESS OF HORTON—Roan, calved Oct. 31, 1875, bred by R. P. Davies, Horton Park, got by Oxford King (34997), out of Kirklevington Duchess of Kent by Lord Liverpool (22168)—Kirklevington 14th by 4th Duke of Oxford (11387) —Kirklevington 7th by Earl of Derby (10177)—Kirklevington 4th by Earl Liverpool (9061)—Kirklevington by Duke of Northumberland (1940)—Nell Gwynne by Belvedere (1706)—Northallerton by Son of 2d Hubback (2683)—a cow of Mr. Bates', from the stock of Mr. Maynard, of Eryholme.

Recorded, A. H. B., vol. 25, p. 455. This heifer was begotten in England, and calved at Vinewood, after the arrival there of her dam. Sold by B. B. Groom's assignees, June 19 and 20, 1878, to Bow Park Association, of Canada, for $2,800.

WILD ROSE 4TH—Red and white heifer, calved Feb. 28, 1875, bred by R. P. Davies, Horton, got by Grand Duke of Clarence (28750), out of Wild Flower Duchess by Lord Oxford 2d (20215)— Wildair 3d by Marquis of Speke (13307)—Wildair by Santiago (12047)—Wild Eyes 8th by Duke of Northumberland (1940)—by Belvedere (1706)—by Emperor (1915)—by Wonderful (700)—by Cleveland (145)—by Butterfly (104)—by Hollon's Bull (313)— by Mowbray's Bull (2342)—by Masterman's Bull (422)—descended from M. Dobison's stock.

Recorded, A. H. B., vol. 16, p. 12389, as the property of Bow Park Association, Brantford, Can., under the name of Wild Eyes of Horton Park. Sold, Oct. 14, 1875, at public auction, at Vinewood, Ky., to W. N. Offutt, of Georgetown, Ky., for $1,700.

ROGUISH EYES—Red and white cow, calved June 28, 1864, bred by Col. Kingscote, Kingscote, Gloucestershire, got by Earl of Walton (19672), out of Red Eyes by Duke of Richmond (7996)—Wild Eyes 23d by 2d Cleveland Lad (3408)—Wild Eyes 9th by Duke of Northumberland (1940)—by Belvedere (1706)—by Emperor (1975)—by Wonderful (700)—by Cleveland (145)—by Butterfly (104)—by Hollon's Bull (313)—by Mowbray's Bull (2342)—by Masterman's Bull (422)—descended from M. Dobison's stock.

Recorded, E. H. B., vol. 18, p. 693, and A. H. B., vol. 15, p. 872. Sold by assignees of B. B. Groom, June 19 and 20, 1878, at public sale, and bought by L. G. B. Cannon, of Burlington, Vt., for $1,000.

POPPY JUICE—Roan cow, calved in November, 1864, bred by Col. Kingscote, Kingscote, Gloucestershire, got by 2d Earl of Walton (19672), out of Peach by General Canrobert (12927)—Poppy by 2d Cleveland Lad (3408)—Place 2d by 2d Earl of Darlington (1945)—Place by Son of 2d Hubback (2683)—a cow of Mr. Bates', Kirklevington.

Recorded, E. H. B., vol. 20, p. 694, and A. H. B., vol. 15, p. 826. Sold for $35, at sale of assignees of B. B. Groom, June 19 and 20, 1878.

GEORGIANA 8TH—Red and white cow, calved July 24, 1865, bred by Mr. T. Bell, Brockton Hall, Eccleshall, Staffordshire, got by 2d Baron Westbury (19288), out of Georgiana 5th by General Canrobert (12926)—Georgiana by St. Bernard (15227)—by Lord George Bentinck (10444)—by King Pippin (14769)—by Earl of Stanhope (5966).

Recorded, E. H. B., vol. 17, p. 507, and A. H. B., vol. 15, p. 575. Sold by assignees of B. B. Groom, June 19 and 20, 1878, at Vinewood, at public auction, and bought by C. C. Chiles, of Independence, Mo., for $60.

GEORGIANA 10TH—Red and white cow, calved March 11, 1868, bred by Mr. T. Bell, got by Northern Light (26984), out of Georgiana 7th by 4th Duke of Oxford (11387)—Georgiana 5th by

35

Gen. Canrobert (12926)—Georgiana by St. Bernard (15227)—by Lord George Bentinck (10444)—by King Pippin (14769)—by Earl Stanhope (5966).

Recorded, E. H. B., vol. 20, p. 537; A. H. B., vol. 15, p. 575, and S. H. R., vol. 6, p. 431. Sold at assignee's sale of B. B. Groom, June 19 and 20, 1878, to C. J. Spillman, Bryantsville, Ky., for $180.

BLOSSOM 4TH—Red heifer, calved March 10, 1874, bred by Mr. G. Allen, Knightly Hall, got by 2d Duke of Wetherby (21618), out of Blossom by 8th Duke of York (28480)—Peach Blossom 2d by Baron Westbury (19287)—Peach by Gen. Canrobert (12927)—Poppy by 2d Cleveland Lad (3408)—Place 2d by 2d Earl of Darlington (1945)—by Son of 2d Hubback (2683)—a cow of Mr. Bates', of Kirklevington.

Sold publicly to J. Collard, Des Moines, Iowa, Oct. 14, 1875, under the name of Duchess of Knightly Hall, for $1,850.

PRIGGISH—Red heifer, calved March 14, 1874, bred by Col. Kingscote, got by Duke of Hillhurst (28401), out of Pride by Northern Duke (22431)—Pride of Fashion by Cherry Duke 4th (17552)—Princess Alice by Gen. Canrobert (12927)—Princess by Earl of Derby (10177)—Poppy by 2d Cleveland Lad (3408)—Place 2d by 2d Earl of Darlington (1945)—Place by Son of 2d Hubback (2683)—a cow of Mr. Bates', of Kirklevington.

Sold at public auction, Oct. 14, 1875, at Vinewood, under the name of Duchess of Kingscote (vol. 20, p. 15605, A. H. B.), to J. V. Grigsby, of Winchester, Ky., for $2,550.

KIRKLEVINGTON DUCHESS 18TH—Roan, calved April 14, 1874, bred by R. P. Davies, got by 2d Duke of Gloster (28392), out of Kirklevington 13th by 4th Duke of Oxford (11387)—Kirklevington 9th by Gen. Canrobert (12926)—Kirklevington 7th by Earl of Derby (10177)—Kirklevington 4th by Earl of Liverpool (9061)—Kirklevington by Duke of Northumberland (1940)—Nell Gwynne by Belvedere (1706)—Northallerton by Son of 2d Hubback (2683)

—a cow of Mr. Bates', descended from the stock of Mr. Maynard, of Eryholme.

Recorded, A. H. B., vol. 17, p. 12952, and E. H. B., vol. 21, p. 665, under dam. Sold, Oct. 4, 1875, at public auction, at Vinewood, to I. R. Craig, Canada, for $5,550.

FENNEL DUCHESS OF KNIGHTLY HALL.—White heifer, calved Feb. 15, 1875, bred by Mr. G. Allen, Knightly Hall, Eccleshall, Staffordshire, got by Grand Duke of York (31304), out of Fennel Duchess 5th by 13th Duke of Oxford (21604)—Fennel 3d by Cherry Duke 2d (14265)—Fennel by 5th Duke of York (10168) —Filbert by 2d Cleveland Lad (3408)—Felicia by 4th Duke of Northumberland (1940)—Fanny by Short Tail (2621)—Fletcher 2d by Belvedere (1706)—Fletcher by a son of Young Wynyard (2859) —descended from J. Brown's Red Bull (97).

Sold, Oct. 14, 1875, at public sale at Vinewood, to T. J. Megibben, for $2,500.

KIRKLEVINGTON LADY 3D—Roan cow, calved Sept. 5, 1871, bred by Mr. G. Allen, of Knightly Hall, got by 8th Duke of York (28480), out of Lady Kirklevington 2d by Earl of Oxford (21651)—Kirklevington Lady by Grand Duke of York (12966)— Kirklevington 9th by Gen. Canrobert (12926)—Kirklevington 7th by Earl of Derby (10177)—Kirklevington 4th by Earl of Liverpool (9061)—Kirklevington by Duke of Northumberland (1940)—Nell Gwynne by Belvedere (1706)—Northallerton by Son of 2d Hubback (2683)—a cow of Mr. Bates', descended from the stock of Mr. Maynard, of Eryholme.

Recorded, under dam, E. H. B., vol. 20, p. 601. Sold, Oct. 14, 1875, at Vinewood, at public sale, to J. V. Grigsby, Winchester, Ky., for $3,000.

KIRKLEVINGTON LADY 6TH—White, calved May 5, 1875, bred by Geo. Allen, Knightly Hall, Staffordshire, got by Grand Duke of York (31304), out of Kirklevington Lady 3d by 8th Duke of York (28480)—Kirklevington Lady 2d by Earl of Oxford (21651)—Kirklevington Lady by Grand Duke of York (12966)—

Kirklevington 9th by Gen. Canrobert (12926)—Kirklevington 7th by Earl Derby (10177)—Kirklevington 4th by Earl of Liverpool (9061)—Kirklevington by Duke of Northumberland (1940)—Nell Gwynne by Belvedere (1706)—Northallerton by Son of 2d Hubback (2683)—a cow of Mr. Bates', descended from the stock of Mr. Maynard, of Eryholme.

Recorded, A. H. B., vol. 16, p. 12129. This heifer was calved a few days after the importation reached Vinewood, and was sold at Mr. Groom's sale, Oct. 14, 1875, to H. Corbin, of Bourbon Co., Ky., for *2,000.

SECOND DUCHESS OF CLARENCE—Roan heifer, calved Aug. 3, 1875, bred by Col. Kingscote, Wotton-under-Edge, got by Duke of Wortley (33759), out of Cosset alias Duchess of Clarence by Duke of Hillhurst (28401)—Comfort by 3d Duke of Clarence (23727)—Clove by Duke of Wharfdale (19648)—Cinnamon by Gen. Canrobert (12927)—Cleveland by Tinsel (13886)—Chapman by Earl Beverly (3674)—Betty by Skipton Bridge Bull (5208)—by 2d Earl of Darlington (1945).

This heifer was begotten in England, and calved after reaching Kentucky, and was sold, Oct. 14, 1875, at public sale, by Mr. Groom, at Vinewood, to J. H. Spears & Sons, of Tallula, Ill., for *3,175.

WILD FLOWER DUCHESS—Red cow, calved Dec. 5, 1863, bred by Mr. T. Atherton, Chapel House, got by Lord Oxford 2d (20215), out of Wildair 3d by Marquis of Speke (13307)—Wildair by Santiago (12047)—Wild Eyes 8th by Duke of Northumberland (1940)—by Belvedere (1706)—by Emperor (1975)—by Wonderful (700)—by Cleveland (145)—by Butterfly (104)—by Hollon's Bull (313)—by Mowbray's Bull (2342)—by Masterman's Bull (422)—descended from M. Dobison's stock.

Recorded, E. H. B., vol. 19, p. 787. Sold at public sale, Oct. 14, 1875, at Vinewood, near Winchester, Ky., and bought by N. G. Pond, of Melford, Conn., for $1,900, and again, June 19 and 20, 1878, by assignees of B. B. Groom, to W. W. Curry, Winchester, Ky., for *280.

WILD ROSE—Red and white cow, calved Jan. 29, 1871, bred by Mr. R. P. Davies, Horton, Chipping Sodbury, Gloucestershire, England, got by Earl of Kirklevington (26064), out of Wild Flower Duchess by Lord Oxford 2d (20215)—Wildair 3d by Marquis of Speke (13307)—Wildair by Santiago (12047)—Wild Eyes 8th by Duke of Northumberland (1940)—by Belvedere (1706)—by Emperor (1975)—by Wonderful (700)—by Cleveland (145)—by Butterfly (104)—by Hollon's Bull (313)—by Mowbray's Bull (2342)—by Masterman's Bull (422)—descended from M. Dobison's stock.

Recorded, A. H. B., vol. 26, p. 1015, and E. H. B., vol. 21, p. 666. Sold, Oct. 14, 1875, at public sale, at Vinewood, near Winchester, Ky., to W. N. Offutt, of Georgetown, Ky., under the name of Wild Eyes Rose, for $3,050.

SIXTH DUKE OF KIRKLEVINGTON 22828 (30982)—Roan bull, calved July 18, 1871, bred by R. P. Davies, got by Duke of Kirklevington (25982), out of Kirklevington 13th by 4th Duke of Oxford (11387)—Kirklevington 9th by Gen. Canrobert (12926)—Kirklevington 7th by Earl of Derby (10177)—Kirklevington 4th by Earl of Liverpool (9061)—by Duke of Northumberland (1940) —by Belvedere (1706)—by Son of 2d Hubback (2683)—a cow of Mr. Bates', descended from the stock of Mr. Maynard, Eryholme.

LORD GARLAND 2d—Red and white bull, calved March 23, 1874, bred by Mr. H. J. Sheldon, Brailes House, Warwickshire, got by 4th Duke of Barrington (30924), out of Lady Florence 2d by 18th Duke of Oxford (25995)—Lady Florence by Duke of Brailes (23724) —Countess by Duke of Cambridge (12742)—Chrysalis by Earl of Dublin (10178)—Garland by Gray Friar (9172)—Fillet by Fawsley (6004)—Marguerite by Marcellus (2260)—Pearl by Rufus (2576)— Ruby by Wellington (683)—by Windsor (698)—by Windsor (698) —by own brother to North Star (459).

Recorded, E. H. B., vol. 22, p. 559, under dam.

GENERAL HILLHURST 3d ALIAS FIRST DUKE OF UNDEREDGE 22963—Red and white, calved July 15, 1874, bred by Col. Kingscote, Wotton-under-Edge, got by Duke of

Hillhurst (28401), out of Georgia Walton by 2d Earl of Walton (19672)—Georgiana 6th by 4th Duke of Oxford (11387)—Georgiana 5th by Gen. Canrobert (12926)—Georgiana by St. Bernard (15227) —by Lord George Bentinck (10444)—by King Pippin (14769)—by Earl Stanhope (5966).

Sold at sale of B. B. Groom & Son, Oct. 14, 1875, to Mrs. Long, Monroe, Iowa, under name of 1st Duke of Underedge, for $1,050.

THIRD DUKE OF UNDEREDGE 43560—Red roan, calved July 3, 1875, bred by Col. Kingscote, got by Duke of Hillhurst (28401), out of Georgiana 13th by Prince Oxford (27172)—Georgiana 8th by 2d Baron Westbury (19288)—Georgiana 5th by Gen. Canrobert (12926)—Georgiana by St. Bernard (15227)—by Lord George Bentinck (10444)—by King Pippin (14769)—by Earl Stanhope (5966).

This bull was calved after his dam reached Vinewood, and was sold, Oct. 14, 1875, to John Collard, of Des Moines, Iowa, for $2,100.

GENERAL HILLHURST 5TH—Red, calved March 15, 1875, bred by Col. Kingscote, got by Duke of Hillhurst (28401), out of Georgiana 8th by 2d Baron Westbury (19288)—Georgiana 5th by Gen. Canrobert (12926)—Georgiana by St. Bernard (15227)—by Lord George Bentinck (10444)—by King Pippin (14769)—by Earl Stanhope (5966).

Sold, Oct. 14, 1875, at public auction, at Vinewood, to W. M. Smith, Lexington, Ill., for $1,650, under name of 2d Duke of Underedge, and so recorded, No. 22964.

July 27, 1875. B. B. GROOM & SON.
By Ship Wisconsin, from Liverpool, July 27, 1875.

EIGHTH DUKE OF GENEVA 26231 (28390)—Roan, calved Nov. 24, 1868, bred by J. O. Sheldon, Geneva, N. Y., got by Baron of Oxford (23371), out of 3d Duchess of Thorndale by Duke of Gloster (11382)—Duchess 66th by 4th Duke of York (10167)— Duchess 55th by 4th Duke of Northumberland (3649)—Duchess 38th by Norfolk (2377)—by Belvedere (1706)—by 2d Hubback

(1423)—by The Earl (646)—by Ketton 2d (710)—by Comet (155) —by Favorite (252)—by Daisy Bull (186)—by Favorite (252)—by Hubback (319)—by J. Brown's Red Bull (97).

This bull had been exported to England by Mr. Sheldon in 1869. Bought at sale of B. B. Groom's assignees, June 19 and 20, 1878, by I. Dawson, Owingsville, Ky., for *200.

EARL OF OXFORD ALIAS OXFORD GENEVA 24221— Roan, calved March 13, 1875, bred by F. Leney & Sons, Orpines, Waterbury, Kent, got by 8th Duke of Geneva (28390), out of 7th Maid of Oxford by 7th Duke of Airdrie (23718)—2d Maid of Oxford by Grand Duke of Oxford (16184)—Oxford 20th by Marquis of Carrabas (11789)—Oxford 5th by Duke of Northumberland (1940) —Oxford 2d by Short Tail (2621)—Matchem Cow by Matchem (2281)—by Young Wynyard (2859).

Sold, Oct. 14, 1875, to D. L. Hughes, of Iowa, for *5,000, who let Mr. G. have him again. Sold again by B. B. Groom's assignees to F. C. Cornell, for *850.

SEVENTH MAID OF OXFORD—Roan, calved July 16, 1866, bred by J. O. Sheldon, Geneva, N. Y., got by 7th Duke of Airdrie (23718), out of Maid of Oxford 2d by Grand Duke of Oxford (16184)—Oxford 20th by Marquis of Carrabas (11789)—Oxford 5th by Duke of Northumberland (1940)—Oxford 2d by Short Tail (2621)—Matchem Cow by Matchem (2281)—by Young Wynyard (2859).

Recorded, E. H. B., vol. 18, p. 604, and A. H. B., vol. 15, p. 732. This cow was exported to England by Mr. J. O. Sheldon. Sold to Bow Park Association, 1878, by assignees.

May 31, 1876. B. B. GROOM & SON,

Of Clark, Co., Ky. Shipped by the Wisconsin, from Liverpool, May 31, 1876; arrived in Kentucky, June 23, 1876.

WINSOME 16TH—Roan heifer, calved Aug. 10, 1873, bred by the Duke of Devonshire, Holker Hall, Lancashire, got by Baron Oxford 4th (25580), out of Bright Eyes 5th by Grand Duke 6th (19876)—Bonny by Oxford Duke (15036)—Beauty by Crusade

(7938)—Bright Eyes by 3d Duke of York (10166)—Wild Eyes 23d by 2d Cleveland Lad (3408)—Wild Eyes 9th by Duke of Northumberland (1940)—by Belvedere (1706)—by Emperor (1975)—by Wonderful (700)—by Cleveland (145)—by Butterfly (104)—by Hollon's Bull (313)—by Mowbray's Bull (2342)—by Masterman's Bull (422)—descended from M. Dobison's stock.

Recorded, A. H. B., vol. 18, p. 13963. Sold by B. B. Groom's assignees, June 19 and 20, 1878, to Bow Park Association, Can., for $2,600.

OXFORD'S ANNETTE—Roan heifer, calved Oct. 27, 1873, bred by Mr. Geo. Fox, Elmhurst Hall, Litchfield, got by 22d Duke of Oxford (31000), out of Annette by Grand Duke of Lightburne 2d (26291)—Athena by Duke of Darlington (21586)—Acacia by Count de Gourey (17632)—Asia by 2d Grand Duke (12961)—Apricot by Fusileer (11499)—Augusta by 3d Duke of York (10166)—Annie by 2d Cleveland Lad (3408)—Annabella by Duke of Cleveland (1937) —by Belvedere (1706)—a cow of Mr. Bates'.

Recorded, under dam, E. H. B., vol. 21, p. 720, and A. H. B., vol. 18, p. 13852. Sold, June 19 and 20, 1878, by B. B. Groom's assignees, to Bow Park Association, for $430.

MISS WILD EYES 3D—Roan heifer, calved August 12, 1875, bred by Mr. R. P. Davies, of Horton, got by Oxford's King (34997), out of Miss Wild Eyes 2d by Grand Prince of Claro (28718)—Miss Wild Eyes by Lord Red Eyes (24459)—Wild Eyes 20th by Lord Barrington 1st (13170)—Wild Eyes 16th by 2d Duke of Oxford (9046)—Wild Eyes 15th by 4th Duke of Northumberland (3949)— Wild Eyes 8th by Duke of Northumberland (1940)—Wild Eyes 2d by Belvedere (1706)—Wild Eyes by Emperor (1875)—by Wonderful (700)—by Cleveland (145)—by Butterfly (104)—by Hollon's Bull (313)—by Mowbray's Bull (2342)—by Masterman's Bull (422)— descended from the stock of M. Dobison.

Recorded, under dam, E. H. B., vol. 22, p. 394, and A. H. B., vol. 19, p. 14716. Sold, June 19 and 20, 1878, by B. B. Groom's assignees, to L. G. B. Cannon, of Burlington, Vt., for $1,900.

ANNETTE—Red and white. calved April 15. 1871, bred by R. B. Hetherington. Park Head. got by Grand Duke of Lightburne 2d (26291), out of Athena by Duke of Darlington (21586)—Acacia by Count de Gourcy (17632)—Asia by 2d Grand Duke (12961)—Apricot by Fusileer (11499)—Augusta by 3d Duke of York (10166)—Annie by 2d Cleveland Lad (3408)—Annabella by Duke of Cleveland (1937)—by Belvedere' (1706)—a cow of Mr. Bates'.

Recorded. E. H. B.: vol. 21, p. 720, and A. H. B., vol. 18, p. 13564. Sold by B. B. Groom's assignees. June 19 and 20, 1878. to Bow Park Association, Can., for $330.

GEORGIE HILLHURST 5th—Red and white heifer, calved June 4, 1874. bred by Col. Kingscote, Kingscote, Gloucestershire, got by Duke of Hillhurst (28401), out of Georgiana 10th by Northern Light (26984)—Georgiana 7th by 4th Duke of Oxford (11387)—Georgiana 5th by Gen. Canrobert (12926)—Georgiana by St. Bernard (15227)—by Lord George Bentinck (10444)—by King Pippin (14769)—by Earl Stanhope (5966).

Recorded. S. H. R., vol. 6, p. 431, under dam, and A. H. B., vol. 18, p. 13686. Sold, June 19 and 20, 1878, by B. B. Groom's assignees. to Mr. Cornell, Ithaca, N. Y., for $300.

FANNY CRAGGS—Red heifer. calved March 16, 1876, bred by Mr. G. Fox, got by Grand Duke of Weston 3d (34079), out of Joan of Arc by Duke of Brailes (23724)—Johanna Southcott by John o' Gaunt (16322)—Duchess of Sussex by Duke of Sussex (12772)—Countess of Beverly by Lord Foppington (10437)—Cowslip Belle by 2d Cleveland Lad (3408)—by Duke of Northumberland (1940)—by Son of 2d Hubback (2683)—a cow of Mr. Bates', from Mr. Maynard's stock, of Eryholme.

E. H. B., vol. 23, p. 443, under dam, and A. H. B., vol. 22, p. 17224, as Lady Craggs. Sold as Lady Craggs, No. 112 in catalogue, by assignees of B. B. Groom, at sale June 19 and 20, 1878, to J. C. Stevens, Kenton, Ohio, for $300.

PEACH BLOSSOM 8th—White cow, calved June 25, 1871, bred by Mr. T. Bell, Brockton House, Eccleshall, got by 8th Duke

of York (28480), out of Peach Blossom 6th by Earl of Gloster
(21644)—Peach Blossom 2d by Baron Westbury (19287)—Peach by
Gen. Canrobert (12927)—Poppy by 2d Cleveland Lad (3408)—Place
2d by 2d Earl of Darlington (1945)—by son of 2d Hubback (2683).

Recorded, E. H. B., vol. 20, p. 683, under dam, and A. H. B.,
vol. 20, p. 16054. Sold by B. B. Groom's assignees, June 19 and
20, 1878, to T. C. Anderson, of Mt. Sterling, Ky., for $300.

FIFTEENTH LADY OF OXFORD—Red and white, calved
May 1, 1873, bred by Mr. E. H. Cheney, Gadesby Hall, Leicester,
got by 9th Duke of Geneva (28391), out of 13th Lady of Oxford
by Baron of Oxford (23371)—7th Lady of Oxford by 6th Duke of
Thorndale (23794)—2d Lady of Oxford by 2d Grand Duke (12961)
—Oxford 13th by 3d Duke of York (10167)—Oxford 5th by Duke
of Northumberland (1940)—Oxford 2d by Short Tail (2621)—
Matchem Cow by Matchem (2281)—by Young Wynyard (2859).

Recorded, E. H. B., vol. 22, p. 358, under dam, and A. H. B., vol.
18, p. 13756. Sold, June 19 and 20, 1878, by assignees of B. B.
Groom, to Bow Park Association, Canada, for $1,000.

LALLY 8TH—Roan, calved Nov. 12, 1867, bred by C. W. Harvey,
Liverpool, got by 7th Duke of York (17754), out of Lally 3d by
4th Duke of Oxford (11387)—Lally by Earl of Derby (10177)—
Olive Leaf 3d by Earl of Liverpool (9061)—Olive Leaf 2d by 2d
Duke of Cambridge (3638)—Olive Leaf by Belvedere (1706)—
Lady Barrington by son of Herdsman (304)—Young Alicia by
Wonderful (700)—Old Alicia by Alfred (23)—by Young Favorite
(6994).

Recorded, E. H. B., vol. 20, p. 619, and A. H. B., vol. 20, p. 15843.
Sold by assignees of B. B. Groom, June 19 and 20, 1878, to Messrs.
Hamilton, Mt. Sterling, Ky., for $1,550.

BARRINGTON LALLY—Red, calved Sept. 8, 1876, bred by W.
Angerstein, got by Duke of Rosedale 2d (33722), out of Lally 8th,
above.

A. H. B., vol. 20, p. 15471.

BLANCHE 10TH—Roan heifer, calved April 17, 1875, bred by Mr. H. D. de Vitre, Charlton House, Wantage, got by Lord Thorndale (31756), out of Blanche 8th by Grand Duke 10th (21848)—Blanche 3d by 10th Duke of Oxford (17739)—Blanche by Dundas (17763)—Sylph by Gloucester (14619)—by Childers (10052)—by Selim (6454)—by Rex (6385)—by Norfolk (2377)—by Belvedere (1706)—by Belvedere (1706)—by Lancaster (360)—by Petrarch (488)—by Major (397)—by Chapman's Son of Punch (122)—by Dickson's Grandson of Punch (213)—by Check (132)—by R. Grimston's Bull (282)—by J. Coate's Bull (148).

Recorded, E. H. B., vol. 23, p. 371, under dam, and A. H. B., vol. 18, p. 13591. Sold, June 19 and 20, 1878, by B. B. Groom's assignees, to Bow Park Association, for $510.

ACOMB BELL—Roan heifer, calved Jan. 8, 1876, bred by Mr. H. D. de Vitre, Charlton House, Wantage, got by Lord Thorndale (31756), out of Alexandria by Grand Duke 3d (17994)—America by Marmaduke (14897)—Asia by 2d Grand Duke (12961)—Apricot by Fusileer (11499)—Augusta by 3d Duke of York (10166)—by 2d Cleveland Lad (3408)—by Duke of Cleveland (1937)—by Belvedere (1706).

Recorded, A. H. B., vol. 18, p. 13551. Sold, June 19 and 20, 1878, to C. H. Andrews, Youngstown, Ohio, for $400.

RASPBERRY 8TH—Roan heifer, calved March 12, 1875, bred by Lord Fitz Hardinge, Berkley Castle, Gloucestershire, got by Oxford's Tony (35000), out of Raspberry 4th by Earl of Gloster (21644)—Raspberry 2d by Baron Westbury (19287)—Raspberry by Gen. Canrobert (12927)—Nightshade by Earl of Derby (10177)—from the stock of Mr. Mason, Kirklevington.

Recorded, A. H. B., vol. 16, p. 12299, as Raspberry 7th.

DUCHESS OF LANCASTER 13TH—Roan, calved Sept. 14, 1871, bred by Lord Penrhyn, Penrhyn Castle, got by 11th Grand Duke (21849), out of Duchess of Lancaster 7th by Marmaduke (14897)—Duchess of Lancaster 4th by Sir Colin Campbell (16961)—Duchess of Lancaster by Duke of York (11393)—Buttercup by

2d Duke of Lancaster (5951)—by Crichton (3516)—by Ploughboy (4726).

Recorded, E. H. B., vol. 20, p. 490, under dam, and A. H. B., vol. 17, p. 12858. Sold to H. P. Thompson, and by him to Gen. Meredith & Son, of Indiana, in 1878.

WILD ROSE—Red and white cow, calved Feb. 2, 1872, bred by M. H. Cochrane, Hillhurst, Compton, Canada, got by 6th Duke of Geneva (30959), out of Wild Eyes 26th by Earl of Walton (17787) —Wild Eyes 24th by 4th Duke of Oxford (11387)—Wild Eyes 22d by Wild Duke (19148)—Wild Eyes 20th by Lord Barrington 1st (13170)—Wild Eyes 16th by 2d Duke of Oxford (9046)—by 4th Duke of Northumberland (3649)—by Duke of Northumberland (1940)—by Belvedere (1706)—by Emperor (1915)—by Wonderful (700)—by Cleveland (145)—by Butterfly (104)—by Hollon's Bull (313)—by Mowbray's Bull (2342)—by Masterman's Bull (422)— descended from M. Dobison's stock.

A. H. B., vol. 12, p. 1295. This cow was exported by Mr. M. H. Cochrane, by Ship Sarmatian, in 1873, and imported by Mr. Groom, in 1876.

OXFORD LOO 2D—Red heifer, calved Feb. 13, 1877, bred by B. B. Groom, of Winchester, Ky., got by Oxford Duke 2d (34994), out of Loo Bell by 3d Lord Lally (24408), &c., as in Loo Bell, above.

A. H. B., vol. 18, p. 13854.

LOO—Red and white cow, calved April 30, 1863, bred by Mr. P. Stevenson, Rainton, Thirsk, Yorkshire, got by Cromwell (17640), out of Lily Bell by Horrox (11591)—Lily by 2d Duke of Oxford (9046)—Harmless by Cleveland Lad (3407)—Hawkeye by by Red Rose Bull (2493)—by Rex (1375)—bred by Mr. Richardson, of Hart.

Recorded, E. H. B., vol. 17, p. 597, and S. H. R., vol. 6, p. 478. Sold by assignees of B. B. Groom, June 19 and 20, 1878, to S. R. Grundy, Springfield, Ky., for $170.

LOO BELL—Red cow, calved May 25, 1871, bred by P. Stevenson, Rainton, Thirsk, Yorks., got by 3d Lord Lally (24408), out of Loo by Cromwell (17640)—Lily Bell by Horrox (11591)—Lily by 2d Duke of Oxford (9046)—Harmless by Cleveland Lad (3407)—Hawkeye by Red Rose Bull (2493)—Hart by Rex—bred by Mr. Richardson, of Hart.

Recorded, A. H. B., vol. 20, p. 15875, and E. H. B., vol. 20, p. 628, under dam. Sold by B. B. Groom's assignees, June 19 and 20, 1878, to Geo. Hamilton, Mt. Sterling, Ky., for $560.

OXFORD LOO—Red and white, calved April 7, 1876, bred by P. Stevenson, Rainton, Thirsk, Yorkshire, got by Oxford Duke 2d (34994), out of Loo by Cromwell (17640)—Lily Bell by Horrox (11591)—Lily by 2d Duke of Oxford (9046)—Harmless by Cleveland Lad (3407)—Hawkeye by Red Rose Bull (2493)—Hart by Rex (1375).

Recorded, A. H. B., vol. 25, p. 1291. Sold by B. B. Groom's assignees, June 19 and 20, 1878, to M. H. Cochrane, Canada, for $605.

KIRKLEVINGTON 13TH—Red and white cow, calved May 1, 1863, bred by Mr. C. W. Harvey, Walton-on-the-Hill, Liverpool, got by 4th Duke of Oxford (11387), out of Kirklevington 9th by Gen. Canrobert (12926)—Kirklevington 7th by Earl of Derby (10177)—Kirklevington 4th by Earl of Liverpool (9061)—Kirklevington by Duke of Northumberland (1940)—Nell Gwynne by Belvedere (1706)—Northallerton by Son of 2d Hubback (2683)—a cow of Mr. Bates', descended from the stock of Mr. Maynard, of Eryholme.

Recorded, E. H. B., vol. 21, p. 665, and S. H. R., vol. 6, p. 450. Died at Vinewood.

KIRKLEVINGTON 14TH—Red and white cow, calved May 24, 1863, bred by Mr. C. W. Harvey, Walton-on-the-Hill, Liverpool, got by 4th Duke of Oxford (11387), out of Kirklevington 7th by Earl of Derby (10177)—Kirklevington 4th by Earl of Liverpool (9061)—Kirklevington by Duke of Northumberland (1940)—Nell

Gwynne by Belvedere (1706)—Northallerton by Son of 2d Hubback (2683)—a cow of Mr. Bates', descended from the stock of Mr. Maynard, of Eryholme.

Recorded, E. H. B., vol. 18, p. 550.

WILD EYES CONNAUGHT 34099—Roan bull, calved April 21, 1876, bred by Col. Kingscote, Walton-under-Edge, got by Duke of Connaught (33604), out of Wild Rose by 6th Duke of Geneva (7933)—Wild Eyes 26th by Earl Walton (17787)—Wild Eyes 24th by 4th Duke of Oxford (11387)—Wild Eyes 22d by Wild Duke (19148)—Wild Eyes 20th by Lord Barrington 1st (13170)—Wild Eyes 16th by 2d Duke of Oxford (9046)—Wild Eyes 15th by 4th Duke of Northumberland (3649)—Wild Eyes 8th by Duke of Northumberland (1940)—Wild Eyes 2d by Belvedere (1706)—by Emperor (1975)—by Wonderful (700)—by Cleveland (145)—by Butterfly (104)—by Hollon's Bull (313)—by Mowbray's Bull (2342)—by Masterman's Bull (422)—descended from M. Dobison's stock.

Sold to T. C. Anderson, Mt. Sterling, Ky., for $460, June 19 and 20, by B. B. Groom's assignees.

GRAND DUKE OF WESTON 3d 28296 (34079)—Roan bull, calved March 26, 1873, bred by Mrs. Dawson, Weston Hall, Otley, Yorkshire, got by Grand Duke of Weston (28768), out of Grand Duchess 19th by Imperial Oxford (18084)—Grand Duchess 9th by Grand Duke 3d (16182)—Grand Duchess 4th by Cherry Duke (12589)—Grand Duchess by Grand Duke (10284)—by Cleveland Lad (3407)—by Belvedere (1706)—by 2d Hubback (1423)—by 2d Hubback (1423)—by The Earl (646)—by Ketton 2d (710)—by Comet (155)—by Favorite (252)—by Daisy Bull (186)—by Favorite (252)—by Hubback (319)—by J. Brown's Red Bull (97).

Sold by assignees to Messrs. Hamilton.

WATER SPRITE—Red heifer, calved Feb. 3, 1874, bred by Mr. Geo. Fox, Elmhurst Hall, Litchfield, got by Grand Duke of Waterloo (28766), out of Water Lass 2d by Gen. Napier (24023)—Water Lass by Oxford 2d (18507)—Water Maid by Marc Antony (14895)—Water Witch by Squire Blanche (12139)—Waterloo 10th by 4th Duke of Northumberland (3649)—by 2d Cleveland Lad (3408)—by Duke of Northumberland (1940)—by Norfolk (2377)—by Waterloo (2816).

Recorded, E. H. B., vol. 21, p. 721, under dam. Sold to J. D. Guthrie, Shelby Co., Ky., by B. B. Groom's assignees, June 19 and 20, 1878.

GRAND DUCHESS OF HAVERING—Roan, calved June 26, 1872, bred by Mr. D. McIntosh, Havering Park, Romford, Essex, got by 3d Duke of Geneva (23753), out of Grand Duchess 21st by Baron of Oxford (23375)—Grand Duchess 17th by Imperial Oxford (18084)—Grand Duchess 10th by Grand Duke 3d (16182)—Grand Duchess 5th by Prince Imperial (15095)—by Grand Duke (10284) —by Cleveland Lad (3407)—by Belvedere (1706)—by 2d Hubback (1423)—by 2d Hubback (1423)—by The Earl (646)—by Ketton 2d (710)—by Comet (155)—by Favorite (252)—by Daisy Bull (186) —by Favorite (252)—by Hubback (319)—by J. Brown's Red Bull (97).

Recorded, under dam, E. H. B., vol. 20, p. 547. Died at Vinewood.

AMELIA—Red, calved Aug. 22, 1865, bred by Mr. H. Thurnall, Royston, Cambridge, got by Surprise (20922), out of Acacia by Count de Gourey (17632)—Asia by 2d Grand Duke (12961)—Apricot by Fusileer (11499)—Augusta by 3d Duke of York (10166) —Annie by 2d Cleveland Lad (3408)—Annabella by Duke of Cleveland (1937)—by Belvedere (1706)—a cow of Mr. Bates'.

Recorded, E. H. B., vol. 17, p. 357, under dam. Died at Vinewood.

OXFORD'S BARONET 27381 (29499)—Roan, calved Oct. 25, 1869, bred by Col. Towneley, Towneley Park, Lancashire, got by Baron Oxford (23375), out of 6th Maid of Oxford by Imperial Oxford (24185)—Maid of Oxford 1st by Grand Duke of Oxford (16184) —Oxford 20th by Marquis of Carrabas (11789)—Oxford 5th by Duke of Northumberland (1940)—by Short Tail (2621)—by Matchem (2281)—by Young Wynyard (2859).

Died at Vinewood.

GRAND DUKE OF GENEVA 23344 (28756)—Roan, calved Jan. 30, 1870, bred by Messrs. F. Leney & Son, Wateringbury, Kent, got by 15th Grand Duke (21852), out of 7th Duchess of Geneva by 3d Lord Oxford (22200)—3d Duchess of Geneva by Oxford Lad (24713)—Duchess of Geneva by 2d Grand Duke (12961)—Duchess 71st by Duke of Gloster (11382)—by 4th Duke of York (10167)—by 4th Duke of Northumberland (3649)—by Norfolk (2377)—by Belvedere (1706)—by 2d Hubback (1423)—by The Earl (646)—by Ketton 2d (710)—by Comet (155)—by Favorite (252)—by Daisy Bull (186)—by Favorite (252)—by Hubback (319)—by J. Brown's Red Bull (97).

Sold at assignees' sale June 19 and 20, 1878, to Vanmeter & Hamilton, Winchester, Ky., for $850.

FIFTH LORD OXFORD 10382 (31738)—Red, calved Aug. 1, 1870, bred by Messrs. Walcott & Campbell, Utica, N. Y., got by 4th Duke of Geneva (30958), out of Countess of Oxford 2d by 2d Duke of Geneva (23752)—Gem of Oxford by 2d Grand Duke (12961)—Romeo's Oxford by Romeo (13619)—Oxford 5th by Duke of Northumberland (1940)—by Short Tail (2621)—by Mutchem (2281)—by Young Wynyard (2859).

Exported by Wolcott & Campbell, imported by B. B. Groom & Son, sold by them to Wm. & B. F. Thompson, of Clark Co., Ky., and sold at their sale Dec. 7, 1877, to Josh Barton, Bourbon Co., Ky., for $750.

GENEVA ROVER ALIAS GENEVA WILD EYES 29656—Red bull, calved Aug. 27, 1875, bred by Mr. E. H. Cheny, Gaddesby, Leicester, got by 9th Duke of Geneva (28391), out of Wild Duchess of York by 7th Duke of York (17754)—Wild Oxford by Lord Oxford 2d (20215)—Wild Eyes 24th by Lord Barrington 3d (16381)—Wild Eyes 16th by 2d Duke of Oxford (9046)—Wild Eyes 15th by 4th Duke of Northumberland (3649)—by Duke of Northumberland (1940)—by Belvedere (1706)—by Emperor (1975) —by Wonderful (700)—by Cleveland (145)—by Butterfly (104)—

by Hollon's Bull (313)—by Mowbray's Bull (2342)—by Masterman's Bull (422)—descended from the stock of M. Dobison.

Sold to Gen. Meredith, of Indiana, as Geneva Wild Eyes 29656.

WATERLOO BARONET 34071 (45761)—Red, calved Jan. 21, 1876, bred by Mr. T. Barber, Sproatley Rise, got by Oxford's Baronet (29499), out of Water Duchess by Grand Duke 6th (19876)—Water Lass by Oxford 2d (18507)—Water Maid by Marc Antony (14895)—Water Witch by Squire Blanche (12139)—Waterloo 10th by 4th Duke of Northumberland (3649)—Waterloo 8th by 2d Cleveland Lad (3408)—Waterloo 6th by Duke of Northumberland (1940)—Waterloo 3d by Norfolk (2377)—by Waterloo (2816)—by Waterloo (2816).

Sold by assignees to I. & B. Renick, Ohio.

UNDERLEY WILD EYES 31312—Roan, calved Nov. 25, 1875, bred by Lord Fitzhardinge, got by Duke of Underley (33745), out of Lady Wild Eyes 7th by Grand Duke of Waterloo (28766)—Lady Wild Eyes 2d by Touchstone (20986)—Lady Wild Eyes by Wethercock (9815)—Wild Eyes 27th by 2d Cleveland Lad(3408)—Wild Eyes 17th by 2d Duke of Northumberland (3646)—Wild Eyes 5th by Short Tail (2621)—by Emperor (1975)—by Wonderful (700)—by Cleveland (145)—by Butterfly (104)—by Hollon's Bull (313), &c.

1824. CHAS. H. HALL.
New York.

REGENT 899—Calved May 10, 1822, bred by Samuel Scotson, Foxheth Park, near Liverpool, got by Regent.

CANADA—Roan, calved about 1821, bred by Thos. Hymers, got by Sir Peter (606), dam by a son of Constellation (163)—by Young Washington.

YOUNG HECTOR—Calved 1825, got by Hector, out of Canada by Sir Peter (606)—by a son of Constellation (163)—by Young Washington.

PRIMROSE—Got by George, out of Leopardess by Sampson.

COMET.

37

1882.   J. H. HALL.
Manitoba.

ROSA LEE—Roan, calved May 9, 1877, bred by A. O. Stevenson, Blairshennoch, Bauff, Scotland, got by Conservative (46111). out of Florence 2d of Careston by Northern Duke (46806)—Florence 2d by Lord Granville (24395)—Florence by Lord Ythan 2d (22238)—Mary Ann by Garibaldi (17916)—Mary by Moss Trooper (11827)—Geraldine by Robin-o'-Day (4973)—by Holkar (4014)—by Jopp's Bull (9256)—Kate of Darlington.

E. H. B., vol. 29, p. 702.

CINDERELLA 2D—Roan, calved Dec. 14, 1879, bred by A. Scott, Towie, Barclay, Scotland, got by Roseberry (39016), out of Cinderella by Royal Duke of Gloster (29864)—Charmer by Champion of England (17526)—Ceremony by The Baron (13833)—Clipper by Billy (3151)—Favorite by Dandy (6918)—Keepsake by Tip Top (7633)—Old Lady, bred by Mr. Mason, of Chilton.

E. H. B., vol. 26, p. 628, under dam.

ROSE OF DALKEITH—Roan, calved January, 1880, bred by the Duke of Buccleuch, Dalkeith Park, got by Cypress (41314). out of Rose of the Valley by Lord Cecil (26621)—Rose of Smeaton by Royal Errant (22780)—Red Rose by Sir James the Rose (15290) —Lady Rose by Robinson Crusoe (13610)—Lady Alicia by Zadig (8796)—by Earl of Durham (5965)—by St. Helena (5055)—by Emperor (3716)—by Satellite (1420)—by Cato (119)—by Jupiter (342)—by George (273)—by Chilton (136)—by Irishman (329)—by B. (45).

JAMES A. HAMILTON.
Dobbs Ferry, Westchester Co., N. Y.

LADY DAY—Roan, calved March 25, 1836, bred by Mr. Sedgewick, England, owned by C. M. Geddings, Cleveland, Ohio. got by Armitage (1655), out of Superior by Gambier (2046)—Splendor by Anson (1639)—Sophia by Cupid (938)—by Merlin (429)—by Windsor (698)—by R. Colling's White Bull (151).

E. H. B., vol. 3, p. 646, under dam.

ABIJAH HAMMOND.
Throg's Neck, N. Y.

OLD WILLEY—Dam of Young Willey. vol. 1.

Oct. 4, 1877. R. F. HARRISON,
Of Stephensburg, Va. By the Australia, to New York, from London.

CYNOSURE—Roan, calved November, 1874, bred by R. Marsh, Little Offley, Hitchin, got by Mantilini Prince (29273), out of Cinderella by May Prince (24572)—Coronet by British Baronet (17453)—Carlotta by Grand Prince (16187)—Borealis by Northern Light (15004)—Strawberry.

A. H. B., vol. 17, p. 12836.

PRIDE OF SPRINGDALE 30611—Roan, calved Dec. 20, 1877, bred by R. F. Harrison, got by M. C. (31898), out of Cynosure by Mantilini Prince (29273), &c., as in dam, above.

WARRIOR KING 31359—Red and white, calved April 1, 1877, bred by R. Marsh, got by Royal Warrior (35415), out of Fairy Queen (vol. 21, p. 841, E.) by Baron Killerby (23364)—Faithful by Sir James (16980)—Faith by Sir Charles (12075)—Fanchette by Petrarch (7329)—Fame by Raspberry (4875)—Farewell by Young Matchem (4422)—Flora by Isaac (1129)—by Young Pilot (4702)—by Pilot (496)—by Julius Cæsar (1143).

GEO. HENTIG.
Marshall, Mich.

COTTINGHAM 46—Roan, bred by John Colling, Danthorpe Hall, got by Invincible (4077), out of a cow bred by John Colling.

1839-40. JOEL HIGGINS and CALVIN C. MORGAN.
Into Fayette Co., Ky.

MARY—Roan, calved in 1838, bred by Mr. Chrisp, of Doddington, Northumberland, Eng., got by Prince Eugene (2463), out of Carnation by Leopold (370)—Red Rose (sister to Countess) by Yorkshireman (2862)—Flora by Mr. Maynard's Eryholme (1980)—Old Flora (bred by Mr. Wetherell, of Field House, near Darlington) by James Brown's Red Bull (97).

E. H. B., vol. 5, p. 135, under dam, without name.

THEODOSIA—Roan, calved in 1838, bred by Mr. Chrisp, of Doddington, got by Prince Eugene (2463), out of Lady Fanny by Young Hector (2104)—by Emperor (1011)—Mr. Jobling's Fanny by Phenomenon (491)—by Colonel (1845)—by a son of Hubback (319).

E. H. B., vol. 5, p. 520, under dam, without name.

HENRIETTA—Roan, calved in 1837, bred by J. Chrisp, of Doddington, Eng., got by Red Prince (2489), out of Nanny by Emperor (1974)—Young Nanny by Snowball (2647)—Nanny by a son of St. Albans (2584)—by Charlton [a son of Lawnsleeves (365)]—by Barnaby (1678)—by Togston (2762).

E. H. B., vol. 5, p. 722, under dam, without name.

PRINCESS, or ANNE—Red, calved in 1838, bred by Mr. Chrisp, of Doddington, got by Captain (3273), out of Young Princess by Young Lancaster (361)—Princess (bred by John Henderson, Dunstan's Square) by St. Albans (2584)—by Lawnsleeves (365)—bred by Messrs. James, of Stamford, Northumberland.

E. H. B., vol. 5, p. 799, under dam, without name.

ELEANOR—Roan, calved in 1837, bred by J. Chrisp, of Doddington, got by Brougham (1746), out of Madame Vestris by Albion (731)—Pekuah by Diomed (974)—Purity by Barmpton (54)—Charity by Wellington (679)—Dairymaid by Sultan (631)—Ruth by a son of Punch (531)—Broadhooks by Hubback (319)—by Dalton Duke (188).

E. H. B., vol. 5, p. 593, under dam, without name.

July, 1882. C. HILLS,

Of Delaware, Ohio. By Ship Sardinian, July 20, 1882. Selected by
F. P. Hills.

GRAND DUKE OF BARRINGTON 2D (46443)—Roan, calved
April 9, 1881, bred by Henry Lovatt, Esq., Low Hill, Bushbury,
Straffordshire, got by Grand Duke 37th (43307), out of Lady Thorndale and Gloster Bates by Grand Duke of Kirklevington (34071)
—Lady Clarence Bates by Grand Duke of Thorndale (31297)—Lady
Thorndale Bates by 4th Duke of Thorndale (17750)—Lady Bates
3d by 4th Duke of Oxford (11387)—Lady Bates 2d by The Buck
(13836)—Lady Bates by Duke of Gloster (11382)—Lady Blanche
by 4th Duke of York (10167)—Lady Barrington 8th by 2d Duke
of Oxford (9046)—Lady Barrington 5th by 4th Duke of Northumberland (3649)—by Cleveland Lad (3407)—by Belvedere (1706)—
by son of Herdsman (305)—by Wonderful (700)—by Alfred (23)—
by Young Favorite (6994).

September, 1882. C. HILLS,

Of Delaware, Ohio. By Ship Circassian, Sept. 7, 1882. Selected by
F. P. Hills.

BUSHBURY COUNTESS OF OXFORD—Roan, calved Dec.
25, 1881, bred by Mr. Lovatt, got by Grand Duke 37th (43307),
out of 11th Maid of Oxford by 9th Duke of Geneva (28391)—Maid
of Oxford 10th by 4th Duke of Geneva (30958)—3d Maid of Oxford by Grand Duke of Oxford (16184)—Oxford 20th by Marquis
of Carrabas (11789)—Oxford 5th by Duke of Northumberland
(1940)—by Short Tail (2621)—by Matchem (2281)—by Young
Wynyard (2859).

E. H. B., vol. 28, p. 485, under dam.

WILD EYES LASSIE 4TH—Roan, calved March 24, 1879,
bred by Mr. S. P. Foster, Killhow, Carlisle, Cumberland, got by
Duke of Ormskirk (36526), out of Wild Eyes Lassie 2d by 22d
Duke of Oxford (31000)—Wild Eyes Lassie by 3d Duke of Claro
(23729)—Wild Eyes 24th by 4th Duke of Oxford (11387)—Wild
Eyes 22d by Wild Duke (19148)—by Lord Barrington 1st (13170)
—by 2d Duke of Oxford (9046)—by 4th Duke of Northumberland

(3649)—by Duke of Northumberland (1940)—by Belvedere (1706)
—by Emperor (1975)—by Wonderful (700)—by Cleveland (145)—
by Butterfly (104)—by Hollon's Bull (313)—by Mowbray's Bull
(2342)—by Masterman's Bull (422)—descended from the stock of
M. Dobison.

E. H. B., vol. 26, p. 431, under dam.

KIRKLEVINGTON LADY 8TH—Roan, calved May 31, 1880,
bred by Earl Beetive, Underley Hall, got by Lord Acomb 2d (43474),
out of Kirklevington Lady 7th by Duke of Clarence 5th (36479)—
Kirklevington Lady 3d by Duke of Wetherby 6th (33756)—Lady
Kirklevington 2d by 2d Earl of Oxford (21651)—Lady Kirklevington by Grand Duke of York (12966)—Kirklevington 9th by
General Canrobert (12926)—Kirklevington 7th by Earl of Derby
(10177)—Kirklevington 4th by Earl of Liverpool (9061)—Kirklevington 1st by Duke of Northumberland (1940)—by Belvedere (1706)
—by Son of 2d Hubback (2683)—descended from the stock of Mr.
Maynard, of Eryholme.

E. H. B., vol. 27, p. 300.

— ———

HARVEY HILL,

*Of New Orleans. Taken to his farm, in Tennessee, and bought by Mark
R. Cockrill, of Nashville, Tenn.*

GENTLE—Roan, calved in 1837, bred by Mr. Pittam, got by
Cupid (7941), out of Spotted Broughton by Crusader (934)—
Bombazine by Regent (544).

Recorded, under dam. E. H. B., vol. 5, p. 962.

LADY LITTLETON—White, calved Oct. 14, 1837, bred by
Mr. Kemp, got by Ranunculus (2479), out of The Kicker by
Memnon (2294)—by Ratify (2481)—by Logic (4248).

E. H. B., vol. 5, p. 1009, under dam, and A. H. B., vol. 22,
p. 17234.

MILD SPRING.

1822 to 1824. H. HILL.

New York.

JENNY—See Cora Creampot.

C. H. B., vol. 1. p. 247.

ECLIPSE.

Aug. 30, 1882. J. J. HILL,

Of St. Paul, Minn. Landed at Quebec, Aug. 30, 1882, from Steamship Manitoban, of Allan Line.

GAMBETTA (49618)—Roan, calved Feb. 4, 1881, bred by Watt Garhetty, Fochabers, Scotland, got by Fitz Henry (41552), out of Marchioness 2d by Baronet (25564)—Marchioness by Emperor (19689)—Averne by Picotee (15063)—Alice by Duke 3d (17697)—Fanny by Garioch Lad (17938)—Averne by Bucephalus (6784)—Young Amazon by Crusader (934)—Amazon by Sultan (1483)—Bellona by Mars (411)—Rolla by North Star (458).

ROSA LEE—Roan, calved May 9, 1877, bred by A. O. Stevenson, Blairshinnock, Banff, Scotland, got by Conservative (46111), out of Florence 2d of Careston by Northern Duke (46806)—Florence 2d by Lord Granville (24395)—Florence by Lord Ythan 2d (22238)—Mary Anne by Garibaldi (17916)—by Moss Trooper (11827)—by Robin-o'-Day (4973)—by Holkar (4041)—by Mr. Jopp's Bull (9256)—Kate of Darlington.

E. H. B., vol. 29, p. 702.

LOUIS BLANC—White, calved May 4, 1883, got by Lord of the North (46687), out of Rosa Lee by Conservative (46111), &c., as in dam, next above.

ROSE OF DALKEITH—Roan cow, calved Jan. 3, 1880, bred by the Duke of Buccleuch, Dalkeith Park, Scotland, got by Cyprus (41314), out of Rose of the Valley (vol. 24, p. 355, E.) by Lord Cecil (26621)—Rose of Smeaton by Royal Errant (22780)—Red Rose by Sir James the Rose (15290)—Lady Rose by Robinson Crusoe (13610)—Lady Alicia by Zadig (8796)—Lady Alice by Earl of Durham (5965)—Lady Ellen by St. Helena (5055)—Lady Mary by Emperor (3716)—Lady Sarah by Satellite (1420)—by Cato (119) —by Jupiter (342)—by George (273)—by Chilton (136)—by Irishman (329)—by B. (45).

Recorded, A. H. B., vol. 26, p. 821.

STAR PRINCESS—Red, little white, cow, calved April 10. 1878, bred by Robt. Arkley, Ethieveaton, Scotland, got by King Hope (36843), out of Annandale Star by Annandale (30386)— Evening Star by Morning Star (29391)—Omega by Harlequin (18026)—Broomley by Hassan (12995)—Flora by Roderick Random (10731)—by Velocipede (5552)—by Burley (1766)—by Sir Alexander (591)—by Marske (418)—from the stock of Mr. Maynard, of Eryholme.

E. H. B., vol. 30.

CINDERELLA 2D—Roan, calved Dec. 14, 1879, bred by A. Scott, Towie, Barclay, Scotland, got by Roseberry (39016), out of Cinderella by Royal Duke of Gloster (29864)—Charmer by Champion of England (17526)—Ceremony by The Baron (13833)—Clipper by Billy (3151)—Favorite by Dandy (6918)—Keepsake by Tip Top (7633)—Old Lady, bred by Mr. Mason, of Chilton.

E. H. B., vol. 26, p. 628, under dam.

CINDERELLA OF NORTH OAKS—Calved April 5, 1883. got by Vienna (45731), out of Cinderella 2d by Roseberry (39016), &c., as in dam, next above.

May, 1883. J. J. HILL.
Per Ship Manitoban.

ANEMONE 3D—Roan, calved May 23, 1880, bred by E. Bowley, got by Duke of Holkar 2d (39749), out of Anemone 2d by 4th Baron Oxford (25580)—Anemone by Duke of Kent (19619)— Acacia by Count de Gourey (17632)—Asia by 2d Grand Duke (12961)—Apricot by Fusileer (11499)—Augusta by 3d Duke of York (10166)—by 2d Cleveland Lad (3408)—by Duke of Cleveland (1937)—by Belvedere (1706)—a cow of Mr. Bates'.

A. H. B., vol. 26, p. 820, and E. H. B., vol. 27, p. 317, under dam.

FAERY COUNTESS 2D—Roan, calved March 11, 1881, bred by A. H. Lloyd, got by 27th Duke of Airdrie (41351), out of Faery Queen by Cherry Emperor (33347)—Fanny by Enterprise (16002)—Emma 3d by Richmond (13592)—Emma by Freetrader (10246)—Nanny 3d by Gainford (2044)—by Magnum Bonum (2243)—by Rob Roy (557)—by son of Houghton (318)—by Sir Stephen (1456)—by Sedbury (1424).

COUNTESS OF WORCESTER—Red, calved May 26, 1881, bred by A. H. Lloyd, got by Duke of Gloster 5th (36494), out of Lady Worcester 19th by 6th Duke of Geneva (30959)—Lady Worcester 2d by Charleston (21400)—by Marton Duke (22307)—by Red Duke (18676)—by 3d Duke of York (10166)—by 2d Cleveland Lad (3408)—by Duke of Northumberland (1940)—by Belvedere (1706)—by Emperor (1975)—by Wonderful (700)—by Cleveland (145)—by Butterfly (104)—by Hollon's Bull (313)—by Mowbray's Bull (2342)—by Masterman's Bull (422)—descended from M. Dobison's stock.

A. H. B., vol. 26, p. 820, and E. H. B., vol. 28, p. 478.

BERKELEY DUKE OF OXFORD 2D 54790 (45973)—Roan, calved April 29, 1881, bred by Lord Fitzhardinge, got by Duke of Connaught (33604), out of Oxford Belle 5th by Grand Duke of Gloster (36721)—Oxford Belle 2d by Duke of Hillhurst (28401)—Countess of Oxford by 7th Duke of Airdrie (23718)—Gem of Oxford by Grand Duke 2d (12961)—by Romeo (13619)—by Duke of Northumberland (1940)—by Short Tail (2621)—by Matchem (2281)—by Young Wynyard (2859).

July 6, 1883.   J. J. HILL.
By Ship Lucerne, to Quebec.

GOLDEN LACE—Roan, calved Aug. 16, 1878, bred by E. Baillie, Dochfour, Inverness, Scotland, got by Abbot of Windsor (32903), out of Golden Link by Flower of the Forest (33948)—Golden Locket by No Mistake (34918)—Golden Rose by Scotch Rose (25099)—Golden Reed by Baronet (15614)—by Lord Raglan (13244)—by Duplicate Duke (6952)—by Robin-o'-Day (4973)—by

Sir Walter (2639)—by Young Jerry (8177)—by Roseberry (567)—
by Roseberry (567)—by Constellation (163)—by Hastings (293)—
by Leopold (372).

A. H. B., vol. 26, p. 820, and E. H. B., vol. 29, p. 309.

GOLDEN MINT—Red, calved July 14. 1883, got by Wilfrid
Benedict (45811), out of Golden Lace by Abbot of Windsor (32903),
&c., as in dam, next above.

FANNY B. 30TH—Red and white, calved Jan. 9, 1882, bred by
Mr. James Bruce, Burnside, Fochabers, Scotland, got by Knicker-
bocker (38510), out of Fanny B. 6th by Earl of March (33807)—
Fanny 5th by Champion (28155)—Fanny 2d by Prince of Worcester
(20597)—Fanny by Signet Seal (18824)—by Western Star (25430)
—by Red Lugs (32263)—by Orbliston (24691).

ROSEBUD 2D—Roan, calved Feb. 20, 1879, bred by Jas. Swan,
got by King Charming, out of Moss Rose by Master Booth (34801)
—by Jacobite (28905)—by White Knight (25437)—by White
Knight (25437)—by Representative.

Roan bull, calved Feb. 1, 1884, got by Red Knight. out of
Rosebud 2d, above.

Aug. 6, 1883.   J. J. HILL.
Per Ship Manitoban.

BELLE OF ALBION—Roan, calved March 5, 1879, bred by
E. Baillie, got by Abbot of Windsor (32903), out of Lady Margaret
by Red Prince (35240)—Annie 2d by Count Bickerstaffe (23630)—
Annie by Baron Melbourne (15621)—Carnation by Saracen (15237)
—Garland 2d by Alfred (6732)—Garland by Brunswick (6814)—
by Lycurgus (7180)—by Ranunculus (2479)—by William (2840)—
by Childers (1824)—by Richard (1376)—by Jupiter (342)—by
Charles (127)—by Windsor (698)—by Chilton (136)—by Colonel
(152).

A. H. B., vol. 26, p. 820, and E. H. B., vol. 28, p. 277.

JENNY LIND 12TH—Roan, calved Dec. 12, 1879, bred by
A. E. Hector, Asbelow, got by British Champion (36273). out of

Jenny Lind by Rajah (32232)—Jenny Lind 6th by Prince Louis (20560)—Jenny Lind 3d by Kelvinside (14756)—Jenny Lind by The Duke (7593)—Cora by The Pasha (7612)—Alice by Mahomed (6170)—Matilda by Sillery (5131)—Young Corridore by Young Western Comet (1575)—by Favorite (256).

VENUS 2D—Red, calved Jan. 4, 1878, bred by A. E. Hector, got by British Champion (36273), out of Venus by Chronometer (33385)—Verbena 2d by Rajah (32232)—Verbena by Golden Eagle (26267)—Violet by Sir Thomas Stanley (25176)—Duchess by Marquis of Bute (18336)—by Count Fairfax (8991)—by South Durham (9676)—by Uptaker (5534)—by Miracle (2320)—by Martin (2280) —by Fitz Remus (2025)—by Cato (119)—by Whitworth (695).

VICEROY—Red bull, calved Dec. 25, 1882, got by Norman (45272), out of Venus 2d by British Champion (36273), &c., as in dam, next above.

Oct. 16, 1883.   J. J. HILL.

By Steamship Hanoverian, landing at Quebec.

SWEET PEA—Red and white, calved Jan. 13, 1878, bred by Mr. E. Baillie, got by Oliver Cromwell (43706), out of Pretty Gold by Flower of the Forest (33948)—Marigold 13th by Gold Digger (24044)—Marigold 6th by Young Pacha (20457)—Marigold 2d by Lord Lorne (18258)—Marigold by Rubens (13641)—by Jemmy (11611)—by Captain (8925)—by Duke of St. Albans (8001)—by Guy Faux (7062)—by Pedestrian (7321)—by Lucian (2228)—by Raby (2473).

A. H. B., vol. 26, p. 821, and E. H. B., vol. 28, p. 278.

RUBY OF NORTH OAKS—Red cow, calved Aug. 24, 1883, bred by Mr. Evan Baillie, calved the property of Jas. J. Hill, got by Wilfrid Benedict (45811), out of imp. Sweet Pea (vol. 26, p. 821) by Oliver Cromwell (43706)—Pretty Gold by Flower of the Forest (33948)—Marigold 13th by Gold Digger (24044)—Marigold 6th by Young Pacha (20457)—Marigold 2d by Lord of Lorne (18258)—Marigold by Rubens (13641)—Rose by Jemmy (11611)—

Rosamond by Captain (8925)—Ruth by Duke of St. Albans (8001) —Ruby by Guy Faux (7062)—Little Red Rose by Pedestrian (7321) —by Lucian (2228)—by Raby (2473).

Recorded, A. H. B., vol. 26, p. 821.

June, 1884.   J. J. HILL,

Of St. Paul, Minn.   By Steamer Austrian, landed at Boston, June 16, 1884.

DUCHESS OF WAPPENHAM—Red and white, calved March 4, 1882, bred by Robert Loder, Whittlebury, Towcester, Northampton, got by Grand Duke 25th (34065), out of Grand Duchess of Geneva 5th by 8th Duke of Geneva (28390)—Grand Duchess of Geneva by Grand Duke 15th (21852)—Duchess of Geneva 7th by 3d Lord Oxford (22200)—Duchess of Geneva 3d by Oxford Lad (24713)—Duchess of Geneva by Grand Duke 2d (12961)—by Duke Gloster (11382)—by 4th Duke of York (10167)—by 4th Duke of Northumberland (3649)—by Norfolk (2377)—by Belvedere (1706) —by 2d Hubback (1423)—by The Earl (646)—by Ketton 2d (710) —by Comet (155)—by Favorite (252)—by Hubback (319)—by J. Brown's Red Bull (97).

E. H. B., vol. 29, p. 560, under dam.

GRAND DUCHESS 43D—Roan, calved July 24, 1880, bred by R. E. Oliver, Sholebroke Lodge, Towcester, got by Grand Duke 30th (38373), out of Grand Duchess 28th by 3d Duke of Clarence (23727) —Grand Duchess 17th by Imperial Oxford (18084)—Grand Duchess 10th by Grand Duke 3d (16182)—by Prince Imperial (15095) —by Grand Duke (10284)—by Cleveland Lad (3407)—by Belvedere (1706)—by 2d Hubback (1423)—by 2d Hubback (1423)—by The Earl (646)—by Ketton 2d (710)—by Comet (155)—by Favorite (252) —by Daisy Bull (186)—by Favorite (252)—by Hubback (319)—by J. Brown's Red Bull (97).

E. H. B., vol. 28, p. 527, under dam. Burned to death in August, 1884.

GRAND DUCHESS 47TH—Red and white, calved March 24, 1882, bred by R. E. Oliver, got by Grand Duke 30th (38373), out of Grand Duchess 38th by Duke of Underly 3d (38196)—Grand Duchess 17th by Imperial Oxford (18084), &c., as in Grand Duchess 43d, next above.

E. H. B., vol. 29, p. 620, under dam.

DUCHESS OF OXFORD 2D—Red, calved Dec. 4, 1880, bred by Mr. D. A. Green, East Donyland, Colchester, got by Kirklevington Lord (45003), out of Duchess of Oxford by Duke of Underly 3d (38196)—Lady of Oxford 16th by 9th Duke of Geneva (28391)—Lady Oxford 13th by Baron of Oxford (23371)—Lady of Oxford 7th by 6th Duke of Thorndale (23794)—Lady of Oxford 2d by Grand Duke 2d (12961)—by 3d Duke of York (10166)—by Duke of Northumberland (1940)—by Short Tail (2621)—by Matchem (2281)—by Young Wynyard (2859).

E. H. B., vol. 27, p. 422, under dam.

NORTH OAKS LADY OF OXFORD—Red, calved June 25, 1884, got in England by Duke of Leicester (43112), out of Duchess of Oxford 2d by Kirklevington Lord (45003), &c., as in next above.

LADY YORK AND THORNDALE BATES 8TH—Red and white, calved May 13, 1882, bred by Mrs. Fawcet, Scalesby Castle, Carlisle, got by Grand Duke of Kirklevington 4th (43317), out of Lady York and Thorndale Bates 4th by Grand Duke of Kirklevington 2d (38379)—Lady York and Thorndale Bates 3d by 8th Duke of York (28480)—Lady Tregunter Bates by 2d Duke of Tregunter (26022)—Lady Thorndale Bates by 4th Duke of Thorndale (17750)—Lady Bates 3d by 4th Duke of Oxford (11387)—Lady Bates 2d by The Buck (13836)—Lady Bates by Duke of Gloster (11382)—by 4th Duke of York (10167)—by 2d Duke of Oxford (9046)—by 4th Duke of Northumberland (3649)—by Clevaland Lad (3407)—by Belvedere (1706)—by son of Herdsman (304)—by Wonderful (700)—by Alfred (23)—by Young Favorite (6994).

E. H. B., vol. 29, p. 355, under dam.

WILD LADY 2D—Roan, calved Sept. 18, 1882, bred by the Countess of Stamford and Warrington, got by 3d Duke of Gloster (33653), out of Wild Duchess of Gloster (vol. 26, p. 623, E.) by 3d

Duke of Gloster (33653)—Wild Oxford by Lord Oxford 2d (20215)—Wild Eyes 24th by Lord Barrington 3d (16382)—Wild Eyes 16th by 2d Duke of Oxford (9046)—by 4th Duke of Northumberland (3649)—by Duke of Northumberland (1940)—by Belvedere (1706)—by Emperor (1975)—by Wonderful (700)—by Cleveland (145)—by Butterfly (104)—by Hollon's Bull (313)—by Mowbray's Bull (2342)—by Masterman's Bull (422)—descended from M. Dobison's stock.

CONISHEAD WILD EYES 2D—Roan, calved March 29, 1879, bred by W. Ashburner, Conishead Grange, got by 2d Duke of Gloster (28392), out of Bright Eyes 5th by Royal Lancaster (29870)—Bright Eyes 4th by 3d Duke of Wharfdale (21619)—Bright Eyes 3d by Beau of Oxford (21254)—by Oxford Duke (15036)—by Crusade (7936)—by 3d Duke of York (10166)—by 2d Cleveland Lad (3408)—by Duke of Northumberland (1940)—by Belvedere (1706)—by Emperor (1975)—by Wonderful (700)—by Cleveland (145)—by Butterfly (104)—by Hollon's Bull (313)—by Mowbray's Bull (2342)—by Masterman's Bull (422)—descended from M. Dobison's stock.

E. H. B., vol. 28, p. 391.

GRAND DUCHESS OF BARRINGTONIA 5TH—Red and white, calved Aug. 31, 1879, bred by R. E. Oliver, got by Grand Duke 30th (38373), out of Grand Duchess of Barringtonia 2d by 3d Duke of Clarence (23727)—Grand Duchess of Barrington by Grand Duke 7th (19877)—Countess of Barrington 2d by 9th Duke of Oxford (17738)—Countess of Barrington by 3d Grand Duke (16182)—by Grand Turk (12969)—by Earl of Derby (10177)—by Earl of Liverpool (9061)—by 2d Duke of Cambridge (3638)—by Belvedere (1706)—by son of Herdsman (304)—by Wonderful (700)—by Alfred (23)—by Young Favorite (6994).

E. H. B., vol. 26, p. 574, under dam.

YOUNG JULIA 3D—Roan, calved April 15, 1876, bred by Lord Lovat, Beaufort Castle, Inverness, N. B., got by Bachelor of Arts (32982), out of Young Julia by King's Seal (26525)—Julia by

Allan (21172)—Josephine by Matadore (11800)—by Fairfax Royal (6987)—by Premier (6308)—by Saturn (5089)—by Favorite (6997) —by Grindon (3942)—bred by Mr. Rennie, of Phantassie.

E. H. B., vol. 25, p. 554.

1821. HUMPHREY HOLLIS.

HART—By Colling's Wellington. See vol. 15, p. 6.

NUDD—By Colling's Wellington, out of Ruddy, owned by Geo. Morley, Boylestone, Eng. See vol. 15, p. 6.

1839. DANIEL HOLLMAN.
New Jersey.

JANE—Red and white, calved in 1836, bred by G. L. Ridley, got by Young Magog (2247), out of Laura by Budget (1759)— Strawberry by Sandoe (2598)—Old Red Rose by Copeland (1871) —Greyhooks by Marquis (407)—by Styford (629)—by Bolingbroke (86)—by a son of Hubback (319)—by Hubback (319).

A. H. B., vol. 15, p. 609, and E. H. B., vol. 3, p. 462.

1874. JOHN HOPE.
From Liverpool to Quebec, Aug. 19, 1874.

TRINKET JEWELER 41146 (35814)—Roan bull, calved April 10, 1874, bred by R. S. Bruere, Braithwaite Hall, got by Booth's Royal Signet (28061), out of Crown Trinket by Royal Booth (22772) —Garnet by King Arthur (13110)—Georgiana by The Silky Laddie (10947)—Rosy by Leonardo (7137)—by Musician (6234)—by Priam (2452).

FENNEL DUCHESS OF LANCASTER—White, calved September, 1874, bred by E. Musgrove, got by Royal Lancaster (29870), out of Fennel Duchess 7th by 13th Duke of Oxford (21604), &c., as in dam, above.

A. H. B., vol. 16, p. 12070.

FENNEL DUCHESS 7TH—Roan, calved May 10, 1870, bred by Mr. Atherton, Chapel House, Liverpool, got by 13th Duke of Oxford (21604), out of Fennel Duchess by Lord Oxford 2d (20215) —Fennel 3d by Cherry Duke 2d (14265)—Fennel by 5th Duke of York (10168)—Filbert by 2d Cleveland Lad (3408)—Felicia by 4th Duke of Northumberland (3649)—by Short Tail (2621)—by Belvedere (1706)—by son of Young Wynyard (2859)—by James Brown's Red Bull (97).

E. H. B., vol. 19, p. 512, under dam, and A. H. B., vol. 14, p. 536.

SUNLIT FLOWER—Roan, calved April 3, 1870, bred by R. S. Bruere, Braithwaite Hall, Middleham, Yorkshire, got by Booth's Kinsman (25658), out of Sunflower by Prince George (13510)— Marigold Flower by King Arthur (13110)—Marigold by Gainford (7029)—Nutty by Rouge (5012)—by Cleveland (3404)—by Burton (3250).

E. H. B., vol. 19, p. 742, under dam.

SAMPHIRE (35465)—Red bull, calved Dec. 31, 1873, bred by R. S. Bruere, Braithwaite Hall, got by Booth's Royal Signet (28061), out of Golden Marigold Flower by Booth's Kinsman (25658)—Marigold Flower by King Arthur (13110)—Marigold by Gainford (7029)—Nutty by Rouge (5012)—by Cleveland (3404)— by Burton (3250).

Oct. 14, 1875. JOHN HOPE.
By Ship Polynesian, from Liverpool, Oct. 14, 1875, to Markham, Canada.

ANEMONE—Red and white cow, calved April 1, 1872, bred by Mr. R. B. Hetherington, got by Grand Duke of Lightburne 2d (26291), out of Athena by Duke of Darlington (21586)—Acacia by Count de Gourcy (17632)—Asia by 2d Grand Duke (12961)—Apricot by Fusileer (11499)—by 3d Duke of York (10166)—by 2d Cleveland Lad (3408)—by Duke of Cleveland (1937)—by Belvedere (1706)—a cow of Mr. Bates'.

Recorded, E. H. B., vol. 21, p. 968, and A. H. B., vol. 16, p. 11960. Sold at sale of John Hope, June 14, 1876, to Emory Cobb, Esq., Kankakee, Ill.

PRINCESS VICTORIA 10TH—Red and white heifer, calved March 21, 1873, bred by Lord Skelmersdale, got by 1st Duke of Oneida (30996), out of Princess Victoria 5th by Lord Oxford 2d (20215)—Princess Victoria 2d by Lord Oxford 2d (20215)—Princess Alice by Gen. Canrobert (12927)—Princess by Earl of Derby (10177)—by 2d Cleveland Lad (3408)—by 2d Earl of Darlington (1945)—by Son of 2d Hubback (2683)—a cow bought of Mr. Bates.

Recorded, E. H. B., vol. 21, p. 931, under dam, A. H. B., vol. 16, p. 12293. Bought at John Hope's sale, June 14, 1876, by Emory Cobb, of Kankakee, Ill.

LADY ACOMB 3D.—Roan heifer, calved June 26, 1873, bred by Mr. T. Gow, Cambo, Newcastle-on-Tyne, got by Oxford Le Grand (29496), out of Annette by Duke of Brailes (23724)—Amy by Duke of Darlington (21586)—Anemone by Duke of Kent (19619)—Acacia by Count de Gourcy (17632)—Asia by 2d Grand Duke (12961)—Apricot by Fusileer (11499)—by 3d Duke of York (10166)—by 2d Cleveland Lad (3408)—by Duke of Cleveland (1937)—by Belvedere (1706)—a cow of Mr. Bates'.

Recorded. E. H. B., vol. 21, p. 741, under dam, and A. H. B., vol. 16, p. 12131. Sold, June 14, 1876, to T. L. McKean, Easton, Pa.

ACOMB'S DUKE 25486—Red, calved Nov. 24, 1875, bred by Sir W. Trevelyan, got by Oxford Beau 4th (34964), out of Anemone by Grand Duke of Lightburne 2d (26291), &c., as in Anemone, above.

DAMSEL—Roan heifer, calved July 9, 1873, bred by J. P. Foster, Killhow, Carlisle, got by 22d Duke of Oxford (31000), out of Daffodil by 11th Grand Duke (21849)—Dorothy by 3d Duke of Wharfdale (21619)—Dora by Duke of Geneva (19614)—Duchess 1st by Master Rembrandt (16545)—Duchess Nancy by Jasper (11609)—by 2d Duke of Oxford (9046)—by 2d Duke of Northumberland (3646)—by Belvedere (1706)—by Son of 2d Hubback (2683)—a cow of Mr. Bates'.

Recorded, E. H. B., vol. 21, p. 715, under dam, and A. H. B., vol. 18, p. 13630. Sold, June 14, 1876, to H. N. More, Red Oak, Iowa, for $2,800.

ACOMB J.—Red heifer, calved Aug. 21, 1873, bred by Sir W. C. Trevelyan, Wallington, Northumberland, got by Baron Acomb 2d (30419), out of Jessy by Duke of Waterloo (21616)—Julia by Ranter (18666)—Lady Jessy by 7th Duke of York (17754)—Lady Jane by Duke of Gloster (11382)—Jardine by Lord Warden (7167) —by Fawsley (6004)—by Warden (5595)—by Javelin (4093)—by Blyth (797)—by Wellington (684)—by Phenomenon (491)—by Favorite (252)—by Favorite (252)—by Favorite (252)—by Hubback (319)—by Snowdon's Bull (612)—by Waistell's Bull (669)—by Masterman's Bull (422)—by Studley Bull (626).

Recorded, E. H. B., vol. 21, p. 969, under dam.

OXFORD WATERLOO 5th—Red heifer, calved November 25, 1873, bred by Mr. R. Lodge, Bishopdale, Bedale, Yorkshire, got by Duke of Athelstane (21562), out of Oxford's Waterloo 4th by 13th Duke of Oxford (21604)—Oxford's Waterloo by Lord Oxford 2d (20215)—Waterloo 19th by 2d Grand Duke (12961)—Waterloo 15th by Matadore (11800)—Waterloo 12th by 3d Duke of York (10166) —Waterloo 4th by Cleveland Lad (3408)—by Norfolk (2377)—by Waterloo (2816)—by Waterloo (2816).

E. H. B., vol. 21, p. 818, under dam, and A. H. B., vol. 16, p. 12272. Sold to Maj. Greig, Toronto, Can., June 14, 1876, for $1,600.

COUNT HILLHURST 26016—Red, calved July 4, 1876, bred by John Foster, got by Duke of Hillhurst (28401), out of Damsel by 22d Duke of Oxford (31000), &c., as in Damsel, above.

SILVER LADY—Roan heifer, calved Feb. 14, 1874, bred by J. P. Foster, Killhow, Carlisle, got by 22d Duke of Oxford (31000), out of Surmise 2d by May Duke (13320)—Surmise by Duke of Gloster (11382)—Silence by Earl of Derby (10177)—Secret 3d by Duke of Sutherland (6945)—by Locomotive (4242)—by Short Tail (2621)—by Gambier (2046)—by Young Wynyard (2859)—Bulls of R. & C. Colling.

Recorded, E. H. B., vol. 21, p. 717, under dam, and A. H. B., vol. 18, p. 13929. Bought, June 14, 1876, by H. N. More, Red Oak, Iowa, for $2,600.

OXFORD QUEEN—Red heifer, calved Nov. 8, 1874, bred by Mr. G. Moore, Whitehall, Cumberland, got by 17th Duke of Oxford (25994), out of Oxford Donna by Didmarton Duke (21546)—Lady Oxford by Imperial Oxford (18084)—Lady Gloster by Harry of Gloster (14674)—Lady Warden by Lord Warden (7167)—Belinda 2d by Lion (9299)—Belinda by Rebel (4882)—by Coxcomb (928)—by Minor (441)—by son of Phenomenon (491)—by Traveler (655)—by Colonel (152)—by R. Colling's son of Broken Horn (95) —by R. Colling's son of Hubback (319).

Recorded, E. H. B., vol. 21, p. 855, under dam. Bought, June 14, 1876, by A. L. Stebbin, Port Huron, Mich., for $430.

ROYAL LANCASTER (29870)—Roan bull, calved June 20, 1870, bred by Mr. D. R. Davies, High Legh Hall, Knutsford, Cheshire, got by 10th Grand Duke (21848), out of Moss Rose by Marmaduke (14897)—Cambridge Rose 6th by 3d Duke of York (10166)—Cambridge Rose 5th by 2d Cleveland Lad (3408)—Cambridge Rose 2d by Belvedere (1706)—by Belvedere (1706)—by 2d Hubback (1423)—by His Grace (311)—by Yarborough (705)—by Favorite (252)—by Punch (531)—by Foljambe (263)—by Hubback (319).

BARON SIDDINGTON 25641—Roan, calved Nov. 11, 1874, bred by J. P. Foster, Killhow, Carlisle, got by 6th Baron Oxford (33075), out of Siddington 12th by 2d Duke of Tregunter (26022) —Siddington 2d by 4th Duke of Oxford (11387)—Kirklevington 7th by Earl of Derby (10177)—Kirklevington 4th by Earl of Liverpool (9061)—Kirklevington by Duke of Northumberland (1940)— Nell Gwynne by Belvedere (1706)—Northallerton by Son of 2d Hubback (2683)—a cow of Mr. Bates'.

Sold, June 14, 1876, to W. W. Pickrell, of Mechanicsburg, Ill.

EARL OF DERBY 26462—Roan, calved Sept. 10, 1875, bred by Lord Skelmersdale, got by Baron Barrington 5th (33007), out of Princess Victoria 10th by 1st Duke of Oneida (30996), &c., as in Princess Victoria, above.

October, 1876. JOHN HOPE.

DESTINY—Roan, calved May 3, 1874, bred by J. P. Foster, Kilhow, Cumberland, got by 22d Duke of Oxford (31000), out of Daffodil by 11th Grand Duke (21849)—Dorothy by 3d Duke of Wharfdale (21619)—Dora by Duke of Geneva (19614)—Duchess 1st by Master Rembrandt (16545)—Duchess Nanny by Jasper (11609)—Duchess Nancy by 2d Duke of Oxford (9046)—Nettle by 2d Duke of Northumberland (3646)—Nell Gwynne by Belvedere (1706)—Northallerton by Son of 2d Hubback (2683)—a cow of Mr. Bates', descended from the stock of Mr. Maynard, of Eryholme.

E. H. B., vol. 21, p. 715, under dam.

THIRD DUCHESS OF THORNDALE—Roan, calved March 20, 1876, bred by W. W. Slye, got by Grand Duke of Thorndale 2d (31298), out of Lady Walton 2d by Earl of Gloster (21644)—Lady Walton by 3d Earl of Walton (19678)—Levity by Lord Thoresby (14856)—Lily Belle by Horrox (11591)—Lily by 2d Duke of Oxford (9046)—Harmless by Cleveland Lad (3407)—by Red Rose Bull (2493)—by Rex (1375)—bought of Mr. Richardson, of Hart.

A. H. B., vol. 17, p. 12861.

September, 1876. JOHN HOPE.

DOCILE—Roan, calved Feb. 22, 1874, bred by J. P. Foster, Kilhow, Cumberland, got by 22d Duke of Oxford (31000), out of Dorothy by 3d Duke of Wharfdale (21619)—Dora by Duke of Geneva (19614)—Duchess 1st by Master Rembrandt (16545)—Duchess Nanny by Jasper (11609)—Duchess Nancy by 2d Duke of Oxford (9046)—Nettle by 2d Duke of Northumberland (3646)—Nell Gwynne by Belvedere (1706)—Northallerton by Son of 2d Hubback (2683)—cow of Mr. Bates', descended from stock of Mr. Maynard, of Eryholme.

E. H. B., vol. 21, p. 715, under dam, and A. H. B., vol. 18, p. 13634. Sold, June 6, 1877, at London, Ont., to Canada West Farm Stock Association.

DUCHESS OF CLARENCE 12TH—Roan, calved Jan. 20, 1875, bred by T. Barber, Sproatley Rise, got by Oxford Baronet

(29499), out of Duchess of Clarence 7th by King of the Roses (22043)—Duchess of Clarence 3d by Grand Duke 6th (19876)—Duchess of Clarence by Duke of Clarence (19611)—Duchess Nan by Romulus (15185)—Duchess Nanny by Jasper (11609)—Duchess Nancy by 2d Duke of Oxford (9046)—Nettle by 2d Duke of Northumberland (3646)—Nell Gwynne by Belvedere (1706)—Northallerton by Son of 2d Hubback (2683), &c.

E. H. B., vol. 22, p. 311, under dam, and A. H. B., vol. 18, p. 13641. Sold, June 6, 1877, at London, Ont., to Canada West Farm Stock Association.

DUCHESS OF CLARENCE 15TH—White, calved March 20, 1877, bred by J. P. Foster, got by 7th Baron Oxford (36199), out of Docile by 22d Duke of Oxford (31000), &c., as in Docile, above.

A. H. B., vol. 18, p. 13641. Sold to Canada West Farm Stock Association, June 6, 1877, at London, Ont.

DUCHESS OF CLARENCE 16TH—Roan, calved June 22, 1877, bred by J. P. Foster, got by Duke of Ormskirk (36526), out of Duchess of Clarence 12th by Oxford's Baronet (29499), &c., as in dam, above.

A. H. B., vol. 18, p. 13641. Sold, at London, Ont., June 6, 1877, to Canada West Farm Stock Association.

WATERLOO 36TH—Red and white, calved Aug. 24, 1872, bred by Sir Wilfrid Lawson, Brayton, got by Royal Cambridge (25009), out of Waterloo 35th by Waterloo Chief (23184)—Waterloo 20th by Cherry Duke 2d (14265)—Waterloo 16th by Red Knight (11976)—Waterloo 14th by Grand Duke (10284)—Waterloo 13th by 3d Duke of Oxford (9047)—Waterloo 9th by 2d Cleveland Lad (3408)—Waterloo 6th by Duke of Northumberland (1940)—Waterloo 3d by Norfolk (2377)—Waterloo Cow by Waterloo (2816)—by Waterloo (2816).

A. H. B., vol. 18, p. 13957, and E. H. B., vol. 20, p. 814, under dam. Sold, June 6, 1877, at London, Ont., to Canada West Farm Stock Association.

WATERLOO 37TH—Red and white, calved Dec. 26, 1876, bred by Sir W. Lawson, got by 6th Baron Oxford (33075), out of Waterloo

36th by Royal Cambridge (25009)—Waterloo 35th by Waterloo Chief (23184), &c., as in dam, above.

A. H. B., vol. 18, p. 13957. Sold, June 6, 1877, at London, Ont., to Canada West Farm Stock Association.

1815. (Supposed) SAMUEL M. HOPKINS.
Moscow, Genesee Valley, N. Y.

MARQUIS (408)—Calved in 1815, bred by Mr. Whitaker, got by Wellington (679), out of Magdalena by Comet (155)—by Cupid (177).

MOSCOW (9413)—Roan, bred by Sir H. Vane Tempest, imported in 1817, got by Wynyard (703), out of Elvira by Phenomenon (491)—Princess by Favorite (252)—own sister to R. Colling's White Bull by Favorite (252)—by Hubback (319)—by Snowdon's Bull (612)—by J. Masterman's Bull (422)—by Waistell's Bull (669) —by Studley Bull (626)—bought of Mr. Pickering by Mr. Hall.

PRINCESS—Yellow red, calved in 1813, bred by Sir Henry Vane Tempest, Wynyard, Durham, got by Wynyard (703), out of a cow bred by Sir H. V. Tempest.

A. H. B., vol. 1, p. 215.

GEO. HOUSEMAN,
Of London, Can.

LORD OF LUNE 4119 (16428) [418]—Roan, calved May 10, 1858, bred by Mr. Houseman, Lune Bank, Lancaster, got by Duke of Buckingham (14428), out of Grace Darling by Rex (6385)— Geneva by Rex (6385)—Elizabeth by Solomon (6513)—by Skelbrook (5206)—by Frickley (3849).

1879. J. HUNTER.
By Ship Quebec, from Liverpool.

SIR EDMUND 53925 (42391)—Roan, calved June 4, 1878, bred by Hugh Aylmer, got by Sir Wilfrid (37484), out of Maid of Lorne by Royal Broughton (27352)—Maid of the Abbey by Prince Christian (22581)—Maid of the Morn by Hildebrand (18068)

—Maid of Orleans by Knight of Windsor (16349)—by Vanguard (10994)—by Banker (11136)—by Samuel (5084)—by Reformer (2512)—by Imperial (2151)—by Favorite (1030)—by Young Dimple (971)—by Brown's White Bull (98).

May 5, 1881. J. HUNTER.
Alma. Ont., Can. Ship Oxenholme, Liverpool.

SOCRATES 58151 (45640)—Red bull, calved May 20, 1880, bred by Hugh Aylmer, West Dereham Abbey, Norfolk, got by Sir Simeon (42412), out of Casseopea by Sir Wilfrid (37484)—Cassandra by Royal Monk (35392)—Celeste by Royal Broughton (27352) —Christine by Prince of the Realm (22627)—Charmian by Valasco (15443)—Lady Hasseltine by British Boy (11206)—Calomel by Hamlet (8126)—Chalk by Leonard (4210)—Bellona by Buckingham (3239)—from the stock of Sir M. W. Ridley.

GOLDEN BELLE—Red heifer, calved March 1, 1879, bred by H. Aylmer, got by Sir Wilfrid (37484), out of Golden Feather by Royal Prince (32404)—Gold Hope by Royal Broughton (27352)— Golden Drop by Prince Christian (22581)—Gold Chain by Fitz Windsor (17860)—Gilt Hope by Hopewell (10332)—Gilt by Vanguard (10994)—Gold Leaf by Leonard (4210)—by Remus (4932)— by Prince Comet (1342)—by Count (170)—by Prince of Waterloo (528)—by Young Favorite (255).

E. H. B., vol. 26, p. 302, under dam, and Brit.-Ami. H. B., vol. 1, p. 482.

1870. J. & R. HUNTER.

LADY FANNY—Red, calved March 1 1870, bred by A. Cruickshank, Sittyton, Aberdeenshire, Scotland, got by Rob Roy (22740), out of Lady Fair by Royal Butterfly (18753)—Lady Like by The Baron (13833)—Lady Isabella by Matadore (11800)—Arabella by Robin-o'-Day (4973)—Picotee by Premier (6308)—Sunflower by Unicorn (8725)—by Monarch (4495)—by Young Satellite (8538)—by Valentine (661)—bred by Mr. Rennie, of Phantassie.

A. H. B., vol. 14, p. 617; C. H. B., vol. 2, p. 551, and Brit.-Am. H. B., vol. 1, p. 482.

June 20, 1871. J. & R. HUNTER.

Sunnyside, Pilkington, Wellington Co., Canada. By Ship Nova Scotia, from Liverpool.

KNIGHT OF WARLABY 20163 (29014) [1634]—Roan bull, calved Jan. 2, 1870, bred by Mr. J. Whyte, Little Clinterty, Aberdeen, Scotland, got by Baron Booth (21212), out of Fan Fan by Sir James (16980)—Faith by Sir Charles (12075)—Fanchette by Petrarch (7329)—Fame by Raspberry (4875)—by Young Matchem (4422)—by Isaac (1129)—by Young Pilot (4702)—by Pilot (496)—by Julius Cæsar (1143).

June, 1872. J. & R. HUNTER.

By Ship Vicksburg, from Liverpool.

ROSE OF SPRING—Red and white, calved April 9, 1873, bred by T. E. Pawlett, Beeston, Bedfordshire, got by Prince Regent (29677), out of Rose of Autumn by Prince Alfred (27107)—Rose of Summer by Prince Hopewell (22592)—Rose of Promise by Heir-at-Law (13005)—Rose of Autumn by Sir Henry (10824)—Pelerine by Buckingham (3239)—Mantilini by Marcus (2262)—Maiden by Matchem (2281)—Lady by Alderman (1622)—Lady Mowbray by Pilot (496)—Sylph by Remus (550)—Matilda by Sir Charles (592)—Alpine by R. Colling's son of Favorite (252)—Young Strawberry by a son of Favorite (252)—Strawberry.

A. H. B., vol. 14, p. 845, and C. H. B., vol. 3, p. 753.

ROSE OF AUTUMN—Red and white, calved Nov. 24, 1870, bred by T. E. Pawlett, got by Prince Alfred (27107), out of Rose of Summer by Prince Hopewell (22592), &c., as in Rose of Spring.

Recorded, Brit.-Am. H. B., vol. 1, p. 485; E. H. B., vol. 19, p. 712, under dam, and A. H. B., vol. 14, p. 841.

HUNTER & VAN CLOVE.

DAISY—Bred by J. Morris, Shuttleworth, Leicester, got by Emperor (1973), out of Cherry by Comet (155)—Beauty, owned by Mr. Bradie, Northallerton, Eng.

1839. REUBEN HUTCHCRAFT.
Bourbon Co., Ky.

BEDA—Red, calved about 1836, bred by Col. Cradock, Hartforth, got by Magnum Bonum (2243), out of Daisy by Gainford (2044)—by Forester (1055)—by Rob Roy (557).

WILD ROSE—Roan, calved in February, 1838, bred by Mr. Watkin, got by Chorister (3378), out of Whitefoot by Harrington (2121)—by Lenny (2197)—by Eclipse (238)—Violet by Western Comet (689)—Best Twin by Favorite (252)—Old Simmon, descended from Studley White Bull (627).

VAN BUREN 1062—Bred by Col. Cradock, got by Magnum Bonum (2243), out of Cherry by Pirate (2430)—by Young Houghton (1119)—by Marshal Blucher (416).

DON JOHN (3603)—Red roan, calved Feb. 24, 1838, bred by Mr. Watkin, Plumpton, Penrith, Eng., got by Chorister (3378), dam by Emperor (1974)—by Margrave (2263)—by Leopold (2199)—by Hector (2103)—by Traveler (655)—by Surley (2715)—by Colonel.

BLOSSOM—Got by Magnum Bonum (2243), dam by Forester (1055)—by Frederick (1060)—by Rockingham.

FATIMA—Roan, bred by Col. Cradock, Eng., got by Magnum Bonum (2243), out of Strawberry by a son of Pirate (2430)—by Frederick (1060)—by Rockingham.

HARRIET—Red cow, calved about 1835, bred by Col. Cradock, Hartford, near Richmond, got by Gainford (2044), out of Young Lofty by Forester (1055)—Dairymaid by Rockingham—by Frederick (1060).

August, 1874. E. ILES.
Springfield, Ill.

FLORA 3D—Roan cow, calved April 25, 1869, bred by J. Gordon, Cluny Castle, Aberdeen, Scotland, got by Masterman (26864), out of Flora by Squire of Bushey (20889)—Tulip by Richard Cœur de Lion (13590)—Opulent by Abraham Parker (9856)—Flounce by Tomboy

(9750)—Flora by Magician (7185)—by Vivian (5575)—by Belvedere 2d (3126)—by Alive O! (2995)—by Young Alive O! (2996)—by Eclipse (236)—by Charge's Gray Bull (872)—by Paddock Bull (477).

E. H. B., vol. 20, p. 523. under dam, and A. H. B., vol. 15, p. 559.

FLORA 7TH—Roan heifer, calved Oct. 21, 1872, bred by J. Gordon, Cluny Castle, Aberdeen, Scotland, got by Watchman 2d (27756), out of Flora 2d by Masterman (26864)—Flora by Squire of Bushey (20889)—Tulip by Richard Cœur de Lion (13590)—Opulent by Abraham Parker (9856)—Flounce by Tomboy (9750) —Flora by Magician (7185)—by Vivian (5575)—by Belvedere 2d (3126)—by Alive O! (2995)—by Alive O! 2d (2996)—by Eclipse (236)—by Charge's Gray Bull (872)—by Paddock Bull (477).

E. H. B., vol. 20, p. 522, under dam, and A. H. B., vol. 15, p. 559.

FLORA BELLE—Red, calved Feb. 22, 1875, bred by J. Gordon, Cluny Castle, Aberdeen, Scotland, got by Windsor Booth (32875), out of Flora 2d by Masterman (26864)—Flora by Squire of Bushey (20889)—Tulip by Richard Cœur de Lion (13590)—Opulent by Abraham Parker (9856)—Flounce by Tomboy (9750)—Flora by Magician (7185)—by Vivian (5575)—by Belvedere 2d (3126)—by Alive O! (2995)—by Young Alive O! (2996)—by Eclipse (236)—by Charge's Gray Bull (872)—by Paddock Bull (477).

A. H. B., vol. 15, p. 559.　Calved after dam reached Illinois.

ORANGE BLOSSOM 18TH—Red, calved Jan. 31, 1873, bred by A. Cruickshank, got by Viceroy (32764), out of Orange Blossom 14th by Knight of the Whistle (26558)—Orange Blossom 12th by Prince Imperial (22595)—Orange Blossom 2d by The Baron (13833) —Orange Blossom by Dr. Buckingham (14405)—Queen of Scotland by Matadore (11800)—Edith Fairfax by Sir Thomas Fairfax (5196) —Fancy by Billy (3151)—Jessie by Sovereign (7539)—Rose by Satellite (1420)—by Baronet (60)—by Cleveland (144)—by Symmetry (646).

A. H. B., vol. 14, p. 764, and E. H. B., vol. 21, p. 657, under dam.

MISSIE 40TH—Red and white, calved Jan. 5, 1873, bred by
W. S. Marr, Aberdeenshire, Scotland, got by Young Englishman
(31113), out of Missie 35th by Prince of Stokesley (27177)—
Missie 5th by Lord of Lorne (18258)—Missie 2d by Augustus
(15598)—Missie by son of 3d Duke (17697)—Countess by Pacha
(7612)—Jessamine by Mahomed (6170)—Rose by Plenipo (4725)—
Thorn by Abbot (2899). ,

A. H. B., vol. 15, p. 774.

DUKE OF RICHMOND 21525—Red, calved March 16, 1873,
bred by Jas. Bruce, Scotland, got by Lord St. Leonard's (29202),
out of Fanny by Royal Errant (22780)—Flora by The Gipsey
Chief (15385)—Red Rose by Captain Balco (12546)—Daisy by Kossuth (11646)—Lucy Neil by Roger (13615)—Snowdrop 2d by Studley
(628)—Snowdrop by Rockingham (2547)—by Wonder (2853)—by
Denton (198)—by Ladrone (353)—by Henry (301).

July, 1857. ILLINOIS IMPORTING ASSOCIATION.
(JAS. N. BROWN, H. C. JOHNS AND H. JACOBY, Agents.)

*By sailing vessel Georgia, in July, 1857, to Philadelphia. Sold at Fair
Grounds, Springfield, Ill., Aug. 27, 1857.*

BULLS.

KING ALFRED (14760)—Red, calved April 4, 1855, bred
by Jonas Webb, Babraham, got by Cheltenham (12588), out of
Heart's Ease by Lord of the North (11743)—Countess of Hardwick
by Percy (6283)—Celia by 3d Duke of Northumberland (3647)—
Cornflower by Bashaw (1692)—by Helmsman (2109)—by Columella
(904)—by Regent (544)—by Palatine (478)—by Palmflower (480)
—by Patriot (486)—by Driffield (223)—by C. Holmes' Bull (314).

Sold to Brown, Jacoby & Co., for $1,300.

DOUBLOON 3833½—Red, calved Feb. 26, 1856, bred by Jas.
Topham, Downston, Ireland, got by Orphan Boy (13429), out of
Splendor 10th by Gen. Lax (12933)—Splendor 4th by Marquis of
Rockingham (10506)—Splendor 2d by Baronet (6763)—Splendor
by Guy Faux (7062)—White Rose by Pedestrian (7321)—Roan

Rosalind by Edrom (1986)—Young Rosalind by Scipio (1421)—Rosalind by Hector (1104)—Rose by Midas (435)—Red Rose by Marquis (407)—by Chilton (136)—by Ben (70).

Sold to Wash. Iles, for $1,075.

MASTER LOWNDS 3140½—Roan, calved May 25, 1855, bred by R. C. Lowndes, Liverpool, Eng., got by Bellerophon (11165), out of Caradorio (vol. 12, p. 303, E.) by Alderman (9882)—Alice by Colonel (8967)—Moss Rose by Locomotive (4242)—Adelaide by Cleveland (3403)—by Young Eryholme (1981)—by Wonderful (700) —by Merlin (429)—by Alfred (23)—by Cupid (177)—by Suwarrow (636).

Sold to J. H. Spears, for $725.

ARGUS (14102)—Roan, calved Dec. 14, 1855, bred by Harvey Combe, Cobham Park, Surrey, got by Beau (12182), out of Annie by Broughton Hero (6811)—Minosa by Duke of Cornwall (5947)—Minna by Fergus (3782)—Starlight by Dandy (1902)—Moonshine by Oliver (2386)—by Blyth (797)—by Midas (435)—by Boughton (90)—by Windsor (698)—by Colling's son of Favorite (252).

Sold to Geo. Barnett, for $2,058.

ROMULUS (13625)—Roan, calved Jan. 9, 1854, bred by William Coppinge, got by Rolla 2d (13618), out of Lovely by a son of Alfred (6732)—Louisa by Clementi (3399)—Duchess by Maximus (2284) —Duchess by Matchem (2281)—by Scipio (1421)—by Stephen (1456)—by Western Comet (689)—by Charge's Gray Bull (872)—by Favorite (252)—by Bartle (777)—descended from Studley White Bull (627).

Died before sale.

DEFENDER (12687)—Roan, calved March 6, 1854, bred by A. Cruickshank, got by Matadore (11800), out of Diadem by Fitz Leonard (7010)—Betty Floss by Cleveland (3406)—Ranunculus by Conservative (1865)—by Liston (2210)—by Wonderful (700)—by Symmetry (643)—by Jupiter (345)—by Phenomenon (491)—by Favorite (252)—by Punch (531).

Sold to A. G. Carle, for $2,500.

GOLDFINDER 29204—Roan, calved Oct. 5, 1856, bred by H. Ambler, Watkinson Hall, got by Grand Turk (12969), out of Gipsey by Charlie (12583)—Urella by Captain Edwards (8929)— Fair Rosamond by Senator (8548)—Emily by Sir Philip (8588)—by Premier (2448)—by Alfred (2985).

Sold to J. C. Bone, for $725.

CULTIVATOR (12670)—Red and white, calved March 18, 1854, bred by R. C. Lowndes, Rice House, Club Moor, got by Harry Lorrequer (12991), out of Belinda by Lord Clarendon (10434)— Annie by Harkaway (9184)—Primrose by Egremont (9075)—Orinda by Marmion (4383)—Adelaide by Cleveland (3403)—by Young Eryholme (1981)—by Wonderful (700)—by Merlin (429)—by Alfred (23)—by Cupid (177)—by Suwarrow (636).

Died before sale.

YOUNG CHELTENHAM (14264)—Red, calved Jan. 10, 1855, bred by John Webb, Horseheath, got by Cheltenham (12588), out of May Day by The Minstrel (8687)—Cherry by Prince Albert (7358)—Miss Cooke by Favorite (3769)—Gulnare by Oliver (2386) —by Albany (13)—by George (273)—by son of Mason's White Bull (421).

Died before sale.

ADMIRAL 2473—Red, calved July 1, 1855, bred by Lord Talbot, Ireland, got by Phœnix (10608), out of Maid of Moynalty by The Beau of Killerby (8126)—Venus by Forrest (10240)—Virtue by Marquis of Chandos (6180)—by Prince Ernest (7366)—by Madcap (7183)—by Paul Pry.

Sold to S. Dunlop & Co., for $2,500.

COWS.

WESTERN LADY—Roan, calved March 2, 1855, bred by H. Ambler, Watkinson Hall, Halifax, got by Grand Turk (12969), out of Wiseton Lady by Humber (7102)—Zeal by Roman (2561)— Roguery by Mercury (2301)—Pageant by Monarch (2324)—No. 3 Mason's Sale by St. Albans (2584)—No. 4 Mason's Sale by Jupiter

(342)—Sir Oliver Cow by Sir Oliver (605)—Raspberry by Trunnell (659)—Strawberry by Favorite (252)—Lily by Favorite (252)—Miss Lax by Dalton Duke (188)—Lady Maynard by Alcock's Bull (19) —by Jacob Smith's Bull (608)—by Jolly's Bull (337).

A. H. B., vol. 4, p. 592, and E. H. B., vol. 12, p. 664, under dam. Sold to J. N. Brown, for $1,325.

EMPRESS EUGENIE—Red and white, calved April 21, 1855, bred by H. Ambler, Watkinson Hall, Halifax, got by Bridegroom (11203), out of Imperial Cherry (vol. 10, p. 403, E. H. B.) by Emperor (6973)—Cherry Ripe by Sir Walter (2639)—Young Cherry by Young Waterloo (8757)—Cherry by Waterloo (2816)—Old Cherry by Waterloo (2816)—by Kitt (2179)—by Kitt (2179)—by Page's Bull (6269)—by Middleton's Bull (438).

Sold to J. Ogle, St. Clair Co., for $675.

LADY HARRIET—Roan cow, calved April 1, 1854, bred by A. Cruickshank, Sittyton, Aberdeen, Scotland, got by Procurator (10657), out of Countess of Lincoln by Diamond (5918)—Nonpareil 3d by Young Frederick (3836)—Nonpareil 2d by Commodore (1850)—by Tathwell Studley (5401)—Twine Tail by Blyth Comet (85).

A. H. B., vol. 4, p. 418. Sold to J. H. Jacoby, for $1,300.

CASSANDRA 2D—Roan, calved Sept. 19, 1854, bred by Rev. Thos. Cator, Shelbrook Park, near Doncaster, Yorkshire, got by Master Charlie (13312), out of Lady Hawke by Norfolk (9442)—Rebecca by Rex (6385)—Fair Maid of Athens by Sir Thomas Fairfax (5196)—Medora by Ambo (1636)—Blossom by Memnon (2295) —own sister to Isabella by Pilot (496)—by Agamemnon (9)—by Burrell's Bull of Burdon.

E. H. B., vol. 11, p. 524, under dam, and A. H. B., vol. 15, p. 476. Sold to H. Owsley, for $675.

FAMA—Red and white, calved June 7, 1856, bred by S. E. Bolden, Springfield Hall, got by 2d Grand Duke (12961), out of Finella 2d by Grand Duke (10284)—Fay by Foig-a-Ballagh (8082)—

Fame by Raspberry (4875)—Farewell by Young Matchem (4422)—
Flora by Isaac (1129)—by Young Pilot (497)—by Pilot (496)—by
Julius Cæsar (1143).

E. H. B., vol. 12, p. 383, under dam.   Sold to J. H. Spears & Co.,
for $1,050.

POMEGRANATE—Roan, calved Nov. 25, 1855, bred by Rev.
Thomas Cator, Shelbrook Park, got by Master Charlie (13312),
out of Cassandra by Norfolk (9442)—Florence by Rex (6385)—
Fair Maid of Athens by Sir Thomas Fairfax (5196)—Medora by
Ambo (1636)—Blossom by Memnon (2295)—own sister to Isabella
by Pilot (496)—White Cow by Agamemnon (9)—Darlington Cow
by Burrell's Bull of Great Burdon.

A. H. B., vol. 5, p. 421, and E. H. B., vol. 14, p. 378, under
dam.   Sold to T. Simpkins, for $975.

LILY—White, calved Jan. 17, 1855, bred by Ed. Bowly, Siddington House, Cirencester, got by Snowstorm (12119), out of
Sunflower by California (10017)—Diana by Thirty-Two (8704)—
Limpid by Viceroy (7678)—Lemon by Marquis (2271)—by Isaac
(1129)—by Blucher (83)—by Cecil (120).

A. H. B., vol. 4, p. 432, and E. H. B., vol. 12, p. 618, under
dam.   Sold to G. Barnet, for $550.

CONSTANCE—Roan, calved Oct. 11, 1854, bred by Ed. Bowly,
Siddington House, Cirencester, got by Snowstorm (12119), out of
Felicity by Sol (8608)—Joyous by Leo (4208)—Playful by Frederick (3837)—Delight by Rival (2534)—Diana by Darlington (956)
—Dahlia by Denton (198)—Gaudy by Rockingham (560)—by Jobling's son of Phenomenon (491)—by Colonel (152)—by Styford
(629).

A. H. B., vol. 4, p. 303, and E. H. B., vol. 11, p. 449, under dam.
Sold to Geo. Barnett, Will Co., Ill., for $700.

EMPRESS—Roan, calved April 20, 1855, bred by Mr. E. Bowly,
Siddington House, Cirencester, got by Tortworth Duke (13892),
out of Flippant (vol. 11, p. 453, E. H. B.) by Bourton Hero (9983)—

La Polka by Monzani (6222)—by The Prince (7615)—by Fitzroy (3808)—by Morpeth (2339)—by Roman (2559)—by Admiral (5)— by a son of Blyth Comet (85)—by a bull of Mr. Foljambe's.

A. H. B., vol. 4, p. 338. Sold to Henry Jacoby, for $1,725.

RACHEL 2D—Roan, calved July 5, 1855, bred by S. E. Bolden, Springfield Hall, Lancaster, got by Duke of Bolton (12738), out of Young Rachel (vol. 10, p. 542, E. H. B.) by Leonard (4210)— Rachel by Young Red Rover (4905)—Rally by Rowton (5019)— Young Carnation by Admiral (5)—Carnation by Pilot (496)— White Rose by Albion (14)—Halnaby by Lame Bull (359)—by Easby (232)—by Suwarrow (636).

A. H. B., vol. 4, p. 529. Sold to J. N. Brown, for $3,025.

MINX—Red, calved Jan. 3, 1856, bred by John Christy, Fort Union, Adare, Limerick, got by Lord Spencer (13251), out of Mabel by Young Shaftoe (9625)—Eva by Shotley 2d (7494)— Actress by Magnum Bonum (2243)—Sylph by Robin—by Regent (544)—by Cecil (120)—by Midas (435)—by Meteor (431)—by Petrarch (488)—by Alexander (20)—by Traveler (655)—by son of Bolingbroke (86).

E. H. B., vol. 12, p. 476, under dam. Sold to J. G. Loose, for $800.

ADELAIDE—Roan, calved March 1, 1856, bred by A. Cruick-shank, Sittyton, Aberdeen, got by Matadore (1180), out of Edith Fairfax by Sir Thomas Fairfax (4196)—by Billy (3151)—Jessy by Sovereign (7539)—Rose by Satellite (1420)—by Baronet (60)—by Cleveland (144)—by Symmetry (641).

A. H. B. vol. 4, p. 251. Sold to R. Morrison, for $825.

EMERALD—Roan, calved March 4, 1856, bred by Thos. Barnes, Westland, Moynalty, Ireland, got by Hopewell (10332), out of Ruby by Royal Buck (10750)—Lady Sarah by Burley Fairfax (6822)— Violet by Monarch (4495)—Julia by Invalid (4076)—Lady Sarah by Satellite (1420)—by Cato (119)—by Jupiter (342)—by George (273)—by Chilton (136)—by Irishman (329)—by B. (45).

A. H. B., vol. 4, p. 336, and E. H. B., vol. 12, p. 594, under dam. Sold to J. C. Bone, for $2,125.

PERFECTION—Red, calved April 20, 1856, bred by A. Cruick-shank, Sittyton, Aberdeen, got by The Baron (13833), out of Model by Matadore (11800)—Brunette by Prince Edward Fairfax (9506)—Zenith by Premier (6308)—Strawberry by Soldier (5239)—by Young Commodore (3453)—by Alamode (725).

E. H. B., vol. 12, p. 510, under dam. Sold to E. B. Hill, Scott Co., for $900.

COQUET—Roan, calved May 16, 1856, bred by Mr. Bowly, got by Economist (11425), out of Caprice by Harold (10299)—Juliet by Sol (8608)—Kate by Leo (4208)—Ada by Treasurer (5513)—Lady Byron by Rupert (2580)—by North Star (460)—by Cripple (173)—by Minor (441)—by Freeman (269)—by Danby (190).

A. H. B., vol. 4, p. 303, and E. H. B., vol. 12. p. 303, under dam. Sold to Geo. Barnett, Will Co., Ill., for $550.

CORONATION—Red, calved June 28, 1856, bred by Jonas Webb, got by Cheltenham (12588), out of Young Celia by Lord of the North (11743)—Celia by 3d Duke of Northumberland (3647)—Cornflower by Bashaw (1692)—Columbine by Helmsman (2109)—by Columella (904)—by Regent (544)—by Palatine (478)—by Palmflower (480)—by Patriot (486)—by Driffield (223)—by C. Holmes' Bull (314).

E. H. B., vol. 12, p. 310, under dam. Sold to J. A. Pickrell, Sangamon Co., Ill., for $500.

GENTILITY—Roan, calved June 1, 1855, bred by Harvey Combe, Cobham Park, Surrey, got by Puritan (9523), out of Graceful by Loyalist (10479)—Gazelle by Noble (4578)—Garland by Hector (4000)—Moss Rose by Emperor (1974)—Rosebud by Margrave (2263)—by Leopold (2199)—by Hector (2103)—by Traveler (655)—by Surley (2715)—by Colonel (152).

E. H. B., vol. 12, p. 404, under dam. Died before sale.

VIOLET—Roan, calved Aug. 19, 1856, bred by Jonas Webb, Babraham, got by Young Scotland (13681), out of Lady Love by

21

Red Roan Kirtling (10691)—Belinda by Ranunculus (2479)—Sylph by Sir Walter (2637)—by Hotspur (1117)—by Coxcomb (928)—by Midas (435)—by Comet (155)—by R. Colling's son of Favorite (252)—by son of Favorite (252)—by Hubback (319).

A. H. B., vol. 4, p. 588, and E. H. B., vol. 12, p. 452, under dam. Sold to J. W. Judy, Menard Co., Ill., for $700.

BELLA—Roan, calved March 27, 1852, bred by E. Bowly, got by California (10017), out of Diana by Thirty-Two (8704)—Limpid by Viceroy (7678)—Lemon by Marquis (2271)—by Isaac (1129)—by Blucher (83)—by Cecil (120).

E. H. B., vol. 11, p. 407, under dam. Sold to J. Ogle, St. Clair Co., Ill., for $750.

CAROLINE—Roan, calved May 16, 1853, bred by L. C. Lowndes, got by Arrow (9906), out of Bell by Harkaway (9184)—Gaudy by Marmion (4383)—Moss Rose by Locomotive (4242)—Adelaide by Cleveland (3403)—by Young Eryholme (1981)—by Wonderful (700)—by Merlin (429)—by Alfred (23)—by Cupid (177)—by Suwarrow (636).

A. H. B., vol. 4, p. 289, and E. H. B., vol. 11, p. 331, under dam. Sold to J. M. Hill, Cass Co., Mich., for $500. Her dam, " Bell," is sometimes called " Bellrage."

STELLA—Roan, calved Aug. 22, 1853, bred by E. Bowly, got by Snowstorm (12119), out of Purity by Hesperus (10321)—Changeling by Leonidas (9291)—by Sol (8608)—by Brunel (7857)—by Ovis (4635)—by Treasurer (5513)—by Rupert (2580)—by North Star (460)—by Cripple (173)—by Minor (441)—by Freeman (269)—by Danby (190).

E. H. B., vol. 11, p. 651, under dam. Sold to Mr. Bonnman, St. Clair Co., Mich., for $925.

JULIUS L. INCHIES.
Fredrickton, New Brunswick.

PRINCESS 2D—Got by Prince Regent (13523), out of Princess by Chilton (11279)—Lucy by Belted Will (9952)—Moss Rose by Duke (7595)—Warren Rose by Bellerophon (3119)—White Rose by Belvedere (1706)—by Waterloo (2816)—by Young Wynyard (2859) —by Irishman (329)—by Styford (103).

See Crown Prince 2d 1119, Brit.-Am. H. B.

1859 or 1860. JULIUS L. INCHIES.
Fredrickton, New Brunswick.

DUKE—Light roan, calved Feb. 25, 1858, bred by Robt. Geikie, Roseneath, Baldowrie, Scotland, got by Oronoco (13425), out of Lady by Viscount (13956)—Jannetta by Chilton (11279)—Susanna by Strathmore (6547)—Susan by Sir Robert (5183)—Rosalind by The Baron (3095)—Ruby by Emperor (3716)—Cicely by Invalid (4076)—Lady Sarah by Satellite (1420)—by Cato (119)—by Jupiter (342)—by George (273)—by Chilton (136)—by Irishman (329)—by B. (45).

No. 1116, in Brit.-Am. H. B.

--- --- ---

June, 1867. GEORGE ISAAC,
Of Haldimand, Can. To Montreal, by Ship Abionia.

MARGARET 3D—Red, calved May, 1866, bred by S. Campbell, Kinellar, Aberdeenshire, Scotland, got by Diphthong 3d (21547), out of Margaret 2d by Scarlet Velvet (16916)—Margaret by The Garioch Boy (15384)—Barbara by Unrivaled (13926)—Isabella by The Pacha (7612)—Crocus by 2d Duke of Northumberland (3646) —Nora by Sillery (5131)—Emily by Sillery (5131)—Eliza by Young Western Comet (1575)—Lady Betty by Diamond (205)—Betty by Favorite (256)—by Charge's Red Bull (1810).

A. H. B., vol. 12, p. 1010, and C. H. B., vol. 2, p. 635.

ISABELLA—Red, calved March 26, 1866, bred by S. Campbell, Kinellar, Aberdeenshire, Scotland, got by Diphthong 3d (21547),

out of Mina by Beeswing (12456)—Crocus by Sir Arthur (12072)—
Bashful by Young Ury (10984)—Likely by The Pacha (7612)—
Helen by 2d Duke of Northumberland (3646)—by Sillery (5131)
—by Carleton (843)—by Diamond (205).

A. II. B., vol. 9, p. 672, and C. H. B., vol. 2, p. 511.

PRINCE CHARLIE (27123)—Roan, calved Sept. 2, 1866, bred
by S. Campbell, got by Prince of Worcester (20597), out of Ury
2d by Diphthong (17681)—Ury Lass by Beeswing (12456)—Miss
Isabella by Mosstrooper (11827)—Donside Lassie by Vice-President (11002)—by The Pacha (7612)—by 2d Duke of Northumberland (3646)—by Sillery (5131)—by Sillery (5131)—by Young
Western Comet (1575)—by Diamond (205)—by Favorite (256)—
by Charge's Red Bull (1810).

CANADIAN PRINCE 7646 [1065]—Red, calved Jan. 12, 1868,
bred by Sylvester Campbell, Kinellar, Aberdeenshire, Scotland,
got by Gladstone (26256), out of Isabella by Diphthong 3d (21547)
—Mina by Beeswing (12456)—Crocus by Sir Arthur (12072)—
Bashful by Young Ury (10984)—Likely by The Pacha (7612)—
Helen by 2d Duke of Northumberland (3646)—by Sillery (5131)
—by Carleton (843)—by Diamond (205)—by Diamond (205).

July, 1870. GEO. ISAAC.
To Montreal, by Allan Steamship Gleniffer.

STATESMAN 15539 (32607) [2317]—Red roan, calved April
9, 1869, bred by S. Campbell, Kinellar, got by Nobleman (26967),
out of Nonpareil 28th by Prince of Worcester (20597)—Nonpareil
25th by Diphthong (17681)—Nonpareil 24th by Lord Sackville
(13249)—Nonpareil 23d by The Baron (13833)—Nonpareil 17th by
Matadore (11800)—Nonpareil 10th by Prince Edward Fairfax
(9506)—Countess of Lincoln by Diamond (5918)—Nonpareil 3d
by Young Frederick (3836)—by Commodore (1858)—Nonpareil
by Tathwell Studley (5401)—by Blyth Comet (85).

WELLINGTON 15692 (32828) [2421]—Red, calved Jan. 6,
1869, bred by S. Campbell, Kinellar, got by Eskdale (26111), out

of Bloom 2d by Diphthong 3d (21547)—Bloom by Mosstrooper (11827)—Thessalonica by Duke of Clarence (9040)—Jewess by Bowmont (3200)—Ruth by a son of Exmouth (3747)—by Prince (4765) —by Wellington (2824).

BLOOM 3D—Red, calved April 23, 1868, bred by S. Campbell, got by Diphthong 3d (21547), out of Bloom (vol. 13, E.) by Mosstrooper (11827)—Thessalonica by Duke of Clarence (9040)—Jewess by Bowmont (3200)—Ruth by a son of Exmouth (3747)—by Mr. Robson's Bull (9562)—by Prince (4765)—by Wellington (2824).

A. H. B., vol. 20, p. 15514, and C. H. B., vol. 2, p. 364.

BLOOM 4TH—Red, calved Feb. 25, 1871, bred by S. Campbell, got by Sir Christopher (22895), out of Bloom 3d, above.

C. H. B., vol. 2, p. 364.

MYRTLE—Roan, calved March 28, 1869, bred by S. Campbell, got by Nobleman (26967), out of Rosebud 2d by Prince of Worcester (20597)—Rosebud by Scarlet Velvet (16916)—Thalia by Earl of Aberdeen (12800)—Myrtle by Balmoral (9920)—Fortune Teller by Dannecker (7949)—Enchantress by Ury (17157)—Ladykirk by Heriot (4017)—by Gray Diomed (2076)—by Jupiter (1144).

C. H. B., vol. 2, p. 678.

GOLDEN DROP 2D—Roan, calved in April, 1868, bred by S. Campbell, Kinellar, got by Gladstone (26256), out of imp. Golden Drop 1st (vol. 14, A. H. B.) by Prince of Worcester (20597)—Golden Drop by Scarlet Velvet (16916)—Bloom by Mosstrooper (11986) —Thessalonica (vol. 11, E.) by Duke of Clarence (9040)—Jewess (vol. 10, E.) by Bowmont (3200)—by son of Exmouth (3747)—by Robson's Bull (9562)—by Prince (4765)—by Wellington (2824).

A. H. B., vol. 24, p. 18545, and C. H. B., vol. 2, p. 495.

GOLDEN DROP 3D—Roan, calved Feb. 10, 1871, bred by S. Campbell, got by Sir Christopher (22895), out of Golden Drop 2d by Gladstone (26256), &c., as in dam, above.

A. H. B., vol. 15, p. 582, and C. H. B., vol. 2, p. 495.

MISS RAMSDEN 3D—Roan, calved in March, 1869, bred by S. Campbell, got by Nobleman (26967), out of Miss Ramsden 2d by Diphthong (17681)—Miss Ramsden 1st by Scarlet Velvet (16916)—Miss Ramsden by Bushranger (12516)—by Donside Fairfax (11365) —by Duke (3630)—by Reveller (2528)—by Grazier (1085)—by Cato (857)—by Atlas (42)—by Favorite (257)—by Robinson's Bull (4974) —by Badsworth (47).

A. H. B., vol. 23, p. 17982, and C. H. B., vol. 2, p. 670.

MISS RAMSDEN 4TH—Roan, calved Jan. 24, 1871, bred by S. Campbell, Kinellar, got by Duke (28342), out of Miss Ramsden 3d by Nobleman (26967), &c., as in Miss Ramsden 3d, above.

A. H. B., vol. 25, p. 826, and C. H. B., vol. 3, p. 656.

GERALDINE—Red, calved March, 1869, bred by S. Campbell, got by Nobleman (26967), out of Ellen by Garibaldi (17916)—Moss Rose by Earl of Windsor (15968)—Mary by Mosstrooper (11827) —Geraldine by Robin-o'-Day (4973)—Angelica by Holkar (4041)— Glendronan by A. Jopp's Bull (9256)—Kate of Darlington.

A. H. B., vol. 20, p. 15694, and C. H. B., vol. 2, p. 491.

HELEN—Red, calved April 18, 1868, bred by S. Campbell, got by Gladstone (26256), out of Miss Helen 2d (vol. 16, p. 583, E.) by Scarlet Velvet (16916)—Miss Helen by California (12528)—Isabella by The Pacha (7612)—Crocus by 2d Duke of Northumberland (3646)—Norah by Sillery (5131)—by Young Western Comet (1575) —by Diamond (205)—by Favorite (256)—by Charge's Red Bull (1810).

A. H. B., vol. 23, p. 17865.

LOUISA—Roan, calved March 24, 1869, bred by S. Campbell, got by Nobleman (26967), out of Mary 1st by Diphthong (17681)— Mary by Cœur de Lion (12611)—Alma by Mosstrooper (11827)— Leda by Donside Fairfax (11365)—Miss Ramsden by Duke (3630) —by Reveller (2528)—by Grazier (1085)—by Cato (857)—by Atlas (42)—by Atlas (257)—by Mr. Robinson's Bull (4974)—by Badsworth (47).

C. H. B., vol. 2, p. 609.

July, 1872. GEORGE ISAAC.
By Steamship Gleniffer, landed at Montreal.

MARGARET 4TH—Red, calved Feb. 20, 1871, bred by S. Campbell, got by Duke (28342), out of Margaret 2d by Gladstone (26256)—Margaret 1st by Diphthong 3d (21547)—Margaret by Garioch Boy (15384)—Barbara by Unrivaled (13926)—Isabella by The Pacha (7612)—Crocus by 2d Duke of Northumberland (3646)—Nora by Sillery (5131)—Emily by Sillery (5131)—Eliza by Young Western Comet (1575)—Lady Betty by Diamond (205)—Betty by Favorite (256)—by Charge's Red Bull (1810).

A. H. B., vol. 20, p. 15925, and C. H. B., vol. 3, p. 620.

HIGH SHERIFF 32841 (34162) [3345]—Roan, calved April 10, 1873, bred by S. Campbell, got by Under Sheriff (32745), out of Margaret 4th by Duke (28342), &c., as in Margaret 4th, above.

INKERMANN 26863 (31414)—Roan, calved March 31, 1872, bred by S. Campbell, got by Duke (28342), out of Claret 2d by Diphthong 3d (21547)—Claret (vol. 15, E.) by Scarlet Velvet (16916)—Barbara by Unrivaled (13926)—Isabella by The Pacha (7612)—Crocus by 2d Duke of Northumberland (3646)—Norah by Sillery (5131)—Emily by Sillery (5131)—Eliza by Young Western Comet (1575)—Lady Betsy by Diamond (205)—Betty by Favorite (256)—by Charge's Red Bull (1810).

BUCHAN LASSIE 2D—Roan, calved February 19, 1871, bred by S. Campbell, got by Sir Christopher (22895), out of Buchan Lassie by Diphthong (17681)—Betsey by Scarlet Velvet (16916)—Buchan Lassie by Narcissus (9430)—Whitehead by Sir Arthur (12072)—Miss Grant Duff by Prince Alfred (8421)—Maraschino by Duplicate Duke (6952)—Miss Chrisp by Bachelor (1666)—Peggy by Emperor (1974)—by Houndilee (2139)—by Togston (5487).

QUEEN OF SCOTS—Roan, calved in February, 1873, bred by S. Campbell, got by Under Sheriff (32745), out of Buchan Lassie 2d by Sir Christopher (22895), &c., as in dam, next above.

C. H. B., vol. 3, p. 713.

RUBY HILL—Died after landing.

PRUDENCE—Red, calved Jan. 20, 1871, bred by Mr. Makie, Pilly of Fyvie, Scotland, bred by Boanerges (25647), out of Mary Anne by Blair Athol (25638)—Tibby by Garibaldi (26218)—Queen by Young Van Dunck (19044)—Mary by Albert (17286)—Julia by Albert (17286)—Madonna by Wonderful (700)—by Cardinal (841) —by Butterfly (104).

C. H. B., vol. 3, p. 708.

ROSE OF AUTUMN—Red, calved Feb. 28, 1873, bred by S. Campbell, got by Under Sheriff (32745), out of Prudence by Boanerges (25647), &c.. as in Prudence, above.

C. H. B., vol. 3, p. 751.

URY 7TH—Roan, calved Jan. 14, 1871, bred by S. Campbell, Kinellar, got by Duke (38342), out of Ury 6th by Gladstone (26256)— Ury 2d by Scarlet Velvet (16916)—Ury Lass by Beeswing (12456)— Miss Isabella by Mosstrooper (11827)—Donside Lassie by Vice-President (11002)—Isabella by The Pacha (7612)—Crocus by 2d Duke of Northumberland (3646)—Nora by Sillery (5131)—Emily by Sillery (5131)—Eliza by Young Western Comet (1575)—Lady Betty by Diamond (205)—Betty by Favorite (256)—by Charge's Red Bull (1810).

C. H. B., vol. 3, p. 790.

QUEEN OF THE OCEAN—Red and white, calved Feb. 28, 1873, bred by S. Campbell, Kinellar, got by Under Sheriff (32745), out of Ury 7th by Duke (28342), &c., as in Ury 7th, next above.

A. H. B., vol. 21, p. 16651, and C. H. B., vol. 3, p. 714.

HELEN 6TH—Red and white, calved April 7, 1871, bred by S. Campbell, got by Sir Christopher (22895), out of Helen 3d by Gladstone (26256)—Helen 2d by Diphthong (17681)—Helen 1st by Scarlet Velvet (16916)—Helen by California (12528), &c., as in Helen, above.

C. H. B., vol. 3, p. 504.

August, 1874. JOHN ISAAC,

Of Bowmanton, Can. To Quebec, by the Texas.

CROCUS—Roan, calved in August, 1872, bred by S. Campbell, got by Sir Christopher (22895), out of Helen by Garibaldi (17916) —Moss Rose by Earl of Windsor (15968)—Mary by Mosstrooper (11827)—Geraldine by Robin-o'-Day (4973)—Angelica by Holkar (4041)—Glendronan by Jopp's Bull (9256)—Kate of Darlington.

C. H. B., vol. 3, p. 408.

FAIR QUEEN—Red, calved Jan. 6, 1871, bred by S. Campbell, got by Sir Christopher (22895), out of Fair Queen 3d by Diphthong (17681)—Fair Queen by Master Gunner (22316)—Lovely by Frolic (16086)—Lovely by Sir Arthur (12072)—Nancy by Dannecker (7949)—Likely by The Pacha (7612)—Helen by 2d Duke of Northumberland (3646)—Mary Anne by Sillery (5131)—by Carleton (843)—by Diamond (205)—by Diamond (205).

C. H. B., vol. 3, p. 460.

MINA 6TH—Roan, calved May 2, 1873, bred by S. Campbell, got by Under Sheriff (32745), out of Mina 1st by Diphthong 3d (21547) —Mina by Beeswing (12456)—Crocus by Sir Arthur (12072)— Bashful by Young Ury (10984)—Likely by The Pacha (7612)— Helen by 2d Duke of Northumberland (3646)—by Sillery (5131) —by Carleton (843)—by Diamond (205)—by Diamond (205).

C. H. B., vol. 3, p. 644.

CLARET 7TH—Roan, calved March 17, 1873, bred by S. Campbell, got by Under Sheriff (32745), out of Claret 6th by Sir Christopher (22895)—Claret 4th by Prince of Worcester (20597)—Claret by Scarlet Velvet (16916)—Barbara by Unrivaled (13926)—Isabella by The Pacha (7612)—Crocus by 2d Duke of Northumberland (3646)—by Sillery (5131)—by Sillery (5131)—by Young Western Comet (1575)—by Diamond (205)—by Favorite (256)—by Charge's Red Bull (1810).

C. H. B., vol. 3, p. 399.

RUBY HILL 7TH—Dark roan, calved Jan. 6, 1873, bred by S. Campbell, got by Scarlet Velvet 1st (29939), out of Ruby Hill 3d

by Prince of Worcester (20597)—Ruby Hill 2d by Diphthong (17681)
—Ruby Hill 1st by Scarlet Velvet (16916)—Ruby Hill by Elphin-
stone (14492)—Hawthorne Hill by Duke (7980)—Red Tibby by Sir
Robert Peel (7512)—Lady Provost by Buchan Laddie (5814)—by
Rob Roy—White Beauty.

C. H. B., vol. 3, p. 766.

NONPAREIL 32D—Roan, calved March 6, 1872, bred by S.
Campbell, got by Sir Christopher (22895), out of Nonpareil 26th by
Scarlet Velvet (16916)—Nonpareil 24th by Lord Sackville (13249)
—Nonpareil 23d by The Baron (13833)—Nonpareil 17th by Mata-
dore (11800)—Nonpareil 10th by Prince Edward Fairfax (9506)—
Countess of Lincoln by Diamond (5918)—Nonpareil 3d by Young
Frederick (3836)—by Commodore (1858)—by Tathwell Studley
(5401)—Blyth Comet (85).

A. H. B., vol. 25, p. 912, and C. H. B., vol. 3, p. 676.

BRITISH ENSIGN 34186—Roan, calved Oct. 25, 1874, bred
by S. Campbell, got by British Prince (33228), out of Nonpareil
32d by Sir Christopher (22895), &c., as in dam, above.

May, 1879. JOHN ISAAC.

To Quebec, by Ship Canadian.

MATILDA—Red, calved April 20, 1877, bred by S. Campbell,
got by Borough Member (33186), out of Matilda 1st by Sir Christopher (22895)—Daisy 2d by Waldemar (21054)—Matilda 2d by
Cunningham (11323)—Matilda by Robin Hood (8494)—Raby by
Mahomed 2d (10492)—Daisy by Billy (3151)—Maria by Belshazzar
(1703)—by Abraham (2905)—by Simon (5134).

Brit.-Am. H. B., vol. 1, p. 489, and C. H. B., vol. 6, p. 431.

MINA 6TH—Red, calved Aug. 7, 1877, bred by S. Campbell, got
by Borough Member (33186), out of Mina 2d by Sir Christopher.
(22895)—Mina 1st by Diphthong 3d (21547)—Mina by Beeswing
(12456)—Crocus by Sir Arthur (12072)—Bashful by Young Ury
(10984)—Likely by The Pacha (7612)—Helen by 2d Duke of

Northumberland (3646)—Mary Anne by Sillery (5131)—by Carleton (843)—by Diamond (205)—by Diamond (205).

Brit.-Am. H. B., vol. 1, p. 489, and C. H. B., vol. 6, p. 431.

STATESMAN 1st (44096)—Red, calved July, 1878, bred by S. Campbell, got by Golden Prince (38363), out of Nonpareil 30th by Royal Duke (35356)—Nonpareil 28th by Sir Christopher (22895)—Nonpareil 24th by Lord Sackville (13249)—Nonpareil 23d by The Baron (13833)—by Matadore (11800)—by Prince Edward Fairfax (9506)—by Diamond (5918)—by Young Frederick (3836)—by Commodore (1858)—by Tathwell Studley (5401)—by Blyth Comet (85).

BRITISH STATESMAN (42847)—Roan, calved in April, 1878, bred by S. Campbell, got by Golden Prince (38363), out of Rosebud 3d by Aberdeen (21142)—Rosebud by Scarlet Velvet (16916)—Thalia by Earl of Aberdeen (12800)—Myrtle by Balmoral (9920)—by Dannecker (7949)—by Ury (17157)—by Heriot (4017)—by Gray Diomed (2076)—by Juniper (1144).

June, 1881. JOHN ISAAC.
To Quebec, by Ship Quebec.

CLEMENTINA—Red, calved Jan. 28, 1880, bred by S. Campbell, Kinellar, Aberdeen, Scotland, got by Golden Prince (38363), out of Clementina 4th by Favorite (33894)—Clementina 3d by Sir Christopher (22895)—Clementina 2d by Diphthong 3d (21547)—Clementina by Lord Ythan (14858)—Empress Eugenie by Guy Fawkes (12981)—Bridesmaid by Sir Arthur (12072)—Crescent by The Pacha (7612)—by 2d Duke of Northumberland (3646)—by Emperor (3716)—by Invalid (4076)—by Magnet (392)—by Palm-flower (480).

Recorded, Brit.-Am. H. B., vol. 1, p. 489.

STATESMAN 2d—Roan, calved Jan. 2, 1882, bred by S. Campbell, got by Gladstone (43286), out of Clementina by Golden Prince (38363), &c., as in dam, above.

No. 2239, Brit.-Am. H. B.

MAGNET—Roan, calved June 25, 1879, bred by S. Campbell, got by Golden Prince (38363), out of Magnet 3d by Aberdeen (21142)—Magnet 2d by Prince of Worcester (20597)—Magnet 1st by Dipthong (17681)—Magnet by Scarlet Velvet (16916)—Leda by Donside Fairfax (11365)—by Duke (3630)—by Reveller (2528)—by Grazier (1085)—by Cato (857)—by Atlas (42).

Recorded, Brit.-Am. H. B., vol. 1, p. 489.

MAGNET 4TH—Red, calved Jan. 2, 1882, bred by S. Campbell, got by Gladstone (43286), out of Magnet by Golden Prince (38363), &c., as above.

Recorded, Brit.-Am. H. B., vol. 1, p. 527.

PRINCE OF NORTHUMBERLAND (46911)— Roan, calved Nov. 27, 1880, bred by A. Cruickshank, got by Perfection (37185), out of Circassia by Champion of England (17526)—Cicely by Lancaster Royal (18167)—Crocus by Jemmy (11611)—Kitty by Somerset (10858)—by Hawthorne (7071)—by The Peer (5455)—by George (2057)—by Togston (5487)—bred by Mr. Laing, of Longhoughton.

June, 1883. JOHN ISAAC,
Of Bowmanton, Ont., Can. By Steamer Coline, at Quebec.

COMET—Red, calved Feb. 3, 1882, bred by S. Campbell, Kinellar, Aberdeenshire, Scotland, got by Gladstone (43286), out of Golden Drop by Gold Prince (38363)—Golden Drop 7th by Sir Christopher (22896)—Golden Drop 4th by Nobleman (26967)—Bloom by Mosstrooper (11827)—Thessalonica by Duke of Clarence (9040)—Jewess by Red Bowmont (3200)—Ruth by son of Exmouth (3747)—by Robson's Bull (9562)—by Prince (4765)—by Wellington (2824).

BUSHRANGER—Red, calved March 19, 1882, bred by S. Campbell, Kinellar, got by Good Hope (44884), out of Magnet 4th by Favorite (33894)—Magnet 2d by Prince of Worcester (20597)—Magnet by Scarlet Velvet (16916)—Leda by Donside Fairfax (11365)—Miss Ramsden by Duke (3630)—by Reveller (2528)—by Grazier (1086)—by Cato (857).

MINA 6TH—Red, calved Jan. 2, 1881, bred by S. Campbell, got by Gladstone (43286), out of Mina by British Prince (33228)—Mina by Dipthong 3d (21547)—Mina by Beeswing (12456)—Crocus by Sir Arthur (12072)—Bashful by Young Ury (10984)—Likely by The Pacha (7612)—Helen by 2d Duke of Northumberland (3646)—Nora by Sillery (5131)—by Young Western Comet (1575)—by Diamond (205)—by Favorite (252)—by Charge's Red Bull (1810).

FAIR QUEEN 3D—Roan, calved Feb. 5, 1881, bred by S.. Campbell, got by Gladstone (43286), out of Fair Queen 2d by Sir Christopher (22895)—Queen by Dipthong 3d (21547)—Fair Queen by Master Gunner (22316)—Lady by Frolic (16086)—Lovely by Sir Arthur (12072)—Nancy by Dannecker (7949)—Likely by The Pacha (7612)—Crocus by 2d Duke of Northumberland (3646)—Nora by Sillery (5131), &c., as in Mina 6th, above.

ROSEBUD—Red, calved Jan. 5, 1882, bred by S. Campbell, got by Gladstone (43286), out of Rosebud by Golden Prince (38363)—Rosebud 1st by Dipthong (17681)—Rosebud by Scarlet Velvet (16916)—Thilia by Earl of Aberdeen (12800), &c., as in Rosebud 8th, below.

GOLDEN DROP—Red, calved June 20, 1882, bred by S. Campbell, got by Gladstone (43286), out of Golden Drop 7th by Sir Christopher (22895)—Golden Drop 4th by Nobleman (26967)—Bloom by Mosstrooper (11827), &c., as in bull Comet, above, in this list.

MISS RAMSDEN—Red, calved Feb. 6, 1882, bred by Mr. Reith, Auchindich, got by Nobleman (38797), out of Miss Ramsden by Dipthong 3d (21547)—Miss Ramsden 3d by Scarlet Velvet (16916)—Miss Ramsden by Bushranger (12516)—Leda by Donside Fairfax (11365), &c., as in Bushranger, in this list.

FLORA 87TH—Red, calved Feb. 13, 1882, bred by W. S. Marr, Upper Mill, got by Sovereign, out of Flora 83d by Freemason (33973)—Flora 62d by Scotchman 3d (32465)—Flora 48th by Young Hero (26385)—Flora 31st by Valiant (23108)—Flora 18th by Prince (16716)—Flora 9th by Raglan (18664)—Flora 6th by Topthorn (71662)—Flora by Mahomed 2d (70492)—by son of Monarch (4495)—bred by Mr. Watson, of Wauldby.

LADY 2D—Red, calved June 14, 1882, bred by Geo. Shepherd, Shethin, got by Carberry (44496), out of Lady Matilda by General Windsor (28701)—Lady Medora by Champion of England (17526) —Lady Marion by Bridegroom (11203)—Lady Mitten by Mitten (7238)—Zuleika by Norfolk (2377)—Medora by Ambo (1636)— Blossom by Memnon (2295)—own sister to Isabella by Pilot (496) —White Cow by Agamemnon (9)—by Burrell's Bull (1768).

CECILIA—Red, calved Feb. 7, 1882, bred by S. Campbell, Kinellar, got by Gladstone (43286), out of Cecilia 1st by Borough Member (33186)—Cecilia by Cæsar Augustus (25704)—Columbine by Sir Walter Scott (32922)—Camelia by Lancaster Comet (11663)—Cactus by Lord Sackville (13249)—Sharon Rose by Plantagenet (11906)— Fancy by Billy (3151)—Jessie by Sovereign (7539)—Rose by Satellite (1420).

NONPAREIL 35TH—Roan, calved March 20, 1882, bred by S. Campbell, got by Gladstone (43286), out of Nonpareil 32d by Borough Member (33186)—Nonpareil 31st by Duke (28342)—Nonpareil 29th by Duke (28342)—Nonpareil 25th by Dipthong (17681) —Nonpareil 24th by Lord Sackville (13249)—Nonpareil 23d by The Baron (13833)—Nonpareil 17th by Matadore (11800)—Nonpareil 10th by Prince Edward Fairfax (9506)—Countess of Lincoln by Diamond (5918)—Nonpareil 3d by Young Frederick (3836)— by Commodore (1858)—Nonpareil by Tathwell Studley (5401)—by Blyth Comet (85).

ROSEBUD 8TH—Red, calved Feb. 2, 1881, bred by S. Campbell, got by Gladstone (43286), out of Rosebud 6th by Borough Member (33186)—Rosebud by Sir Christopher (22895)—Rosebud 1st by Dipthong (17681)—Rosebud by Scarlet Velvet (16916)—Thilia by Earl of Aberdeen (12800)—Myrtle by Balmoral (9920)—Fortuneteller by Dannecker (7949)—Enchantress by Ury (17157)—Ladykirk by Heriot (4017)—by Gray Diomed (2076)—by Juniper (1144).

MINA—Red and white, calved in March, 1881, bred by S. Campbell, got by Gladstone (43286), out of Mina by Luminary (34715)—Mina by Dipthong (21547)—Mina by Beeswing (12456)—Crocus by Sir Arthur (12072), &c., as in Mina 6th, above.

BELLA 3D—Red, calved Feb. 28, 1881, bred by Mr. Bruce, Heatherwick, got by Statesman (45659), out of Bella 2d by Heir Apparent (31352)—Bella by John Bright (31441)—Mary by British Prince (23470)—Flora 2d by The Challenge Cup (23022)—Flora by Sam Johnson (15234)—Victoria by Principal Fairfax (10656)—Beauty 2d by Enterprise (10202)—Beauty by Chancellor (6046).

SPOTTIE 2D—Red and white, calved Jan. 26, 1881, bred by Mr. Bruce, got by Statesman (45659), out of Dottie by Socrates (39144)—Spottie by John Bright (31441)—Mayflower by British Prince (23470)—Moss Rose by The Challenge Cup (23022)—Mary by Earl of Windsor (15968)—Mary Ann by Sam Johnson (15234)—Mayflower 8th by Duke of Kent (21596)—Mayflower 7th by Filbert (14548)—Mayflower 5th by Jemmy (11611).

October, 1883. JOHN ISAAC,
Of Bowmanton, Ont., Can. By Ship Buenos Ayrean.

BEAUTY 6TH—Roan, calved March 1, 1882, bred by S. Campbell, Kinellar, got by Lord Granville 5th (43517), out of Beauty 4th by Sir Windsor Broughton (27507)—Beauty 2d by Royal Hope (32302)—Beauty by Baron Colling (25560)—Duchess 7th by Michigan (24504)—Duchess 4th by West Australian (23202)—Duchess 3d by Prince Arthur (16728)—Duchess 2d by Magnum Bonum (13277)—Duchess by Bloomsbury (9972)—Juno by Monsieur Vestris (6220), &c.

QUEEN BESS 7TH—Red, calved March 5, 1882, bred by S. Campbell, got by Lord Granville 5th (43517), out of Queen Bess 5th by Sir Windsor Broughton (27507)—Queen Bess 2d by Windsor's Bridegroom (30525)—Queen Bess by Prince of Wales (20583)—Bessy Booth by Richard Booth (18699)—Bessie by Zetland (14048)—Red Princess by Leader (11674)—Princess by Duke of Richmond (14453)—Lady Vesper by Young Hector (2104).

CLARET 7TH—Roan, calved June 20, 1882, bred by S. Campbell, got by Gladstone (43286), out of Claret 4th by Luminary (34715)—Claret 2d by Novelist (34929)—Claret by Prince of Worcester (20597)—Claret 1st by Duke (28342)—Claret by Scarlet Velvet (16916)—Barbara by Unrivaled (13926)—Isabella by The Pacha (7612)—Crocus by 2d Duke of Northumberland (3646)— Nora by Sillery (5131)—by Young Western Comet (1575)—by Diamond (205)—by Favorite (252)—by Charge's Red Bull (1810).

MINA 10TH—Roan, calved July 6, 1882, bred by S. Campbell, got by Gladstone (43286), out of Mina 5th by Golden Prince (38363) —Mina 3d by Duke (28342)—Mina by Dipthong 3d (21547)—Mina by Beeswing (12456)—Crocus by Sir Arthur (12072)—Bashful by Young Ury (10984)—Likely by The Pacha (7612)—Helen by 2d Duke of Northumberland (3646)—Nora by Sillery (5131), &c., as in Claret 7th, above.

CLEMENTINA—Red, calved June 5, 1882, bred by S. Campbell, Kinellar, got by Gladstone (43286), out of Clementina by British Prince (33228)—Clementina by Sir Christopher (22895)—Clementina by Master Goldschmidt (20305)—Clementina by Lord Ythan (14850)—Empress Eugenie by Guy Fawkes (12981)—Bridesmaid by Sir Arthur (12072)—Crescent by The Pacha (7612)—by 2d Duke of Northumberland (3646)—by Emperor (3716)—by Invalid (4076).

CLEMENTINA—Roan, calved March 4, 1883, bred by S. Campbell, Kinellar, got by Gladstone (43286), out of Clementina by Golden Prince (38363)—Clementina by Sir Christopher (22895), &c., as above.

MARY—Red, calved March 10, 1883, bred by S. Campbell, got by Gladstone (43286), out of Mary 3d by Borough Member (33186) —Mary 2d by Novelist (34929)—Mary by Sir Christopher (22895) —Mary by Diphthong (17681)—Mary by Cœur de Lion (12611)— Alma by Mosstrooper (11827)—Leda by Donside Fairfax (12365) —Miss Ramsden by Duke (3630)—by Reveller (2528)—by Grazier (1086)—by Cato (857).

LADY YTHAN—Roan, calved March 20, 1883, bred by S. Campbell, got by Gladstone (43286), out of Lady Ythan by Novelist (34929)—Lady Ythan by Royal Duke (35356)—Lady Ythan by Marmaduke (20284)—Duchess 5th by Lord Ythan (14852)—Duchess 2d by Prince of Cobourg (15100)—Duchess by Duke of Clarence (9040)—Agnes by Angus Hero (6745)—Rosebud by Darlington.

THE MEMBER—Red, calved March 20, 1883, bred by S. Campbell, got by Gladstone (43286), out of Rosebud by Borough Member (33186)—Rosebud by Novelist (34929)—Rosebud 1st by Dipthong (17681)—Rosebud by Scarlet Velvet (16916)—Thilia by Earl of Aberdeen (12800)—Myrtle by Balmoral (9920)—Fortune-teller by Dannecker (7949)—Enchantress by Ury (17157)—Lady-kirk by Heriot (4017)—by Gray Diomed (2076)—by Juniper (1144).

COMET—Roan, caived April 21, 1883, bred by S. Campbell, got by Gladstone (43286), out of Nonpareil 30th by Royal Duke (35356)—Nonpareil 28th by Sir Christopher (22895)—Nonpareil 24th by Lord Sackville (13249)—Nonpareil 23d by The Baron (13833)—Nonpareil 17th by Matadore (11800)—Nonpareil 10th by Prince Edward Fairfax (9506)—Countess of Lincoln by Diamond (5918)—Nonpareil 3d by Young Frederick (3836)—by Commodore (1858)—by Tathwell Studley (5401)—by Blyth Comet (85).

CLARET—Roan, calved Jan. 5, 1881, bred by S. Campbell, got by Sir Christopher (22895), out of Claret by Prince of Worcester (20597)—Claret by Scarlet Velvet (16916)—Barbara by Unrivalled (13926)—Isabella by The Pacha (7612)—Crocus by 2d Duke of Northumberland (3646)—Nora by Sillery (5131)—by Young Western Comet (1575)—by Diamond (205)—by Favorite (252)—by Charge's Red Bull (1810).

ELIZABETH—Roan, calved March 22, 1881, bred by S. Campbell, got by Nobleman (38797), out of Elizabeth by Lord Oxford [a son of Sir Christopher (22895)]—Elizabeth by Wizard (25467)—Messalina by Domain 1760—Roan Strawberry by Kelvinside (14756)—Red Strawberry by Cecil (12571)—Rose by Commander (8976)—by Jerry.

43

JILT 7TH—Red, calved April 10, 1883, bred by S. Campbell. got by Gladstone (43286), out of Jilt by Baron Colling (25560)—Jilt 3d by Felix Booth (23925)—Jilt by Prince Arthur (16723)—Flirt by Magnum Bonum (13277)—Queen by The Pacha (7612)—Strawberry by 2d Duke of Northumberland (3646)—Margaret by Mahomed (6170)—Mary Ann by Sillery (5131)—Miss Gibson by Carleton (843)—Dora by Diamond (205)—by Diamond (205).

ROSEBUD 7TH—Red, calved March 22, 1883, bred by S. Campbell, got by Gladstone (43286), out of Rosebud 7th by Golden Prince (38363)—Rosebud by Dipthong (17681), &c., as in Rosebud 8th, above.

CLARET 8TH—Roan, calved June 30, 1883, bred by S. Campbell, got by Gladstone (43286), out of Claret 4th by Luminary (34715)—Claret 2d by Novelist (34929), &c., as in Claret 7th, above.

RAW JACKSON and JOHN HODGESON.
Ohio.

SOVEREIGN 995—Red and white, calved in May, 1837, got (in England) by Young Remus (2523), out of Strawberry by Magnum Bonum (2243). &c., as above.
Imported in 1837.

STRAWBERRY—Red and white, calved April 2, 1835, got by Magnum Bonum (2243); out of Old Clara by Marske (4401)—Milkmaid by Wyvill's son of Comet (155)—by Sedbury (1424)—by Smurthwaite's Bull (5219)—by Jolby (4115)—by Burton (3249).
A. H. B., vol. 2. p. 567. Imported in 1837.

1833. WILLIAM JACKSON.
Into New York.

DUCHESS—Roan, calved in 1830, bred by Samuel Scotson, Toxteth Park, got by Ebor (996), dam by Barmpton (54)—by Young Wynyard (704)—by Northumberland (464)—by a son of Comet (155).

A. H. B., vol. 1, p. 171. Brought to Kentucky by Jos. Wasson and Mr. Shropshire.

ROSE—Roan, bred by John Smith, Dashford, near Northallerton, got by Skipton, dam by a son of Parrington's Cleveland.

Brought to Kentucky by N. L. Lindsey, of Bourbon Co., Ky.

MAGNET—Calved May 29, 1832, bred by the Earl of Derby, got by a son of Ebor (996), dam by Viscount—by Young Wynyard (704)—by Northumberland (464), a son of Comet (155).

A. H. B., vol. 16, p. 11606.

BULL WINDLE.

MISS SCOTSON—Roan, calved May 4, 1838, bred by Samuel Scotson, Toxteth Park (imported in 1840), got by Henwood 2d (4012), out of Jenny Wren by a son of Beauchamp (781)—Avonia by Wharfdale (1578).

A. H. B., vol. 1, p. 206.

DIMPLES 421—Roan, bred by Mr. Pilkington, got by a grandson of Fitz Favorite (1042), out of Princess by a son of Young Sir Dimple (1442).

Brought to Kentucky in 1835, by T. Y. Brent, of Paris, Ky.

1872. MR. JEFFS.
Into Canada.

DIADEM 2D—Red, calved Jan. 16, 1870, bred by R. Blackwell, Tansley, got by Jupiter (24228), out of Diadem by Sir Charles (16948)—Diana 2d by Great Duke (12973)—Dinah by Governor (10282)—Dipthong by Red Duke (7619)—Dillicot by Chediston 1st

(6858)—Miss Harrison by Northumberland (1286)—Miss Fortune by Hotspur (1117)—Lily by Northumberland (1286)—Jasmine by Rob Roy (556)—Pink by Simon (590)—by Colonel (152)—by Styford (629)—by a son of Hubback (319).

C. H. B., vol. 3, p. 426.

DIADEM 4TH—Red, calved May 16, 1873, bred by R. Blackwell, got by Byron 21434, out of Diadem 2d by Jupiter (24228), &c., as in Diadem 2d, above.

A. H. B., vol. 14, p. 489.

August, 1881. ARTHUR JOHNSTON,
Of Greenwood, Ont. From Liverpool, by Ship Lake Manitoba, to Quebec.

LEWIS ARUNDEL 46433—Red, calved March 20, 1880, bred by F. Leney & Son, Wateringbury, Kent, England, got by 44th Duke of Oxford (39774), out of Lady Louisa's Duchess 4th by 2d Duke of Tregunter (26022)—Lady Louisa's Duchess 2d by Cambridge Duke 3d (23503)—Lady Louisa by Arch Duke 2d (15588)—Lucy Long by Duke of Gloucester (11382)—Louise by Cramer (6907)—Lady Bird by Cato (6836)—Luna by Helicon (2107)—Lavender by Matchem (2281)—Cora by Sir Alexander (591)—Mary by Stephen (1456)—by Western Comet (689)—by Charge's Gray Bull (872)—by Favorite (252)—by Bartle (777)—descended from Studley White Bull (627).

July, 1883. ARTHUR JOHNSTON.
From Glasgow.

CAPTAIN ERRANT (47547)—Red, calved April 3, 1882, bred by the Duke of Buccleuch, got by King Errant (36839), out of Twin Cherry by Lord Cecil (26621)—Cherry Blossom by Royal Errant (22780)—Cherry by Viscount (15471)—Violet by Kossuth (11646)—by Roger (13615)—by son of Thorp (2757)—by Thorp (2757)—by Albion (731)—by Wellington (679)—by Sultan (631)—by Signior (588).

BOLD BUCCLEUCH—Red, calved Jan. 4, 1883, bred by the Duke of Buccleuch, got by King Errant (36839), out of Lady Madeline by Job (31438)—Lady Heraldine by Grand Herald (26301)—Lady Warlaby by Prince of Warlaby (20593)—Twinness by Harbinger (10297)—Twinna by Fitz Leonard Junior (14553)—Polly by Young Rufus (13649)—by Constitution (12634)—by Young Comet (1853)—descended from Jolly's Bull (4115).

September, 1883. ARTHUR JOHNSTON,

Of Greenwood, Ont. By Steamship Hanoverian, September, 1883, from Liverpool to Quebec.

STATIRA DUCHESS 2D—Red, calved July 16, 1875, bred by Mr. A. Robotham, Drayton, Bassett, got by 4th Duke of Grafton (28396), out of Statira 8th by 12th Duke of Oxford (19633)—Statira 6th by Britannicus 2d (19349)—Statira by Duke of Gloster (11382)—Stately by Balco (9918)—Statice by Sir Launcelot (5166)—Shepherdess by Major (4345)—by Ganthorpe (2049)—by Don Juan (1923)—by Shylock (2622).

E. H. B., vol. 27, p. 494.

STATIRA DUKE 12TH—Red, calved March 28, 1883, bred by H. Lovatt, Low Hill, Wolverhampton, England, got by Lightburne Duke of Oxford 2d (38564), out of Statira Duchess 2d by Duke of Grafton 4th (28396), &c., as in dam, above.

1853. CHARLES KELLY.

Kellyville, Pa.

LIBERATOR 6394 (13153)—Roan, calved June 28, 1852, bred by J. S. Tanqueray, Hendon, got by Lord Marquis (10459), out of Janetta by Lycurgus (7180)—Jocasta by Friar Tuck (3848)—Junta by Warden (5595)—Joyance by Javelin (4093)—Joy by Blyth (797)—Janette by Wellington (684)—by Phenomenon (491)—by Favorite (252)—by Favorite (252)—by Favorite (252)—by Hubback (319)—by Snowdon's Bull (612)—by Waistell Bull (669)—by Masterman's Bull (422)—by Studley Bull (626).

DENNIS KELLY.

Philadelphia.

SALLY WALKER—From the herd of Mr. Walker, Donegal, Ireland.

TRAVELER COW—Bred by Mr. Parrington, England, got by Traveler (1525).

PEACH—Bred by Jonas Whitaker. See 2142.

1854. KENTUCKY IMPORTING COMPANY,

(JAMES BAGG AND WESLEY WARNOCK, Agents.)

Of Scott County, Ky. Sold at farm of C. W. Innes, Oct. 19, 1854.

EMIGRANT 472—Red, calved Aug. 27, 1853, bred by Wm. Wright, Sheriff Hutton, got by Smiling Willy (13758), out of Curiosity 2d by Guy Mannering (3957)—Eliza by Prince Albert (4781) —Fiddle by Marcus (2262)—by Liston (4230)—by White Comet (1582).

Sold to S. Corbin, for $205.

SIRIUS (13737)—Roan, calved Oct. 11, 1852, bred by E. Ackroyd, Denton Park, got by Concord (11302), out of She's Coming Again by Laudable (9282)—Fairy Tale by Sir Thomas Fairfax (5196)— Thomasine by Stillington (5327)—by Young Rockingham (2547)— by Driver (1928)—by Richard (1376).

Sold to R. A. Alexander, for $3,500.

MACGREGOR (13270)—Roan, calved Oct. 1, 1853, bred by F. H. Fawkes, Farnley Hall, got by Bridegroom (11203), out of Lady Milton by Milton (7238)—Zuleika by Norfolk (2377)—Medora by Ambo (1636)—Blossom by Memnon (2295)—by Pilot (496)—by Agamemnon (9)—by Burrell's Bull, of Burdon.

Sold to John Hill, for $600.

EARL DE GREY (12795)—Roan, calved Oct. 22, 1852, bred by R. Cattley, Barndsby, got by De Grey (11346), out of Lavender by Sir Charles Napier (10816)—Sweet Pea by Liberator (7140)—

Sweet Maid by Prince Albert (4791)—Vestris by Marton Comet (4409)—Vesta by Plato (2433)—Venus by Bedford Jr. (1701)—Vesta by Isaac (1129)—by Northern Light (1281)—by White Comet (1582)—by Cattley's Gray Bull (1798).

Sold to T. W. Goodloe, for $250.

OAKUM 763 (13402)—Red, calved March 12, 1852, bred by Mr. Ackroyd, Denton Park, got by Star (9687), out of Oakleaf by Sir Launcelot (5166)—Old Love by Orville (4625)—by Tomboy (2765) —by Vesper (1547).

Sold to James Bagg.

CAPTAIN STOUFFER 311—Red, calved Aug. 30, 1854, bred by E. Ackroyd, Denton Park, got by Oakum (13402), out of She's Welcome by Beaufort (9943)—She's Coming Again by Laudable (9282)—Fairy Tale by Sir Thomas Fairfax (5196)—Thomasine by Stillington (5327)—by Young Rockingham (2547)—by Driver (1928)—by Richard (1376).

Sold to J. McMeekin, for $167.50.

IRENE—Roan, calved Sept. 4, 1851, bred by John Kirkham, Hagnaby, near Spilsby, got by Sheldon (8557), out of Interlude by Sailor (10769)—Imogene by Prince Albert (4781)—Isabella by Commodore (3445)—by Ormsby (4621)—by Major (398)—by Cossack (925)—by Captain (108).

A. H. B., vol. 2, p. 405, and E. H. B., vol. 10, p. 404, under dam. Sold to John Hill, for $530.

INDUSTRY—Roan, calved March 30, 1855, bred by John Kirkham, got by Usurer (9763), out of Irene by Sheldon (8557), &c., as in dam, above.

A. H. B., vol. 2, p. 405, under dam.

POMEGRANATE—Roan, calved winter of 1854-5, bred by Mr. Renton, calved the property of Willis F. Jones, Woodford Co., Ky., got by Bridegroom (11203), out of Pine Apple, &c., as above.

PINE APPLE—Roan, calved Dec. 2, 1850, bred by John Renton, near Farnley, got by Lord Morpeth (13205), out of Christmas

Lady by The Stuart (7623)—Princess by Rockingham (2550)—Queen by Norfolk (2377)—by Follyfoot (3818)—by Hero (1110).

A. H. B., vol. 20, p. 16066, and E. H. B., vol. 11, p. 632. Sold to W. F. Jones, for $510.

AMAZON—Roan, calved June 20, 1851, bred by John Kirkham. Hagnaby, near Spilsby, got by Newmarket (10563), out of Alice Hawthorne by Neptune (7273)—Alice Gray (vol. 9, p. 247, E.) by Prince Albert (4781)—Audleby by Reformer (4917)—by Toneham (2767)—Graceful by a son of Alpha (3004)—by Cossack (925).

A. H. B., vol. 5, p. 203. Sold to H. Clay, for $225.

BESSY HOWARD—Red roan, calved Oct. 11, 1852, bred by F. H. Fawkes, Farnley Hall, got by Fitz Walter (10232), out of Lady Milton (vol. 10, p. 436, E.) by Milton (7238)—Zuleika by Norfolk (2377)—Medora by Ambo (1636)—Blossom by Memnon (2295)—own sister to Isabella by Pilot (496)—by Agamemnon (9) —by Burrell's Bull of Burdon.

Sold to R. A. Alexander, for $650.

RUBY—Roan, calved May 4, 1850, bred by Wm. Linton, of Sheriff Hutton, near York, got by Gen. Fairfax (11519), out of Sybil by Liberator (7140)—Vesper by Prince Albert (4791)—Nunthorpe by Marcus (2262)—Nun by Luck's-All (2230)—(bred by Mr. Simpson) by a descendant of Alfred (23)—by Equinox (245).

A. H. B., vol. 2, p. 549. Sold to R. A. Gano, for $215.

COMMERCE 2D—Roan, calved March 25, 1852, bred by E. Ackroyd, Denton Park, got by Concord (11302), out of Free Trade by The Stuart (7623)—The Pet by Norfolk (2377)—Favorite by Bright (1739)—Campanula by Shylock (2622)—by Percy (1314)—by Blucher (84)—by Jupiter (344)—by Marshal Beresford (415)—by Crocus (932)—by a bull of Mr. Colling's.

E. H. B., vol. 11, p. 463, under dam, and A. H. B., vol. 2, p. 334. Sold to J. McMeeken, for $415.

PEERLESS—Roan, calved Oct. 13, 1853, bred by E. Ackroyd, Denton Park, got by Treasurer (13899), out of The Pearl by

Cotherstone (6903)—Tidy by Plenipo (4724)—by Velocipede (5552) —by Francisco (2032)—by Marske (418)—from the stock of Mr. Maynard, of Eryholme.

E. H. B., vol. 11, p. 725, under dam, and A. H. B., vol. 2, p. 502. Sold to Mr. Gaines, for $275.

WINNY—White, calved April 8, 1853, bred by Mr. Maynard, of Eryholme, got by Udolpho (13907), out of Zara by Jerveaux (13082) —Matilda by Lord Stanley (4269)—Red Rose by Velocipede (5552) —Red Rose by Priam (2452)—by Jerry (4097)—from the stock of Mr. Booth.

E. H. B., vol. 11, p. 763, under dam, and A. H. B., vol. 3, p. 701. Sold to A. Allen, for $300.

MARY—Roan, calved July 19, 1851, bred by Mr. Wright, of Sheriff Hutton, got by Sweet William (9701), out of Mimmy by Liberator (7140)—Modesty by Guy Mannering (3957)—Fanny by Prince Albert (4791)—Fiddle by Marcus (2262)—by Liston (4230).

E. H. B., vol. 11, p. 570, and A. H. B., vol. 2, p. 465. Sold to W. E. Simms, for $240.

SHE'S WELCOME—Roan, calved Sept. 13, 1851, bred by E. Ackroyd, Denton Park, got by Beaufort (9943), out of She's Coming Again by Laudable (9282)—Fairy Tale by Sir Thomas Fairfax (5196)—Thomasine by Stillington (5327)—by Young Rockingham (2547)—by Driver (1928)—by Richard (1376).

E. H. B., vol. 10, p. 577, under dam, and A. H. B., vol. 2, p. 555. Sold to J. McMeeken, for $505.

SHEPHERDESS—Roan, calved Nov. 26, 1853, bred by E. Ackroyd, Denton Park, got by Bridegroom (11203), out of She's Coming Again by Laudable (9282)—Fairy Tale by Sir Thomas Fairfax (5196)—Thomasine by Stillington (5327)—by Young Rockingham (2547)—by Driver (1928)—by Richard (1376).

E. H. B., vol. 11, p. 696, under dam, and A. H. B., vol. 2, p. 555. Sold to R. Innis, for $505.

44

MATILDA—Roan, calved April 24, 1854, bred by W. Wright, Sheriff Hutton, got by Villiers (13959), out of Downhorn by Liberator (7140)—Curiosity 2d by Guy Mannering (3957)—Eliza by Prince Albert (4791)—Fiddle (vol. 5, E.) by Marcus (2262)—by Liston(4230)—by White Comet (1582).

A. H. B., vol. 2, p. 471. Sold to S. Corbin, for $205.

GRACE DARLING—Roan, calved June 12, 1849, bred by W. Wright, Sheriff Hutton, got by Liberator (7140), out of Modesty 2d by Guy Mannering (3957)—Eliza by Prince Albert (4791)—Fiddle by Marcus (2262)—by Liston (4230)—by White Comet (1582).

Died.

DOWNHORN—Roan, calved July 31, 1850, bred by W. Wright, Sheriff Hutton, got by Liberator (7140), out of Curiosity 2d by Guy Mannering (3957)—Eliza by Prince Albert (4791)—Fiddle (vol. 5, E.) by Marcus (2262)—by Liston (4230)—by White Comet (1582).

A. H. B., vol. 3, p. 373. Sold to J. T. McClelland, for $405.

LIZZY—Red, calved March 25, 1853, bred by F. H. Fawkes, of Farnley Hall, got by Marquis of Carrabas (11789), out of Lady Lauretta by Laudable (9282)—Laurel by Petrarch (7329)—Fair Vetch by Sir Thomas Fairfax (5196)—Vervain by Colossus (1847)—Verbena by Burley (1766)—Young Alexina by Pilot (496)—Alexina by Warlaby (672)—Agnes by Albion (14)—by Lame Bull (359)—by Shipton (587)—by a son of Suwarrow (636)—by Booth's Son of Twin Brother to Ben (88)—by Twin Brother to Ben (660).

E. H. B., vol. 11, p. 527, under dam. Sold to A. J. Alexander, for $600.

1883. KENTUCKY IMPORTING COMPANY.

(B. F. VANMETER AND LESLIE COMBS, Agents.)

From Liverpool, January 24, 1883. Landed in New York, February, 1883, by Steamer Lake Huron. Arrived in Kentucky, February, 1883.

LADY WILD EYES 5TH—Red and white, calved March 3, 1879, bred by Earl of Lathom, Lathom House, Ormskirk, Lancashire, got by Duke of Rosedale 6th (38176), out of Lady Wild Eyes 2d by Duke of Gloster 2d (28392)—Lady Wild Eyes by Duke of Geneva 8th (28390)—Wild Eyes 24th by Duke of Oxford 4th (11387)—Wild Eyes 22d by Wild Duke (19148)—Wild Eyes 20th by Lord Barrington 1st (13170)—Wild Eyes 16th by Duke of Oxford 2d (9046)—Wild Eyes 15th by Duke of Northumberland 4th (3649)—Wild Eyes 8th by Duke of Northumberland (1940)—Wild Eyes 2d by Belvedere (1706)—Wild Eyes by Emperor (1975)—by Wonderful (700)—by Cleveland (145)—by Butterfly (104)—by Hollon's Bull (313)—by Mowbray's Bull (2342)—by Masterman's Bull (422) —descended from the stock of M. Dobison.

E. H. B., vol. 26, p. 643, under dam. Sold to A. J. Alexander, for $510.

LADY WILD EYES 7TH—Roan, calved Jan. 16, 1881, bred by the Earl of Lathom, got by Baron Oxford 4th (25580), out of Lady Wild Eyes 2d by Duke of Gloster 2d (28392), &c., as above.

E. H. B., vol. 28, p. 469, under dam. Sold to A. J. Alexander, for $1,000.

ROWFANT KIRKLEVINGTON 2D—Roan, calved March 8, 1877, bred by Sir Curtis Lampson, Bart., Rowfant, Crawley, Sussex, got by Duke of Oxford 22d (31000), out of Siddington 12th by Duke of Tregunter 2d (26022)—Siddington 2d by Duke of Oxford 4th (11387)—Kirklevington 7th by Earl of Derby (10177)—Kirklevington 4th by Earl of Liverpool (9061)—Kirklevington 1st by Duke of Northumberland (1940)—Nell Gwynne by Belvedere (1706) —Northallerton by Son of 2d Hubback (2683)—a cow, the property of Mr. Bates, descended from the stock of Mr. Maynard, of Eryholme, Eng.

E. H. B., vol. 26, p. 515. Sold to George L. Danforth, for $425.

SIDDINGTON DUCHESS—White, calved Dec. 6, 1882, bred by H. Lovatt, Low Hill, Wolverhampton, got by Grand Duke 37th (43307), out of Rowfant Kirklevington 6th by Duke of Ormskirk (36526)—Siddington 12th by Duke of Tregunter 2d (26022), &c., as above.

E. H. B., vol. 29, p. 566, under dam. Sold to A. J. Alexander, for $690.

MARCHIONESS KIRKLEVINGTON—Red, calved July 10, 1882, bred by H. Lovatt, got by Marquis Oxford 2d (37055), out of Kirklevington Duchess 16th by Duke of Gloster 2d (28392)—Duchess of Kent by Lord Liverpool (22168)—Kirklevington 14th by Duke of Oxford 4th (11387)—Kirklevington 7th by Earl of Derby (10177), &c., as above.

E. H. B., vol. 29, p. 564, under dam, and A. H. B., vol. 26, p. 529. Sold to J. M. Bigstaff, for $210.

DUCHESS KIRKLEVINGTON—Red, calved March 27, 1882, bred by H. Lovatt, got by Grand Duke 37th (43307), out of Kirklevington Duchess 26th by Duke of Gloster 3d (33653)—Kirklevington Duchess 11th by Duke of Gloster 2d (28392)—Kirklevington Duchess 5th by Duke of Claro 2d (21576)—Duchess of Kent by Lord Liverpool (22168), &c., as above.

E. H. B., vol. 29, p. 565, under dam. Sold to A. J. Alexander, for $950.

SIDDINGTON KIRKLEVINGTON—Roan, calved April 16, 1882, bred by H. Lovatt, got by Grand Duke 37th (43307), out of Rowfant Kirklevington 2d by Duke of Oxford 22d (31000)—Siddington 12th by Duke of Tregunter 2d (26022)—Siddington 2d by Duke of Oxford 4th (11387)—Kirklevington 7th by Earl of Derby (10177), &c., as above.

E. H. B., vol. 29, p. 566, under dam. Sold to A. J. Alexander, for $425.

GENEROSITY—Roan, calved Feb. 25, 1882, bred by A. Cruickshank, got by Barmpton (37763), out of Gratitude (vol. 23, p. 407, E.) by Breadalbane (28073)—Golden Princess by Lord Raglan (13244)

—Gold Leaf by Lord Cardigan (13177)—Pure Gold by Young 4th Duke (9037)—The Star Pagoda by Duplicate Duke (6952) —The Mint by Robin-o'-Day (4973)—Brawith Bud by Sir Walter (4639)—by Jerry (4097)—by Roseberry (567)—by Constellation (163)—by Hastings (293)—by Hastings (293)—by Leopold (372).

Sold to B. B. Veech, for $250.

VERBENA—Red, calved March 18, 1882, bred by A. Cruickshank, Sittyton, got by Lamlash (45025), out of Veronica by Pride of the Isles (35072)—Violante by Champion of England (17526)— Violette by Lorenzo (20235)—by Dannecker (7949)—by The Chief (5425)—Eliza by Billy (3151)—Princess by Sovereign (7539)— Queen, bred by Mr. Robertson, of Ladykirk.

A. H. B., vol. 26, p. 1168. Sold to George L. Danforth, for $375.

EVANGELINE 4TH—Red, calved Jan. 27, 1880, bred by W. Duthie, Collynie, Aberdeenshire, Scotland, got by Earl of Derby 2d (31061), out of Evangeline 2d by Dipthong (17681)—Evangeline by Hotspur (21960)—Pride of the Dairy by Guy Fawkes (12981)— Bashful by Young Ury (10984)—Likely by The Pacha (7612)— Helen by Duke of Northumberland 2d (3646)—Mary Ann by Sillery (5131)—Miss Gibson by Carleton (843)—Dora by Diamond (205)—Kitty by Diamond (205).

E. H. B., vol. 27, p. 378, under dam. Sold to B. B. Veech, for $660.

EVANGELINE 5TH—Red, calved Feb. 9, 1882, bred by W. Duthie, got by Earl of March (33807), out of Evangeline 2d (vol. 27, p. 378, E.) by Dipthong (17681)—Evangeline by Hotspur (21960), &c., as in Evangeline 4th, above.

A. H. B., vol. 26, p. 1167. Sold to George L. Danforth, for $225.

MARY ANNE 35TH—Red, calved March 23, 1879, bred by G. Marr, Cambrogie, got by Brabagon (37881), out of Mary Anne 34th by Bromley (36289)—Mary Anne 24th by Scotsman 3d (32465)— Mary Anne 13th by Grand Prince (26308)—Mary Anne 8th by Baron Sebastopol (21241)—Mary Anne by Allathan (25508)—Anne

2d by Van Dunck (10992)—by Billy 2d (5794)—by son of **Emperor** (3716)—by Inkhorn (6091).

E. H. B., vol. 26, p. 398, and A. H. B., vol. 26, p. 1168. Sold to Geo. L. Danforth, for $360.

DAINTY DAME—Roan. calved March 25, 1880, bred by J. Cran, Kirkton, Inverness, Scotland. got by Commandant (39610), out of Dainty 8th (vol. 26, p. 384, E.) by Young Hero (26385)—Dainty 7th by Grand Prince (26308)—Dainty 4th by Baron Sebastopol (21241)—Dainty 3d by Duke of Bolton 4th (15915)—Young Dainty by Fairfax Hero (9106)—Dainty by Duke of Northumberland 2d (3646)—Magnet by Emperor (3716)—Young Magnet by Invalid (4076)—Magnet by Magnet (392)—by Palmflower (480).

A. H. B., vol. 26, p. 1167. Sold to Geo. L. Danforth, for $220.

QUEEN MARGARET—Roan, calved May 11, 1882, bred by G. Stephenson, Mains of Dum, Portsoy, Scotland, got by Champion (47565), out of Queen Mary by Cæsar Augustus (25704)—Queen of Scots by Forth (17866)—Queen of the Isles by The Baron (13833)—Queen of Scotland by Matadore (11800)—Edith Fairfax by Sir Thomas Fairfax (5196)—Kirton by Billy (3151)—Jessie by Sovereign (7539)—Rose by Satellite (1420)—by Baronet (60)—by Cleveland (144)—by Symmetry (641).

E. H. B., vol. 29, p. 703, under dam, and A. H. B., vol. 26, p. 1168. Sold to Geo. L. Danforth, for $150.

PANSY 6TH—Roan, calved April 5, 1882, bred by W. Duthie, got by Good Hope (44883), out of Pansy 3d by Rosethorn (32345)—Pansy 2d by Prince of Worcester (20597)—Pansy by Signet Seal (18824)—Rosa by Lennox (31593)—Kinaldie by Corporal (6899)—Missie by Commander (8976)—by Young Duke, bred at Ury.

Sold to B. B. Veech, for $410.

DUKE OF GREYHOLT—Roan, calved June 29, 1883, bred by Wm. Duthie, got by Cayhurst (47560), out of Evangeline 4th by Earl of Derby 2d (31061), &c., as in Evangeline 4th, above.

WILD EYES LADY—Roan, calved July 24, 1883, bred by Earl Lathom, got by Earl of Kirklevington (46302), out of Lady Wild Eyes 7th by Baron Oxford 4th (25580), &c., as in dam, above.

LORD WELLINGTON 52508 (45171)—Red, calved June 19, 1880, bred by Sir Henry Allsopp, Bart., Hindlip Hall, Worcester, got by Duke of Hillhurst 3d (30975), out of Waterloo 37th by Oxford Beau (29485)—Waterloo 30th by Duke of Wharfdale 3d (21619)—Waterloo 25th by Duke of Geneva (19614)—Waterloo 17th by Red Knight (11976)—Waterloo 14th by Grand Duke (10284)—Waterloo 13th by Duke of Oxford 3d (9047)—Waterloo 9th by 2d Cleveland Lad (3408)—Waterloo 6th by Duke of Northumberland (1940)—Waterloo 3d by Norfolk (2377)—Waterloo Cow by Waterloo (2816)—by Waterloo (2816).

Sold to W. W. Estill, for $600.

AURORA—Roan, calved Feb. 8, 1880, bred by W. A. Mitchell, got by Duke of Chamburgh (36052), out of Alma by Prince Alfred (27107)—Adeline by Cæsar Augustus (25704)—Anemone by Forth (17866)—Avalanche by Sir Samuel (16302)—Angerona by Lemnos (13146)—Amy by Earl Stanhope (5966)—Augusta by True Blue (5522)—Albina by Miracle (2321)—Alice by Sir Henry (1446)—Young Madam by Count (170)—Young Venus by Bracken (61)—Venus by Badsworth (47)—by Driffield (223)—bred by Sir George Strickland, Bart.

E. H. B., vol. 27, p. 518, under dam, and A. H. B., vol. 26, p. 1167. Sold to G. L. Danforth, for $375.

PRINCESS ROYAL 6TH—Red, calved Jan. 28, 1881, bred by W. Duthie, got by Border Chief (37874), out of Princess Royal 2d by Heir of Englishman 2d (34128)—Princess Royal by Grand Prince (26308)—Annie by Picotee (15063)—Fanny by Garioch Lad (17938)—Averne by Bucephalus (6784)—Young Amazon by Crusader (934)—Amazon by Sultan (1485)—Bellona by Mars (411)—Rolla by North Star (458).

A. H. B., vol. 26, p. 1168. Sold to G. L. Danforth, for $425.

MARCHIONESS OF MARCH—Roan, calved Feb. 19, 1882, bred by W. Duthie, got by Earl of March (33807), out of Marchioness 3d (vol. 27, p. 379, E. H. B.) by Gold Digger (24044)—Marchioness 2d by Nelson (22401)—Marchioness by Garibaldi

(21795)—Miss Nightingale by Guy Fawkes (12981)—Bride by Sir Arthur (12072)—Aurora by Earl of Liverpool (9061)—Young Amazon by Crusader (934), &c., as in Princess Royal 6th, above.

Sold to W. Warfield, for $325.

FLIRTATION—Roan, calved Feb. 20, 1881, bred by the Duke of Richmond, got by Arthur Benedict (40986), out of Flirt 10th by Royal Hope (32392)—Flirt by Magnum Bonum (13277)—Romp by Bloomsbury (9972)—Queen by The Pacha (7612)—Strawberry by Duke of Northumberland 2d (3646)—Margaret by Mahomed (6170)—Mary Ann by Sillery (5131)—Miss Gibson by Carleton (843)—Dora by Diamond (205)—Kitty by Diamond (205).

A. H. B., vol. 26, p. 1167. Sold to G. L. Danforth, for $250.

WIMPLE 21st—Roan, calved March 13, 1881, bred by Duke of Richmond, got by Arthur Benedict (40986), out of Wimple 17th by Royal Hope (32392)—Wimple 8th by Baron Colling (25560), &c., as in Lady Evelyn Hope 2d, above.

A. H. B., vol. 26, p. 1169, and E. H. B., vol. 29, p. 657. Sold to G. L. Danforth, for $450.

MAID OF MARCH—Red roan, calved Feb. 25, 1882, bred by W. Duthie, got by Earl of March (33807), out of Maid of Windsor by Frederick Fitz-Windsor (31196)—Maid of Honor by Hotspur (21960)—Minerva by General Simpson (14608)—Miss Nightingale by Guy Fawkes (12981)—Bride by Sir Arthur (12072), &c., as in Marchioness of March, above.

A. H. B., vol. 26, p. 1168. Sold to Geo. L. Danforth, for $280.

DAISY OF THE LEA—Roan, calved March 19, 1880, bred by Mr. A. Davidson, Mains of Cairnbrogie, Old Meldrum, Scotland, got by Titus (40822), out of Daisy 10th by King John (31494)—Daisy 4th by Prince (16716)—Daisy by Sir Arthur (12072)—Likely by The Pacha (7612)—by Duke of Northumberland 2d (3646)—by Sillery (5131)—by Carleton (843)—by Diamond (205)—by Diamond (205).

E. H. B., vol. 28, p. 357, under dam. Sold to I. D. Corwin, Ohio, for $325.

LADY EVELYN HOPE 2D—Red and white, calved Feb. 26, 1881, bred by Duke of Richmond, got by Arthur Benedict (40986), out of Lady Evelyn Hope by Royal Hope (32392)—Wimple 8th by Baron Colling (25560)—Wimple 5th by Duke of Oxford 15th (23776) —Wimple 3d by Whipper-In (19139)—Wimple by Prince Arthur (16723)—Anna by Magnum Bonum (13277)—Amelia by Blooms-bury (9972)—Lady Cecilia by Duke 3d (17697)—Sugar Candy by The Peer (5455)—Miss Crisp by Bachelor (1666)—Peggy by Em-peror (1974)—by Houndilee (2139)—by Togston (5487).

A. H. B., vol. 25, p. 915. Sold to D. H. James, for $400.

MARY ANNE 36TH—Roan, calved March 3, 1883, bred by G. Marr, got by Mountain Chief (38767), out of Mary Ann 35th by Brabazon (37881), &c., as in dam, above.

Sold to G. L. Danforth.

WHITE CHIEF—White, calved Feb. 13, 1883, bred by A. Davidson, got by Mountain Chief (38767), out of Daisy of the Lea by Titus (40822), &c., as in dam, above.

Sold to S. Vanmeter.

ROSE OF CORIOLE—Roan, calved March 11, 1878, bred by A. Davidson, got by Coriolanus (33446), out of Rosemary by Em-peror Maximillian (26100)—Moss Rose by Young Freedom (21777) —White Rose by Statesman (18927)—Melody by Richard Cœur de Lion (13590)—Young Rosebud by Brilliant (7851)—Rosebud by Young Hector (7074)—by Bachelor (1665)—by Wallace (5588)— by Leopold (2199)—by Sir Harry (5155)—by Traveler (655)—by Colonel (152)—by son of Hubback (319).

E. H. B., vol. 26, p. 398, under dam, and A. H. B., vol. 26, p. 483. Sold to James Anderson, Ohio, for $250.

CAYHURST 2D—Roan, calved March 13, 1883, bred by W. Duthie, got by Cayhurst (47560), out of Princess Royal 6th by Border Chief (37874), &c., as in dam, above.

Sold to John Shanton, Pennsylvania.

BRUNSWICK—Red roan, calved March 20, 1883, bred by A. Davidson, got by Mountain Chief (38767), out of Rose of Coriole by Coriolanus (33446), &c., as in dam, above.

Sold to J. R. Anderson, Ohio.

LOVELY 25TH—Roan, calved Dec. 26, 1877, bred by A. Cruickshank, Sittyton, Aberdeenshire, Scotland, got by General Windsor (28701), out of Lovely 14th by Master of Arts (26867)—Lovely 10th by Duke of Bedford (23722)—Lovely 6th by Bosquet (14183)—Lovely 3d by Hero (10934)—Lovely by Kelly 2d (9264)—Lady Ythan by Robin-o'-Day (4973)—Lady by Favorite (9116)—Marion by Anthony (1640)—Miranda by Anthony (1640)—Merino by Edgecott (1953)—Matilda by Son of Merlin (6522)—White Cow by Acton (1607).

A. H. B., vol. 26, p. 1168, and E. H. B., vol. 27, p. 379. Sold to Geo. L. Danforth, for $300.

LOVELY 28TH—Roan, calved April 23, 1878, bred by A. Cruickshank, got by Pride of the Isles (35072), out of Lovely 12th by Scotch Rose (25099)—Lovely 9th by Windsor Augustus (19157)—Lovely 8th by Bosquet (14183)—Lovely by Kelly 2d (9265), &c., as in dam, above.

A. H. B., vol. 26, p. 700, and E. H. B., vol. 28, p. 353. Sold to W. W. & R. C. Estill, for $240.

LOVELY 38TH—Red roan, calved Jan. 12, 1883, bred by A. Cruickshank, got by Cumberland (46144), out of Lovely 28th by Pride of the Isles (35072), &c., as in Dam, above.

A. H. B., vol. 26, p. 1120. Sold to John Shanton, Pennsylvania, for $155.

LOVELY BOY—Red, calved March 1, 1883, bred by A. Cruickshank, got by Earl of March (33807), out of Lovely 25th by Gen. Windsor (28701), &c., as in dam, above.

Sold to Geo. L. Danforth.

ACANTHUS—Red, calved Oct. 14, 1881, bred by A. Cruickshank, got by Barmpton (37763), out of Amaryllis by Lord Lancaster (26666)—Azalea by Cæsar Augustus (25704)—Anemone by Forth (17866)—Avalanche by Sir Samuel (15302)—Angerona by

Lemnos (13146)—Amy by Earl Stanhope (5966)—Augusta by True Blue (5522)—Albinia by Miracle (2321)—Alice by Sir Henry (1446) —Young Madame by Count (170)—Young Venus by Bracken (91) —Venus by Badsworth (47)—by Driffield (223)—bred by Sir George Strickland.

E. H. B., vol. 28, p. 353, under dam. Sold to B. B. Veech, for $425.

VIOLET BUD—Red, calved Dec. 26, 1881, bred by A. Cruickshank, got by Barmpton (37763), out of Rose of Knowlmere by Knight of Knowlmere (22055)—Red Violet by Allan (21172)— Violet by Lord Bathurst (13173)—Roseate by Matadore (11800)— China Rose by Hudson (9228)—Carmine Rose by Fairfax Royal (6987)—Red Rose by Inkhorn (6091)—Moss Rose by Grazier (1085) —Cicely by Sampson—Marion by Wallace (1560).

A. H. B., vol. 26, p. 1169, and E. H. B., vol. 28, p. 354, under dam. Sold to Geo. L. Danforth, for $350.

VICTORIA 73D—Red, calved Feb. 4, 1882, bred by A. Cruickshank, got by Roan Gauntlet (35284), out of Victoria 58th (vol. 26, p. 393, E.) by Pride of the Isles (35072)—Victoria 43d by Champion of England (17526)—Victoria 36th by Baronet (15614) —Victoria 31st by Master Butterfly 2d (14918)—Victoria 29th by Red Knight (11976)—Victoria 19th by Lord John (11731)—Victoria 4th by Prince Albert (11933)—Victoria 2d by Belzoni (783)—Victoria by Satellite (1420)—No. 1 Mason's Sale by Cato (119)—Pope Cow by Pope (514)—Flora by Favorite (252)—Nymph by White Bull (421)—Lily by Favorite (252)—Miss Lax by Dalton Duke (188)—Lady Maynard by Alcock's Bull (19)—by Jacob Smith's Bull (608)—by Jolly's Bull (337).

Sold to B. B. Veech, for $500.

LUSTRE 23D—Roan, calved April 15, 1879, bred by the Duke of Richmond, Gordon Castle, Fochabers, Scotland, got by White Duke (32849), out of Lustre 15th (vol. 23, p. 617, E.) by Royal Hope (32392)—Lustre 12th by Baron Colling (25560)—Lustre 10th by Michigan (24594)—Lustre 5th by Prince Arthur (16723)— Lustre 3d by Magnum Bonum (13277)—Lustre 1st by Bloomsbury

(9972)—Lustre by Duke of Northumberland 2d (3646)—Bluebell by Bachelor (1666)—Bellflower by Sultan (1485)—Rolla by North Star (458).

Sold to Hugh A. Moran, for $300.

LITTLE LUSTRE—Roan, calved Nov. 29, 1882, bred by the Duke of Richmond, got by Good Hope (44883), out of Lustre 23d by White Duke (32849), &c., as in dam, above.

Sold to M. F. Arbuckle, for $100.

LUSTROUS—Red and white, calved March 11, 1881, bred by the Duke of Richmond, got by Arthur Benedict (40986), out of Lustre 15th by Royal Hope (32392), &c., as in Lustre 23d, above.

A. H. B., vol. 26, p. 1168. Sold to G. L. Danforth, for $275.

PEACH BLOSSOM 14th—Red, calved Jan. 8, 1880, bred by the Duke of Richmond, got by Chief Officer (36359), out of Rose Blossom (vol. 23, p. 617, E.) by Royal Hope (32392)—Autumn Blossom by Baron Colling (25560)—Red Rose by Magnum Bonum (13277)—Profit by Lochnagar (9303)—Almond Flower by Holkar (4041)—Eglantine by Brougham (1746)—by a bull of Mr. C. Mason's —bred by Mr. Weir, of Goswick.

A. H. B., vol. 25, p. 1166. Sold to M. B. Robertson & Son, of Ohio, for $615.

NEW-YEAR'S MORN 57211—Red and white bull, calved Jan. 1, 1883, bred by the Duke of Richmond, got by Arthur Benedict (40986), out of Peach Blossom 14th by Chief Officer (36359), &c., as in dam, above.

Sold to M. B. Robertson & Son, for $105.

1835. KENYON COLLEGE.

A Present to Mrs. McIlvaine, wife of Bishop McIlvaine, by Thos. Bates, Kirklevington, Yarm, Eng.

SKIPTON BRIDGE (5208)—Roan, calved in 1833, bred by Mr. Clark, of Skipton Bridge, Yorkshire, got by a grandson of Governor (1077), out of Rosebud 1st by a grandson of Sir Dimple (594) —Lady by Young Hector (1107)—Rosebud by Maynard's grandson of Laird (1158)—by Gray Bull—by a son of Simon (590)—by a son of Suwarrow (636).

HONORABLE MISS BARRINGTON—Roan, calved Jan. 19, 1835, bred by T. Bates, got by Belvedere (1706), out of Lady Barrington by a son of Mason's Herdsman (304)—Young Alicia by Wonderful (700)—Old Alicia by Alfred (23)—by a son of Favorite (252).

E. H. B., vol. 3, p. 449, under dam.

Aug. 4, 1871. W. S. KING,

Of Minneapolis, Minn. Shipped by the Germany, from Liverpool, Aug. 4, 1871. Selected by R. Gibson.

GARLAND 2D—Red and white heifer, calved Jan. 1, 1870, bred by Lord Penrhyn, Penrhyn Castle, North Wales, got by 11th Grand Duke (21849), out of Graceful by Duke of Geneva (19614)—Gracious by Marmaduke (14897)—Graceful by Royalist (10479)—Gazelle by Noble (4578)—by Hector (4000)—by Emperor (1974)—by Margrave (2263)—by Leopold (2199)—by Hector (2103)—by Traveler (655)—by Surly (2715)—by Colonel (152).

Recorded, under dam, E. H. B., vol. 19, p. 535. Bought at public sale of W. S. King's, May 21, 1874, at Dexter Park, Chicago, for $2,100, by T. J. Megibben, Cynthiana, Ky.

DOUBLE BUTTERFLY 2D—Roan heifer, calved Sept. 28, 1867, bred by R. Eastwood, Thorneyholme, Clitheroe, got by The Hero (20958), out of Double Butterfly by Royal Butterfly (16862)—Alice Butterfly by Master Butterfly (13311)—Alice 2d by Duke of Athol (10150)—Madaline by Marcus (2262)—Landlady by Matchem (2281)—Landlady by Pilot (496)—Daphne by Young Albion (15)—Gaudy by Albion (14)—Old Gaudy by Suwarrow (636)—by Son of Twin Brother to Ben (88)—by Twin Brother to Ben (660).

Recorded, A. H. B., vol. 11, p. 588, and E. H. B., vol. 19, p. 480, under dam.

DUCHESS OF TOWNELEY—Red, calved July 23, 1866, bred by G. Garne, Churchill Heath, Oxon., got by Duke of Towneley (21615), out of Donna Inez by Gondomar (17985)—Damask by A-1 (15538)—Damsel by Enterprise (11443)—Blond by Patriot (10595)—by son of Elevator (6969)—by No Mistake (8357)—by Young Consul (6893)—by Fairfax (1023).

Recorded, E. H. B., vol. 19, p. 488.

Aug. 26, 1871. W. S. KING,

Of Minneapolis, Minn.　By Steamer Austrian, from Liverpool, Aug. 26, 1871.　Selected by Richard Gibson.

BUTTERFLY'S MEMENTO—Roan heifer, calved Dec. 20, 1869, bred by Col. Towneley, Towneley Park, Lancashire, got by Baron Oxford (23375), out of Duchess of Lancaster 7th by Inglewood (20006)—Duchess of Lancaster 2d by Precedent (11918)—Lancaster Belle by Louis Napoleon 2d (13259)—Duchess of Lancaster by Duke of Lancaster (10929)—Honeycomb by North Star (9447)—Bessy by Thick Hock (6601)—Barmpton Rose by Expectation (1988)—by Belzoni (1709)—by Comus (1861)—by Denton (198).

Recorded, E. H. B., vol. 19, p. 486, under dam.

NOTE.—In this pedigree, as recorded in vol. 11, p. 535, A. H. B., Duchess of Lancaster 7th is given as by Baron Ribblesdale (21235), out of Duchess of Lancaster 5th. The English Herd Book gives it as above. Both Duchess of Lancaster 5th and 7th are by Inglewood. See E. H. B., vol. 16, p. 428, under dam, Duchess of Lancaster 5th; E. H. B., vol. 17, p. 464, Duchess of Lancaster 7th, under dam, and E. H. B., vol. 19, p. 486.

MAID OF THE ABBEY—Roan heifer, calved Sept. 30, 1868, bred by Mr. H. Aylmer, West Dereham Abbey, Norfolk, got by Prince Christian (22581), out of Maid of the Morn by Hildebrand (18068)—Maid of Orleans by Knight of Windsor (16349)—Joan of Arc by Vanguard (10994)—Joan by Bumper (10005)—Milkmaid by Samuel (5084)—Cora by Reformer (2512)—by Imperial (2151)—by Favorite (1030)—by Young Dimple (971)—by Brown's White Bull (98).

Recorded, E. H. B., vol. 18, p. 605, under dam.

COUNTESS OF OXFORD—Roan heifer, calved Jan. 11, 1869, bred by Messrs. Hosken & Son, Hoyle, Cornwall, got by 2d Earl of Oxford (23844), out of Countess by Prince Frederick (16734)—Jocund by Brigadier (14193)—Jilt by Usurper (13929)—Jacinth by Fawsley (6004)—Jacquette by Javelin (4093)—Jonquille by Blyth (797)—Jeanette by Wellington (684)—by Phenomenon (491)—by Favorite (252)—by Favorite (252)—by Favorite (252)—by Hubback (319)—by Snowdon's Bull (612)—by Waistell's Bull (669)—by Masterman's Bull (422)—by Studley Bull (626).

Recorded, A. H. B., vol. 11, p. 569, and E. H. B., vol. 19, p. 456.

HENRIETTA—Roan, calved May 20, 1867, bred by D. Ainslie, Costerton, Edinburg, got by Duke of Tyne (17751), out of Ferooza by Knight Errant (18154)—Barbary by Cardigan (12556)—Barmaid by Lord Fanny (13187)—Beauty by Noble (4578)—Betsey by Newton (2367)—by Baronet (1686)—by Reformer (2502)—by Margrave (2263)—by Leopold (2199)—by Hector (2103)—by Surly (2715)—by Traveler (655)—by Colonel (152).

Recorded, A. H. B., vol. 11, p. 693, and E. H. B., vol. 19, p. 546.

BARON HUBBACK 2D 13199 (27947)—Red, calved Feb. 3, 1870, bred by Col. Towneley, Towneley Park, got by Baron Oxford (23375), out of Duchess 7th by Grand Duke of Lancaster (19883)—Duchess 6th by Royal Duke (16865)—Duchess 5th by Brennus (8902)—Duchess by Duke of Norfolk (5952)—by Cleveland Lad (3407)—by Red Highflyer (2488)—by Sir Charles (5146)—by Harry Lorrequer (3985)—by Blucher (84)—by Magnum Bonum (4322)—by Styford (103)—by son of T. Wetherell's Bull (690).

NOTE.—This bull is recorded in the English Herd Book as Baron Hubback 3d (27947).

LADY BROUGH—Roan heifer, calved Oct. 15, 1869, bred by F. Robinson, Catterrack, Yorkshire, got by Baron Killerby (27949), out of Necklace by Lord Stanley (16452)—by Rifleman (15163)—by Young Hopewell (14719)—by Sir Harry (15287)—by Fitz William (14555).

Recorded thus in A. H. B., vol. 11, p. 760. In E. H. B., vol. 19, p. 647, under dam, Necklace is given as by Fitz Arthur (26161).

SYLPH—Red and white heifer, calved July 2, 1869, bred by Lord Penrhyn, Penrhyn Castle, Bangor, North Wales, got by Cherry Duke (25752), out of Sylvia by Vampire (19043)—Duchess by Penrhyn (13463)—Snowdrop by Gentleman (12937)—Virgin by Charles 1st (8947)—Vestris by Belvedere 4th (3130)—by Panton Favorite (4646)—by Plenipotentiary (2436)—by Mameluke (2257) —by Prime Minister (2454)—by Surprise (2716).

E. H. B., vol. 19, p. 749, under dam, and A. H. B., vol. 11, p. 1107.

GEM OF LYNDALE—Red and white, calved March 8, 1872, bred by D. Ainslie, Costerton, got by Banner Bearer (27907), out of Henrietta by Duke of Tyne (17751), &c., as in dam, above.

A. H. B., vol. 12, p. 813.

FLORENCE BARTON—Roan, calved Sept. 15, 1871, bred by Messrs. Hoskin & Son, Hoyle, got by Towneley Oxford (30170), out of Countess of Oxford by 2d Earl of Oxford (23844), &c., as in dam, in this list.

A. H. B., vol. 11, p. 659.

1871. W. S. KING.

LADY MARGARET—Roan heifer, calved Feb. 17, 1869, bred by H. Aylmer, West Dereham Abbey, got by Prince Christian (22581), out of Lady Fannie (vol. 18, p. 564, E.) by Prince Leopold (20557)—Lady Margaret by Red Knight (16808)—Ladybird by First Fruits (16048)—Laura by Chilton (10054)—Lady Jane by Augustus (8848)—Lady Maynard by Lord Stanley (4269)— Julia by Velocipede (5552)—by Francisco (2032)—by Sir Thomas (2636)—by Sir Alexander (591)—by Marske (418)—Favorite (252).

Recorded, A. H. B., vol. 11, p. 776.

FRILL—Roan heifer, calved Nov. 23, 1869, bred by H. Aylmer, West Dereham Abbey, got by Gen. Hopewell 2d (24021), out of Fame by Norfolk Thorndale Duke (24666)—Flirt by Hildebrand (18068)—Flounce by Alderman 2d (17292)—Filigree by First Fruits (16048)—Flirt by Kirklevington (11639)—Flounce by Broughton Hero (6811)—Frill by Rockingham (2550)—Fancy by

Remus (2524)—White Rosette by Juniper (1144)—Rosette by White Comet (1582)—Young Rose by Grandson of Favorite (2073) —Old Rose by Cattley's Gray Bull (1798).

Recorded, A. H. B., vol. 11, p. 667, and E. H. B., vol. 19, p. 506, under dam.

CHRISTINE—Red and white, calved Feb. 22, 1869, bred by G. K. Cooper, Bowback House, Suffolk, got by Hogarth 2d (24148), out of Christabel by Mountain Chief (20383)—Christiana by Sir Roderick Dhu (18862)—St. Crispen's Daughter by War Eagle (15483)—Flower of May by Vanguard (10994)—Lenny by Leonard (4210)—by Buckingham (3239)—by Raspberry (4875).

Recorded, E. H. B., vol. 19, p. 444, under dam, and A. H. B., vol. 11, p. 555.

A. KNIGHT.
Stanstead, Can.

TOPEKA 13046—Roan, calved April 2, 1868, bred by S. Stowell & Bro., Favordale, Durham, Eng., got by Earl of Oxford (21651), out of May Duchess by Grand Duke of York (12966)—Maydew by Forester (8084)—Martha by Melmoth (6200)—Pretty by Colossus (1847)—Nan by Norfolk (2377)—Lady by Somerville (2659).

NOTE.—(26885) is May Duke, with same pedigree, but calved in April, 1867. Possibly own brother to above.

1821. MR. LAW,
Of Baltimore.

ASSURANCE—Calved in 1816, bred by Mr. Curwen, Working-ton Hall, Cumberland, got by Harlequin (289), out of Scut by Sir Oliver (605)—Fanny by Chilton (136)—by Cleveland (144).

Recorded, E. H. B., vol. 1, p. 492, under dam.

ROSEMARY—Roan, calved in 1816, bred by J. C. Curwen, England, got by Flash (261), out of Red Rose by Petrarch (488)—

Bright Eyes by Alexander (20)—by Traveler (655)—by son of Bolingbroke (86).

E. H. B., vol. 1, p. 458, under dam, and A. H. B., vol. 21, p. 16666.

VIRGINIA—White, calved in 1821, bred by J. C. Curwen, got by General (272), out of Rosemary by Flash (262), &c., as in dam, above.

E. H. B., vol. 2. p. 595, and A. H. B., vol. 21, p. 16690.

BISHOP (73)—Calved in 1818, bred by Mr. Curwen, got by Rockingham or Wellington (560) or (683), out of Arbutus by son of Harlequin (289)—Strawberry by Yarborough (705)—Snowdrop by Duke (224)—Darling by Jobling's Traveler (655)—Flowery by Bolingbroke (86)—a cow bought of Mr. Boazman, near Darlington.

1838. JOSIAH LAWRENCE,
Of Cincinnati, Ohio.

COWS.

LADY ANN—Roan, calved April 8, 1836, got by Magnum Bonum (2243), dam by Linton (2207)—by Jupiter (343)—by Easby (232).

A. H. B., vol. 1, p. 189.

JUNO—Roan, calved in January, 1836, bred by R. Crofton, got by Scrip (2604), out of Whitworth by Miracle (2320)—Daisy by Volunteer (2806)—by Fitz Remus (2025)—bought of Mr. Mason.

E. H. B., vol. 3, p. 694, under dam.

FORTUNA—Roan, calved Aug. 26, 1835, bred by Robert Smith. Givendale, got by Reformer (2512), out of Florence by Lindrick (1170)—Florence by Sir Alexander (591)—by Star (618)—by Snowball—by Sir Dimple (594).

Recorded, E. H. B., vol. 5, p. 387. Under her dam, on page 395, vol. 3, E., and on page 384, vol. 5, E., she is recorded as by Reformer, without a number.

ADELAIDE—Red and white, calved Oct. 13, 1838, begotten in England by Sir Walter (2639), out of Juno by Scrip (2604)—Whitworth by Miracle (2320)—Daisy by Volunteer (2806)—by Fitz Remus (2025)—bought of Mr. Mason.

EMPRESS—White, calved Dec. 11, 1838, begotten in England by Barforth (3085), out of imp. Lady Ann by Magnum Bonum (2243)—by Linton (4227)—by Jupiter (343)—by Easby (232).

VERBINA—Roan, calved in January, 1836, bred by Mr. Houldsworth, got by Rinaldo (4949), out of Wharfdale Lady by 2d Hubback (1423).

NOTE.—See A. H. B., vol. 4, p. 481, in Lile by Napoleon. This should be the Napoleon imported by Mr. Lawrence, *not* 733, as given in Herd Book. This cow was certainly not imported by Mr. Lawrence. She may have been imported by Mr. Neff or Mr. Mahard. See, for Verbina, E. H. B., vol. 5, p. 1047.

BULLS.

BERRYMAN (3143)—Roan, calved in 1835, bred by Rev. Henry Berry, got by Henwood (2114), out of Minikin by Wharfdale (1578)—Minna by Nestor (452)—Minerva by Harold (291)—Mary by Meteor (432)—Magdalena by Comet (155)—by Cupid (177).

NAPOLEON—Red and white, calved Oct. 9, 1838, got in England by Fantastical (3759), out of Fortuna by Reformer (2512)—Florence by Lindrick (1170)—Florence by Sir Alexander (591)—by Star (618)—by Snowball—by Sir Dimple (594).

Oct. 5, 1875. PROF. G. LAWSON.

For Central Board of Agriculture, Nova Scotia. By Ship Nova Scotian.

ROSE GWYNNE 4TH—Roan cow, calved June 22, 1866, bred by J. J. Hetherington, Barmpton, Carlisle, got by Duke of Cumberland (21584), out of Rose Gwynne 2d by Gen. Jackson 2d (17954)—Rose Gwynne by General Jackson (14604)—Rosebud by Mango (4359)—Cowslip by Wallace (5586)—by Tom Gwynne (5498)—by Marmion (406)—bred by Mr. Matthews, of Durham.

E. H. B., vol. 18, p. 704, under dam.

POLLY VAUGHAN—Roan, calved Jan. 30, 1872, bred by G. Moore, Whitehall, Carlisle, got by 17th Duke of Oxford (25994), out of Rose Gwynne 4th by Duke of Cumberland (21584), &c., as in dam, above.

E. H. B., vol. 20, p. 738, under dam.

CAWOOD'S ROSE—Roan cow, calved Feb. 4, 1867, bred by Mr. W. S. Cragg, Arkholme, Lancashire, got by Lord Cawood 3d (24368), out of White Cow by Golden Eclipse (14625)—by Reindeer (15150)—by Horton Boy (13050).

E. H. B., vol. 19, p. 432.

PRINCESS MARY—White, calved Feb. 22, 1872, bred by M. T. Lamb, Hay Carr House, Lancaster, got by Golden Duke (26266), out of Empress by Majestic (20264)—Regina by Constantine (14318)—Lucy Long by the Duke of Lancaster (10929)—Eliza by Gainford 2d (6030)—Primrose by Wharton (2833)—by Count (1883)—by Baronet (1686)—by Young Rockingham (2549)—by Wellington (2824)—by Northumberland (464)—by Styford (103)—by Lame Bull (358)—by Bolingbroke (86).

BARON LIGHTBURNE 2D (36191)—Roan, calved Feb. 23, 1873, bred by A. Brogden, Ulverstone, got by Baron Oxford 4th (25580), out of Winsome 7th by Grand Duke 10th (21848)—Winsome by Oxford 2d (18507)—Beauty by Crusade (7938)—Bright Eyes by 3d Duke of York (10166)—by 2d Cleveland Lad (3408)—by Duke of Northumberland (1940)—by Belvedere (1706)—by Emperor (1975)—by Wonderful (700)—by Cleveland (145)—by Butterfly (104)—by Hollon's Bull (313)—by Mowbray's Bull (2342) —by Masterman's Bull (422).

GWYNNE OF THE FOREST (34100)—Roan bull, calved Aug. 22, 1873, bred by G. Moore, got by 17th Duke of Oxford (25994), out of Flighty Gwynne by Grand Duke of Lightburne (26290)—Fairy Gwynne by 5th Grand Duke (19875)—Fortuna Gwynne by Duke of Leinster (17724)—Frances Gwynne by Captain Hardinge (10023)—by St. Thomas (10777)—by Prime Minister

(2456)—by Marmion (406)—by Merlin (430)—by Layton (366)—by Phenomenon (491)—by Favorite (252), &c.

CAPTAIN CAWOOD (33286)—Roan, calved Nov. 3, 1873, bred by W. S. Cragg, Arkholme, got by Captain Tregunter (28136), out of Double Cawood by 3d Lord Cawood (24368)—by Lord Cawood (15943)—by Grazier (28788)—by Horton Boy (13050).

KENT GWYNNE (34300)—Roan, calved July 10, 1874, bred by W. Ashburner, Netherhouse, got by Grand Duke of Kent 2d (28759), out of Double Gwynne by Rufus (27397)—Dolly Gwynne by Duke of York (14461)—Young Dowager Gwynne by St. Thomas (10777)—Dowager Gwynne by Prime Minister (2456)—by Wallace (5586)—by Marmion (406)—by Merlin (430)—by Layton (366)—by Phenomenon (491)—by Favorite (252)—by Favorite (252)—by Hubback (319), &c.

VISCOUNT OXFORD (35902)—Roan, calved Aug. 21, 1874, bred by G. Moore, got by 6th Baron Oxford (33075), out of Graceful Duchess by Baron Oxford 4th (25580)—Duchess by Grand Duke 15th (21852)—Countess by Knightley Grand Duke (24268)—Chorus by 4th Duke of Thorndale (17750)—by Marmaduke (13289)—by Cardinal (11246)—by White Friar (9827)—by Little John (4232) —by Caliph (1774)—by Sir Walter (2637)—by Hotspur (1117)— by Coxcomb (928)—by Midas (435)—by Comet (155)—by R. Colling's son of Favorite (252)—by same son of Favorite (252)—by Hubback (319).

SKIDDAW—Roan bull, calved Jan. 17, 1875, bred by G. Moore, got by 17th Duke of Oxford (25994), out of Sweetheart 31st by Patrician (24728)—Sweetheart 10th by The Baron (13833)—Sweetheart 5th by Mameluke (13289)—Sweetheart 3d by Daybreak (11338)—Sweetheart by Accordion (5708)—by Little John (4232)— by Caliph (1774)—by Sir Walter (2637)—by Hotspur (1117)—by Coxcomb (928)—by Midas (435)—by Comet (155)—by R. Colling's son of Favorite (252)—by same son of Favorite (252).

GRACEFUL—Roan, calved Nov. 12, 1863, bred by John Clayden, Littleborough, Eng., got by Marquis of Cornwallis (18337), out of Gaiety (vol. 6, p. 468, E.) by Lord Althorpe (14800)—Tit by Horatio (10335)—Titmouse by George (6037)—Cowslip by Pickwick (4698)—Young Spot by Broken Horn (3224)—Prim by Regent (1366)—Sopt by Western Comet (689)—by Viscount (666).

C. H. B., vol. 7, p. 486.

Sept. 19, 1876. PROF. G. LAWSON.

For Dalhousie University, Halifax, N. S. By Ship Hibernian.

CAMBRIDGE WITCH—Roan, calved June 6, 1869, bred by G. Moore, Whitehall, Cumberland, got by Royal Cambridge (25009), out of Oxford Witch by Imperial Oxford (18084)—Lancashire Witch by John O'Gaunt (16322)—Lady Warden by Lord Warden (7167)—Belinda 2d by Lion (9299)—Belinda by Rebel (4882)—by Coxcomb (928)—by Minor (441)—by son of Phenomenon (491)— by Traveler (655)—by Colonel (152)—by R. Colling's son of Broken Horn (95)—by R. Colling's son of Hubback (319).

E. H. B., vol. 19, p. 658, under dam.

FORTUNE TELLER—Red and white heifer, calved March 8, 1874, bred by Sir W. Lawson, Bart., Brayton, Carlisle, got by Wellington (32825), out of Cambridge Witch by Royal Cambridge (25009), &c., as above.

E. H. B., vol. 22, p. 479, under dam.

MAID OF OXFORD 4TH—Roan, calved July 26, 1870, bred by T. G. Curtler, Bevere House, Worcester, got by Lord Waterloo 2d (26755), out of Maid of Oxford 2d by 7th Duke of York (17754)— Maid of Oxford by 4th Duke of Oxford (11387)—Fancy by Avalanche (12418)—Caprice by Harold (10299)—Juliet by Sol (8608)— by Leo (4208)—by Treasurer (5513)—by Rupert (2580)—by North Star (460)—by Cripple (173)—by Minor (441)—by Freeman (269)— by Danby (190).

E. H. B., vol. 19, p. 614, under dam.

LADY MARY—Red and white, calved Oct. 11, 1872, bred by T. G. Curtler, got by Grand Duke of Clarence (28750), out of Lovely by Wild Boy (23219)—Lady by Sir James (16980)—Loyalty by The Corsair (15378)—Lucy Locket by Usurer (9763)—Lavender by Dan O'Connell (3557)—Lily by Brutus (1752)—by Frederick (1060)—by Cato (1794)—by son of Wellington (679)—bred by Mr. Robertson, of Ladykirk.

E. H. B., vol. 20, p. 630.

LORD OF BRAEMAR—Red bull, calved Jan. 20, 1875, bred by Earl of Dunmore, Stirling, Scotland, got by 3d Duke of Hillhurst (30975), out of Red Rose of Braemar by 11th Duke of Thorndale (31024)—Red Rose of Breadalbane by Duke Frederick (30910)—Grace by Airdrie (30365)—Ophelia by John O'Gaunt (11621)—Duchess by Buena Vista (30623)—Red Rose by Prince Charles 2d (32113)—Thames by Shakespeare (12062)—Lady of the Lake by Reformer (2505)—by Belvedere (1706)—by 2d Hubback (1423)—by His Grace (311)—by Yarborough (705).

E. H. B., vol. 23, p. 429, under dam.

WETHERBY STAR (37665)—Roan bull, calved May 11, 1875, bred by Duke of Devonshire, Holkar Hall, got by 5th Duke of Wetherby (31033), out of Evening Star by Baron Oxford 4th (25580)—Bright Star by Red Duke (18676)—Bright Eyes by 3d Duke of York (10166)—Wild Eyes 23d by 2d Cleveland Lad (3408)—by Duke of Northumberland (1940)—by Belvedere (1706)—by Emperor (1975)—by Wonderful (700)—by Cleveland (145)—by Butterfly (104)—by Hollon's Bull (313)—by Mowbray's Bull (2342)—by Masterman's Bull (422)—descended from M. Dobison's stock.

FIFTH DUKE OF LORN (36517)—Red, calved August 1, 1875, bred by E. Hall, Shallcross Hall, got by Duke of Lorn (25985), out of Lancaster 20th by Chilton Hero (17564)—Lancaster 7th by Priam (15079)—Lady Lancaster by The Queen's Roan (7389)—Lancaster by Will Honeycomb (5660)—by Spectator (2688)—by Albion (1619)—by Lancaster (360)—by son of Windsor (698)—by Comet (155).

KINGSTON—Red bull, calved Sept. 11, 1875, bred by C. A. Barnes, Charleywood, got by Barrington Duke (27985), out of Kirkee 8th by Lord Wallace (24473)—Red Hawthorne by Duke of

Darlington (21586)—Pride Bushey by Cock of the Walk (15782)—
Kirkee 5th by Master Butterfly 2d (14918)—Kirkee by Young 4th
Duke (9037)—Jenny Lind by Duke of Richmond (7996)—by Sir
Walter (2639)—by Young Jerry (8177)—by Roseberry (567)—by
Roseberry (567)—by Constellation (163)—by Hastings (293)—by
Hastings (293)—by Leopold (372).

Nov. 13, 1880. PROF. G. LAWSON.

Central Board of Agriculture, Halifax, N. S. By Ship Brooklyn.

EARL OF SURREY—Red and white, calved April 10, 1879,
bred by Captain Moir, Manor House, Reigate, got by Fugleman
(36670), out of Wild Eyes Gwynne by Baron Wild Eyes (19290)—
Rebecca Gwynne by Knight of Distington (18158)—Ruth Gwynne
by Exquisite (14524)—Young Dowager Gwynne by St. Thomas
(10777)—Dowager Gwynne by Prime Minister (2456)—by Wallace
(5586)—by Marmion (406)—by Merlin (430)—by Layton (366)—
by Phenomenon (491)—by Favorite (252)—by Favórite (252)—by
Hubback (319)—by Snowdon's Bull (612)—by Waistell's Bull (669)
—by Masterman's Bull (422)—by Studley Bull (626).

E. H. B., vol. 27, p. 519, under dam.

YOUNG EBOR—Roan bull, calved July 3, 1879, bred by W. G.
Garne, Broadmoor, Gloucestershire, got by Ebor (41499), out of Lady
Like by Ranger (35203)—Lady Day by Duke of York (23804)—
Village Miss by Captain Cherry (21363)—Village Girl by Sheriff
(18822)—Bonhill Lassie by Booth (14180)—Orphan by Lord Rag-
lan (13246)—Maid of Aln 2d by Crusade (7938)—Maid of Aln by
Regent (2517)—by Borderer (3191)—by Eclipse (1949)—by Togston
(5487)—by Bolingbroke (3184)—by son of Midas (435)—by Twin
Brother to Ben (660).

E. H. B., vol. 27, p. 408, under dam.

LORD BRETT—Roan bull, calved Sept. 13, 1879, bred by
W. G. Garne, got by Sir Robert Frogmore (40719), out of Windsor
Beauty by Lord Chief Justice (34507)—Windsor Butterfly by
Royal Butterfly 20th (25007)—Misdeeds by England's Glory (23889)

—Peggy by British Prince (14197)—Miss Folly by Prince Alfred (13494)—Folly by Paris (7314)—by Vanguard (5545)—by Robin Hood (4970)—by Anticipation (750)—by Emperor (1014)—by Young Windsor (699)—by Windsor (698).

E. H. B., vol. 26, p. 442, under dam.

ROWLAND (43928)—Roan, calved Sept. 18, 1879, bred by Col. R. Lloyd Lindsay, Lockinge Park, Wantage, Berkshire, got by Earl of Horton 11th (36588), out of Ronda by Rob Roy (29806)—Rosetta by Costa (21487)—Rosette by Prince of Prussia (16752)—Red Rose by Horatio (10335)—Monia by 3d Duke of Northumberland (3647)—Modesty by Velocipede (5552)—by Sir Thomas (2636) —by Marske (418)—by Comet (155)—by Tom (652)—by Favorite (1033)—by Hutton's Bull (323)—by Barningham (56).

LORD RANDOLPH—Roan bull, calved Oct. 19, 1879, bred by W. G. Garne, got by Sir Robert Frogmore (40719), out of Ranunculus by Piratical (33870)—Red Lass by Marksman (26814).

E. H. B., vol. 27, p. 408, under dam.

CABUL (42862)—Roan, calved Nov. 14, 1879, bred by Col. Lloyd Lindsay, got by Earl of Horton 11th (36588), out of Clotilda Rock by Lord Rockville (34568).

ROSE GWYNNE—Red, calved Dec. 20, 1863, bred by John Cloyden, got by Prince Gwynne (20547), out of Rosette (vol. 6, p. 673, E.) by Prince of Prussia (16752)—Red Rose by Horatio (10335) Maria by Duke of Northumberland (3647)—Modesty by Velocipede (5552)—Crocus by Sir Thomas (2536)—by Marske (418)—Laurestina by Comet (155)—Laura by Tom (652)—Cleasby Lady by Favorite (1033)—Lucinda by Hutton's Bull (322)—Lucy by Barningham (56).

C. H. B. vol. 7, p. 487.

DUCHESS OF WARWICK 3D—Red, calved June 30, 1876, bred by W. G. Garne, got by Grand Duke of Geneva 2d (31288), out of Duchess of Warwick by Earl of Warwickshire 3d (28524)—Butterfly's Duchess by Royal Butterfly 20th (25007)—Delicacy by

47

The Druid (20948)—Destiny by Progression (16770)—Damsel by Enterprise (11443)—Blonde by Patriot (10596)—by son of Elevator (6969)—by No Mistake (8357)—by Young Consul (6893)—by Fairfax (1023)—by Speculation (1472).

E. H. B., vol. 23, p. 450, under dam.

MERRY FACE—Roan heifer, calved Jan. 28, 1878, bred by H. Bettridge, East Hanney, Berks., got by Rockville 2d (37356), out of Medora by Masterpiece (24561)—Miss Peel by Cynric (19542)—Miss Ambler by Royal Oak (16870)—Miss Mitford by Bashaw (12449)—Mitford by Lord George (9314)—by Manager (8271)—by Raffler (7391)—by Gazer (7030)—by a bull of Mr. Champion's, of Blyth.

E. H. B., vol. 25, p. 339, under dam.

ROSELEAF—Red, calved Feb. 20, 1878, bred by H. Bettridge, got by Rockville 2d (37356), out of Rose of Poughley by Baron Booth (27915)—Red Heart Rose by Artemus Ward (23326)—Rynil Rose by A-1 (15538)—Rosette by Royal (13636)—Ringlet by Lord George (9314)—Rosebud by Fitz Hardinge (8073)—Red Rose by Augustus (6751)—by Consul (1868)—by Fairfax 2d (8050).

E. H. B., vol. 25, p. 340, under dam.

1821 or 1822. MR. LEE, or MR. ORR,
Of Boston. Mass.

HARRIET—Roan, calved in 1820, bred by Mr. Wetherell, got by Denton (198), out of Henrietta by Comet (155)—Hannah by Henry (301)—by Danby (190)—by a grandson of Favorite (252).

E. H. B., vol. 1, p. 417, and A. H. B., vol. 1, p. 82.

1836. EDWARD A. LE ROY and THOS. H. NEWBOLD.
Livingston Co., N. Y.

WINDLE 185 (5667)—Roan, calved Nov. 21, 1835, bred by Mr. Pilkington, Windle Hall, Lancashire, Eng., imported in 1836, by Le Roy & Newbold, into Livingston Co., N. Y., got by Hopewell (2135), out of Moss Rose by Waterloo (2816)—by Young Wynyard (2859)—by Irishman (329)—by Styford (629).

DIONE—Yellow red, calved in 1833, bred by Mr. Denton, Harrowby, Lancashire, Eng., imported by Messrs. Le Roy & Newbold, into Livingston Co., N. Y., in 1836, got by Monarch (4494), dam by a son of Comet (155)—by Cupid (177)—by Favorite (252)—by Hubback (319).

Recorded, A. H. B., vol. 1, p. 72.

LADY MORRIS—Red, calved in 1836 or 1837, bred by T. H. Newbold, got in England and calved in America, got by Priam (4758), out of Dione, above.

A. H. B., vol. 1, p. 87.

NETHERBY—Imported in 1836 by Messrs. Le Roy & Newbold, of Avon, Livingston Co., N. Y., got by Monarch (4494), out of Sweetbriar by Barmpton (54)—Roseberry by Western Comet (698) —by Comet (155)—by son of Favorite (252)—by Cupid (177)—by Favorite (252).

NETHERBY—Roan, calved in 1837, bred by T. H. Newbold, got in England and calved in America, got by Gambler (2047), out of Netherby, above.

A. H. B., vol. 1, p. 98.

VENUS—Roan, calved in 1833, bred by Richard Pilkington, Windle Hall, Lancashire, Eng., imported by Messrs. Le Roy & Newbold, in 1836, got by Magnum Bonum (2244), out of Ruby by grandson of Sir Dimple (1442)—Ruby by Marshal Beresford (415) —Miss Champion by Charles (127)—by Prince (521)—by Neswick (453).

A. H. B., vol. 1, p. 233.

1840-43. JAMES LENOX.
New York.

KING CHARLES 2D 84 (4154)—White, calved May 29, 1840, bred by Jonas Whitaker, Otley, Yorkshire, Eng., got by Sir Thomas Fairfax (5196), out of Lingflower by Ellerton's Bull (3701) —Gillyflower by Young Colling (1843)—by Allison's Bull of Danby (2998)—by Pink Bull—bred by Mr. Leonard Carter, of Applegarth, England.

PRINCE ALBERT 133 (4809)—Roan, calved July 10, 1840, bred by Jonas Whittaker, Otley, Yorkshire, got by Sir Thomas Fairfax (5196), out of Paulina by Son of Matchem (2678)—by Falstaff (1993)—by Richard (1376)—by Jupiter (342).

DAFFODIL—Roan, calved in May, 1836, bred by Jonas Whitaker, Otley, Yorkshire, Eng., got by Sampson (5081), out of Young Daisy by Danby (1900)—by Mrs. Wilkinson's Bull (2838)— by Greathead's Gray Bull (3936)—by Ellerton's Bull (3708)—by son of Newby's Bull (4562)—by Charge's Gray Bull (872).

Recorded, E. H. B., vol. 5, page 231, and A. H. B., vol. 1, page 69.

RED LADY—Red, calved in 1834, bred by Jonas Whitaker, Otley, Yorkshire, Eng., got by Hubback (2142), dam by Don Juan (1923)—by Woodhouse's Bull—by Woodhouse's Bull.

Recorded, E. H. B., vol. 5, page 841, and A. H. B., vol. 1, page 221.

GAYLY—Red roan, calved April 3, 1840, bred by Jonas Whitaker, Otley, Yorkshire, Eng., got by Sir Thomas Fairfax (5196), out of Graceville by Hubback (2142)—Germanville by son of Young Warlaby (2812)—by Imperial (2151)—(supposed) by Young Comet (905).

A. H. B., vol. 1, page 80, and E. H. B., vol. 5, page 425, under dam.

B. LETTON,

Of Tennessee.

BEAUTY—Calved in May, 1837.

SPOT—Red and white, calved in October, 1837, got by ———, out of Blossom by Magnum Bonum.

COWSLIP—Roan, calved in 1836, got by Colonel, out of Catherine by Planet—Blyth by Wellington.

Roan heifer, calved in August, 1838, got by Charley, out of Snowdrop.

Red and white heifer, calved Nov. 20, 1838, bred by Mr. Lynn.

NEPTUNE—Red and white, calved at sea in spring of 1840, got by Favorite, out of Spot, above.

AQUA—White, calved at sea in 1840, got by Charley, out of Beauty.

1839. JAMES LETTON.

Bourbon Co., Ky.

MISS SEVERS—Calved Jan. 1, 1837, bred by Leonard Severs, got by Reformer (2510), out of Folly by Linton (4227)—by Jupiter (343)—by Easby (232).

LADY DUNDAS—Calved April 30, 1837, got by Reformer (2510), dam by Streatlam (5338)—by North Star (460)—by Wonder (2853)—by Mr. Parkinson's Bull.

IANTHE—Red roan, calved March 1, 1838, bred by John Hunter, East Park, Branspeth, Eng., got by Barforth (3085), dam by Rowland (2571)—by Snowdrop (2653)—by Henry (301)—by St. John (572)—by George (273).

A. H. B., vol. 2, p. 403, and E. H. B., vol. 10, p. 401.

ADELIA—Light roan, calved Feb. 18, 1838, bred by Thomas Fain, Frenchfield, near Penrith, Cumberland, Eng., got by Mr. Crofton's Majesty (2250), out of Snowdrop by Baronet (1686)—by Young Rockingham (2547)—by Wellington (678)—by Young Rob Roy.

COUNTESS—Red roan, calved in March, 1838, bred by Thomas Fain, Frenchfield, Penrith, Cumberland, Eng., got by Archibald (1652), out of Empress by Baronet (1686)—Nonpareil by Wellington (2824)—by Thorpe (1515)—by Styford (629)—by Lame Bull (358).

E. H. B., vol. 5, p. 326, under dam.

CONVOY—Light roan, calved Nov. 4, 1839 (imported in her dam, Lady Dundas), calved the property of her importer, J. E. Letton, got by Lord Lieutenant (4260), out of Lady Dundas by Reformer (2510), &c., as in Lady Dundas, above.

MAGNUM BONUM JR. 30225—Red and white, calved Jan. 25, 1838, bred by Wm. Rains, of Gainford, near Darlington, Eng., got by Magnum Bonum (2243), out of Kate by Rockingham (2547) —by Rob Roy (557)—by Denton (198)—by Ladrone (353)—by Henry (301).

NOTE.—Mr. Rains called this bull Mozart (4518). The Herd Book says calved in March, 1838. Mr. Rains' certificate to Mr. Letton says Jan. 25, 1838. In many pedigrees in early volumes of the A. H. B. (see, for example, Goodness, vol 3, p. 427) the sire of this bull, Magnum Bonum (2243), is used instead of this bull himself.

LOCOMOTIVE 92 (4245)—Roan (very light), calved Oct. 5, 1838, bred by Thomas Bates, Kirklevington, Eng., got by Duke of Northumberland (1940), out of Oxford Premium Cow by Duke of Cleveland (1937)—Matchem Cow by Matchem (2281)—by Young Wynyard (2859).

1879. WM. LINTON,

Of Aurora, Ontario, Can. By Steamer Dominion, from Liverpool to Quebec.

RACHEL—Red and white, calved April 11, 1876, bred by Wm. Linton, Sheriff Hutton, York, Eng., got by Lord Rose (34659), out of Ruth by Sergeant Major (29957)—Laurel by Earl Marcus (23819)—White Rose by Magnus Troil (14880)—Miss Henderson by Magnus Troil (14880)—Eliza by Bates (12451)—Duchess by Lord Warden (20233)—by Snyders (7525)—by Duke of York (9049).

Brit.-Am. H. B., vol. 1, p. 507.

BRITISH HERO (39506)—Red and white, calved Jan. 19, 1877, bred by Wm. Linton, Sheriff Hutton, got by Sir Arthur Ingram (32490), out of Fanny by Sergeant Major (29957)—Louise by White Windsor (27803)—Mushroom by Earl of Windsor (17788)—Beauty 2d by Magnus Troil (14880)—by Bates (12451)—by General Fairfax (11519)—by Liberator (7140)—by Prince Albert (4791)—by Young Matchem (4422)—by Young Red Rover (4904) or Rockingham (2551)—by Whisker (1579)—by Pilot (496).

SHERIFF HUTTON ROSE—Roan, calved Jan. 27, 1879, bred by William Linton, Sheriff Hutton, Yorkshire, Eng., got by Sir Arthur Ingram (32490), out of Fame by Sergeant Major (29957)—Emily by Earl Torquil (23854)—Josephine by Earl Windsor (17788)—Mint by Magnus Troil (14880)—Clara by 3d Duke of Athol (12734)—Lobelia by Bates (12451)—Eva by Ingram (9236)—by Liberator (7140)—by Prince Albert (4791)—by a descendant of Mars (1199).

E. H. B., vol. 26, p. 523, under dam.

1883. WM. LINTON.

SNOWDROP—Roan, calved Feb. 19, 1879, bred by Wm. Linton, got by Paul Potter (38854), out of Cowslip by Sergeant Major (29957).

FAME 2D—Calved Jan. 26, 1880, bred by John Linton, Sheriff Hutton, got by Arthur Victor (39380), out of Fame (vol. 26, p. 523, E.) by Sergeant Major (29957)—Emily by Earl Torquil (23854), &c., as in Sheriff Hutton Rose, above.

ARTHUR VICTOR 2D—Red and white, calved Oct. 22, 1881, bred by John Linton, got by Arthur Victor (39380), out of Fame 2d by Arthur Victor (39380), &c., as in Sheriff Hutton Rose, above.

FAME 3D—Red and white, calved in 1883, bred by John Linton, Sheriff, Hutton, got by Arthur Victor (39380), out of Fame 2d by Arthur Victor (39380), &c., as in dam, Fame 2d.

1854. LIVINGSTON COUNTY (N. Y.) ASSOCIATION.

(DAVID BROOKS AND S. L. FULLER, Agents.)

PHŒNIX 2D—Red roan, calved in April, 1852, bred by Wm. Ladds, Ellington, got by Horatio (10335), out of Phœnix 1st by Mahomed (6170)—Picole by Jeremy (2157)—Belle by Red Rover (4902)—by Beppo (1712)—by Red Robin (2491)—by Emperor (1014).

A. H. B., vol. 4, p. 513.   Bought by J. H. Bennett, Avon, New York.

MEDORA—Calved in 1853 or 1854, bred by Mr. Berchal or J. S. Tanqueray, got by Horatio (10335), out of Minna (vol. 10, p. 486, E.) by Luke (7179)—Dorothy Gwynne by Conservative (3472)—Cripple by Marmion (406)—Daphne by Merlin (430)—Nell Gwynne by Layton (366)—Nell Gwynne by Phenomenon (491)—Princess by Favorite (252)—by Favorite (252)—by Hubback (319)—by Snowdon's Bull (612)—by Waistell's Bull (669)—by Masterman's Bull (422)—by Harrison's Bull (292)—by Studley Bull (626). See Medora 4th, vol. 8, A. H. B.

MISS DOWLY—Red, calved Aug. 26, 1853, bred by Mr. J. S. Tanqueray, Hendon, got by Fusileer (11499), out of Smile by Humber (7102)—Laughter by Sweet William (5638)—Onion by Roman (2561)—Sage by William (2840)—Goose by Firby (1040)—by brother to Kalmia.

E. H. B., vol. 11, p. 698, under dam.

FALLACY—Roan, calved Dec. 9, 1852, bred by Charles Barnett, Stratton Park, got by Fancy Man (11463), out of Sybil by Sweet William (7571)—Lady East (vol. 9, p. 425, E.) by Earl of Essex (6955)—Verbena by Van Amburg (5543)—Pine Apple by Plenipo (4724)—Jonquil by Young Matchem (4425)—Crocus by Sir Thomas (2636)—by Marske (418)—Laurestina by Comet (155)—Laura by Tom (652)—Cleasby Lady by son of Favorite (252)—Lucinda by Hutton's Bull (322)—Lucy by Barningham (56).

A. H. B., vol. 5, p. 275.

HOPELESS—Red and white, calved Dec. 15, 1852, bred by Charles Barnett, Stratton Park, Bedfordshire, got by Horatio (10335), out of Lady Elizabeth (vol. 9, E.) by Earl of Essex (6955)—White Rose by Senator (1610)—Red Rose by Columella (904)—by Shakespeare (1429)—by Blyth Comet (85)—by Neswick (1266)—by R. Colling's Son of Favorite (1033).

A. H. B., vol. 4, p. 382. Bought by Gen. J. S. Wadsworth, Geneseo, N. Y.

LADY ELLINGTON—Red, bred by Mr. Ladd, Ellington, got by Broughton Hero (6811), out of Fancy by Sweet William (7571)—Fancy by Mahomed (6170)—Faith by Helmsman (2109)—by Columella (904)—by Shakespeare (1429)—by Magnet (392)—by Blyth Comet (85)—by Neswick (1266)—by R. Colling's Son of Favorite (1033).

A. H. B., vol. 3, p. 717. Bought by Gen. J. S. Wadsworth, Geneseo, N. Y.

MUSIC—Roan, calved Dec. 8, 1853, bred by J. S. Tanqueray, Hendon, got by Balco (9918), out of Minstrel (vol. 10, p. 481, E.) by son of Irishman (13076)—Melody by Sir Thomas Fairfax (5196)—Magic by Wallace (5586)—by Wellington (2824)—by Marmion (406)—by Merlin (430)—by Layton (366)—by Phenomenon (491)—by Favorite (252)—by Favorite (252)—by Hubback (319)—by Snowdon's Bull (612)—by Waistell's Bull (669)—by Masterman's Bull (422)—by Studley Bull (626).

A. H. B., vol. 3, p. 718. Bought by Gen. J. S. Wadsworth, Geneseo, N. Y.

HOPE—Red and white, calved April 8, 1855, bred by J. S. Wadsworth, Geneseo, N. Y., got by Usurper (13928), out of Hopeless by Horatio (10335), &c., as in Hopeless, above.

A. H. B., vol. 4, p. 382.

AUSTRALIA—Red and white, calved Feb. 8, 1853, bred by Mr. Marjoribanks, England, got by Lord Foppington (10437), out of Adonea by Harkaway (9184)—Cherry by Henry (4010)—Velveteen by Baron (3095)—Splendid by Matchem 3d (4420)—by Young

48

Eryholme (1981)—by Belzoni (1709)—by Comus (1861)—by Denton (198)—by Henry (301).

A. H. B., vol. 4. p. 264. Bought by Gen. James S. Wadsworth, Geneseo, N. Y.

DAMSEL—Got by Upstart (9760), out of Darling (vol. 10, p. 330. E.) by Robin Hood (8492)—Drusilla by Leander (4199)—by a son of Ivanhoe (1131)—by Turnell's Major (6623)—by George (9151).

CAMILLA—Red roan, calved Dec. 13, 1853, bred by J. S. Tanqueray, Hendon, got by Fusileer (11499), out of Young Sall Gwynne by St. Thomas (10777)—Sall Gwynne by Prime Minister (2456)—Cripple by Marmion (406)—Daphne by Merlin (430)—Nell Gwynne by Layton (366)—by Phenomenon (491)—by Favorite (252)—by Favorite (252)—by Hubback (319)—by Snowdon's Bull (612)—by Waistell's Bull (669)—by Masterman's Bull (422)—by Studley Bull (626).

E. H. B.. vol. 11. p. 691. under dam, and A. H. B.. vol. 5. p. 233.

DORINDA—Red and white, calved Nov. 24, 1853, bred by J. S. Tanqueray, got by Balco (9918), out of Dorcas by North Star (9447)—Sarah by Ribblesdale (7422)—Dorothy Gwynne by Conservative (3472)—Cripple by Marmion (406). &c.. as in Camilla. above.

USURPER 3522 (13928)—Roan, calved March 22, 1853. bred by T. R. B. Cartwright. Aynhoe. got by Upstart (9760). out of Flash by Mowbray (7260)—Flounce by Berryman (3143)—Fancy by Remus (2524)—White Rosette by Juniper (1144)—Rosette by White Comet (1582)—Young Rose by Wright's Grandson of Favorite (2073)—Old Rose by Cattley's Gray Bull (1798).

Bought by Judge Carroll.

BLETSOE 2548 (9970)—Red and white, calved April 27, 1849, bred by Mr. Beauford Bletsoe, Bedfordshire, got by Diamond (5918), out of Fatima by 3d Duke of Northumberland (3647)—Formosa by Sir Thomas (2636)—by Sir Alexander (591)—by

Marske (418)—by North Star (459)—by Wellington (680)—by Favorite (252)—by Ben (70).

Bought by Sachett, Barber & Co.

GOVERNOR 2922 (12957)—Roan, calved in April, 1854, bred by J. C. Adkins, Milcote, Stratford-on-Avon, got by Daybreak (11338), out of Garland by Brunswick (6814)—Graceful by Lycurgus (7180) —Marcia by Ranunculus (2479)—Sackbut by William (2840)— Clarion by Childers (1824)—No 25 Chilton Sale by Richard (1376) —by Jupiter (342)—by Charles (127)—by Windsor (698)—by Chilton (136)—by Colonel (152).

MR. LONGLEY,
Of Maitland, Can.

BETSY—See Lily. C. H. B.. vol. 1, p. 358.

1875. LOWMAN & SMITH.

Importation of Messrs. D. Lowman and J. & J. Smith, of Toulon, Ill. By Ship Corinthian, June, 1875, from Glasgow, to Montreal; arrived at Toulon, July 19, 1875.

LOVELY 18TH—Roan heifer, calved March 4, 1874, bred by Mr. A. Cruickshank, Sittyton, Aberdeenshire, got by Honeycomb (28866), out of Lovely 11th by Allan (21172)—Lovely 10th by Duke of Bedford (23722)—Lovely 6th by Bosquet (14183)—Lovely 3d by The Hero (10934)—Lovely by Kelly 2d (9265)—by Robin-o'-Day (4973)—by Favorite (9116)—by Anthony (1640)—by Anthony (1640)—by Edgecott (1953)—by Son of Merlin (6522)—by Acton (1607).

Recorded, E. H. B., vol. 22, p. 386, under dam, and A. H. B., vol. 15, p. 712. Sold, April 13, 1876, at public sale, at Galesburg, Ill., to A. J. Dunlap, of same place, for $1,010.

BUTTERFLY 45TH—Red heifer, calved Jan. 26, 1874, bred by Mr. A. Cruickshank, Sittyton, Aberdeenshire, Scotland, got by Viceroy (32764), out of Butterfly 37th by Champion of England (17526)—Butterfly 19th by Allan (21172)—Butterfly 4th by Lord

Raglan (13244)—Butterfly by Matadore (11800)—Buttercup by Report (10704)—Bounty by The Pacha (7612)—by 2d Duke of Northumberland (3646)—by Mahomed (6170)—by Sillery (5131)—by Carleton (843)—by Diamond (205)—by Diamond (205).

Recorded, A. H. B., vol. 15, p. 466. Owned by John Bond, Abingdon, Ill.

BUTTERFLY 46TH—Red heifer, calved Feb. 14, 1874, bred by Mr. A. Cruickshank, Sittyton, Aberdeenshire, Scotland, got by Cæsar Augustus (25704), out of Butterfly 12th by St. Clair (25078) —Butterfly 5th by Lord Raglan (13244)—Butterfly by Matadore (11800)—Buttercup by Report (10704)—Bounty by The Pacha (7612)—by 2d Duke of Northumberland (3646)—by Mahomed (6170) —by Sillery (5131)—by Carleton (843)—by Diamond (205)—by Diamond (205).

Recorded, A. H. B., vol. 15, p. 466. Sold, April 13, 1876, by public sale, at Galesburg, Ill., to A. J. Dunlap, same place, for $850.

MISSIE 35TH—Roan, calved July 2, 1870, bred by W. S. Marr, Upper Mill, Tarves, Aberdeenshire, Scotland, got by Prince of Stokesley (27177), out of Missie 5th by Lord of Lorn (18258)— Missie 2d by Augustus (15598)—Missie by son of Duke 3d (17697) —Countess by The Pacha (7612)—by Mahomed (6170)—by Plenipo (4725)—by Abbot (2899).

Recorded, A. H. B., vol. 15, p. 774. Sold, April 13, 1876, at public sale, at Galesburg, Ill., to Edward Isles, Springfield, Ill., for $625.

GOLDIE 18TH—Red heifer, calved April 1, 1874, bred by W. S. Marr, Upper Mill, Tarves, Aberdeenshire, Scotland, got by Young Englishman (31113), out of Goldie 12th by Macduff (26773)— Goldie 2d by Lord Privy Seal (16444)—Goldie by Goldsmith (14632) —Ruby Hill by Elphinstone (14492)—Hawthorne Hill by Duke (7980)—by Sir Robert Peel (7512)—by Buchan Laddie (5814).

Recorded, A. H. B., vol. 15, p. 582, as Goldie 17th. Owned by John Bond, Abingdon, Ill.

RED LADY 3D—Red heifer, calved Feb. 14, 1874, bred by W. S. Marr, Upper Mill Tarves, Aberdeenshire, Scotland, got by Young Englishman (31113), out of Red Lady 2d by Heir of Englishman (24122)—Red Lady by Young Pacha (20457)—Roan Lady by Young Ury (10984)—Red Lady by Van Dunck (10992)—Patience by Duke 7th (7984)—by Teetotaller (5411)—by Ivanhoe (1131)—by Blyth Comet (85)—by Atlas (42)—by Colling's son of Favorite (252).

E. H. B., vol. 23, p. 556, under dam, and A. H. B., vol. 15, p. 856. Sold, at public auction, April 13, 1876, at Galesburg, Ill., to W. W. Pickrell, of Mechanicsburg, Ill., for $1,200.

HEIR OF YOUNG ENGLISHMAN 26801—Roan, calved Jan. 22, 1876, bred by W. S. Marr, got by Young Englishman (31113), out of Missie 35th by Prince of Stokeley (27177), &c., as in Missie 35th, above. (Calved in America.)

GERALDINE 7TH—Roan heifer, calved April 21, 1874, bred by Mr. J. Cochrane, Little Haddo, Scotland, got by Prince Frederick of Cambridge (29621), out of Geraldine 4th by Prince Louis (27158)—Geraldine 1st by Lord Buckingham (20151)—Robina by Liberator (14793)—Geraldine by Robin-o'-Day (4973)—Angelica by Holkar (4041)—Glendronach by Mr. Jopp's Bull (9256)—Kate of Darlington.

Recorded, A. H. B., vol. 15, p. 576. Sold, April 13, 1876, at public sale, at Galesburg, Ill., to W. Scott, of Wyoming, Ill., for $530.

1876. LOWMAN & SMITH,

Of Toulon, Stark Co., Ill. By Ship Phœnician, from Glasgow, June 8, 1876, to Quebec; arrived at Toulon, July 6, 1876.

CACTUS 3D—Red heifer, calved April 3, 1875, bred by Mr. T. Lister, Groby, near Leicester, got by Cambridge Duke 4th (25706), out of Cactus 2d by Alaska (37718)—Cactus by Duke of Belmont (38123)—Camelia by Duke (38111)—Frantic 2d by Halton (11552)—Frantic by 4th Duke of York (10167)—Faith by 4th

Duke of Northumberland (3649)—Fidget by 2d Earl of Darlington (1945)—Fletcher by a son of Young Wynyard (2859)—descended from J. Brown's Red Bull (97).

Recorded, E. H. B., vol. 23, p. 553, under dam, and A. H. B., vol. 19, p. 14422. Sold, Nov. 16, 1876, to John Bond, Abingdon, Ill., at public sale, for $435. Afterward the property of W. Scott, Wyoming, Ill.

ORANGE BLOSSOM 25TH—Red and white heifer, calved April 22, 1875, bred by Mr. A. Cruickshank, Sittyton, got by Ben Wyvis (30528), out of Orange Blossom 8th by Sir Walter Scott (22922)—Orange Blossom by Dr. Buckingham (14405)—Queen of Scotland by Matadore (11800)—by Sir Thomas Fairfax (5196)—by Billy (3151)—by Sovereign (5285)—by Satellite (1420)—by Baronet (60)—by Cleveland (144)—by Symmetry (641).

Recorded, A. H. B., vol. 17, p. 13091. Sold at public sale, Nov. 16, 1876, to L. Hanna, Waveland, Ind., for $705. Afterward owned by Aaron Plumley, West Liberty, Iowa.

BLANCHE 11TH—Red heifer, calved Aug. 26, 1875, bred by W. H. Salt, Maplewell, Loughborough, got by 5th Lord Oxford (31738), out of Baroness Blanche by Baron Waterloo (27977) —Bantling by Britannicus (17452)—Blanchette by Magistrate (13274)—Blanche 5th by Antinous (12401)—Blanche 1st by Killjoy (14759)—Blanche by Diamond (5918)—Blanche 2d by Norfolk (2377)—Blanche by Belvedere (1706)—Lupin by Belvedere (1706) —Tulip by Lancaster (360)—Ruby by Petrarch (488)—Miss Hutchison by Major (397)—Mr. Hutchison's Stranger by Chapman's Son of Punch (122)—Old Roany by Dickson's Grandson of Punch (213)—Roan Heifer by Checks (132)—Red Sall by Grimston's Bull (282)—Sockburn Sall by J. Coates' Bull (148).

Recorded, E. H. B., vol. 22, p. 550, under dam, and A. H. B., vol. 19, p. 14407. Sold to W. Scott, Wyoming, Ill.

www.ingramcontent.com/pod-product-compliance
Lightning Source LLC
Chambersburg PA
CBHW062347220526
45472CB00008B/1727